André B. de Haan, H. Burak Eral, Boelo Schuur
Industrial Separation Processes

Also of Interest

Process Technology.
An Introduction
De Haan, 2015
ISBN 978-3-11-033671-9, e-ISBN 978-3-11-033672-6

Product and Process Design.
Driving Innovation
Harmsen, de Haan, Swinkels, 2018
ISBN 978-3-11-046772-7, e-ISBN 978-3-11-046774-1

Product-Driven Process Design.
From Molecule to Enterprise
Zondervan, Almeida-Rivera, Carmada, 2020
ISBN 978-3-11-057011-3, e-ISBN 978-3-11-057013-7

Process Engineering.
Addressing the Gap between Study and Chemical Industry
Kleiber, 2020
ISBN 978-3-11-065764-7, e-ISBN 978-3-11-065768-5

Basic Process Engineering Control
Agachi, Cristea, Makhura, 2020
ISBN 978-3-11-064789-1, e-ISBN 978-3-11-064793-8

André B. de Haan, H. Burak Eral,
Boelo Schuur

Industrial Separation Processes

Fundamentals

2nd Edition

DE GRUYTER

Authors
Prof. Dr. Ir. André B. de Haan
Delft University of Technology
Department of Chemical Engineering
Section Transport Phenomena
Van der Maasweg 9
2629 HZ Delft
The Netherlands

Assoc. Prof. Dr. H. Burak Eral
Delft University of Technology
Process & Energy Department
Leeghwaterstraat 39
2628 CB Delft
The Netherlands

Prof. Dr. Ir. Boelo Schuur
Sustainable Process Technology group
Faculty of Science and Technology
University of Twente
Meander building 221
De Horst 2
7522 LW Enschede
The Netherlands

ISBN 978-3-11-065473-8
e-ISBN (PDF) 978-3-11-065480-6
e-ISBN (EPUB) 978-3-11-065491-2

Library of Congress Control Number: 2020932665

Bibliographic information published by the Deutsche Nationalbibliothek
The Deutsche Nationalbibliothek lists this publication in the Deutsche Nationalbibliografie;
detailed bibliographic data are available on the Internet at http://dnb.dnb.de.

© 2020 Walter de Gruyter GmbH, Berlin/Boston
Cover image: RonFullHD/iStock/Getty Images
Typesetting: Integra Software Services Pvt. Ltd.
Printing and binding: CPI books GmbH, Leck

www.degruyter.com

Preface

This textbook has originally been written to support the bachelor course "Separation Technology" at the University of Twente and has been later also used at Eindhoven University of Technology, Delft University of Technology, Hogeschool Rotterdam, and other institutions of higher education. Our main objective is to present an overview of the fundamentals underlying the most frequently used industrial separation methods. We focus on their physics principles and the basic computation methods that are required to assess their technical and economic feasibility for a given application. Thus, design calculations are limited to those required for the development of conceptual process schemes. To keep computational time within the limits of our course structure, most examples and exercises of homogeneous mixtures are limited to binary mixtures.

The textbook is organized into three main parts. Separation processes for homogeneous mixtures are treated in the parts on equilibrium-based molecular separations (Chapters 2–5) and rate-controlled molecular separations (Chapters 6–8). Mechanical separation technology presented in Chapters 9–11 provides an overview of the most important techniques for the separation of heterogeneous mixtures. In each chapter a short overview of the most commonly used equipment types is given. Only for gas–liquid contactors Chapter 4 goes into more detail about their design and operation because they are the most commonly used industrial contactors. Chapter 12 has a unique position as this chapter considers the selection of an appropriate separation process for a given separation task.

The design of separation processes can only be learned by an active hands-on approach. The most important aspects are the combination of the right material balances with the thermodynamic equilibrium relations and mass transport equations. To support the reader in learning and applying the presented material, we have extended the number of exercises included at the end of each chapter. Short answers are given at the end of this book; detailed solutions are given in a separate solution manual that is available at the website (http://dx.doi.org/10.1515/9783110306729_suppl.).

In the preparation of 2nd edition of this book, we greatly appreciate Hans Bosch, coauthor of the 1st edition, for allowing us to use his material from the 1st edition. Furthermore Johan Smit (Hogeschool Rotterdam) and Frederico Marques Penha (TU Delft) provided valuable comments and suggestions.

André B. de Haan

H. Burak Eral

Boelo Schuur

https://doi.org/10.1515/9783110654806-202

Contents

Chapter 3
Absorption and Stripping — 57

Chapter 4
General Design of Gas/Liquid Contactors — 81

Chapter 9
Sedimentation and Settling —— 289

Chapter 12
Separation Method Selection —— 385

Chapter 1
Characteristics of Separation Processes

1.1 Significance of separations

When ethanol is placed in water, it dissolves and tends to form a solution of uniform composition. There is no simple way to separate the ethanol and the water again. This tendency of substances to form a mixture is a spontaneous, natural process that is accompanied by an increase in entropy or randomness. In order to separate the obtained mixture into its original species, a device, system or process must be used that supplies the equivalent of thermodynamic work to induce the desired separation. The fact that naturally occurring processes are inherently mixing processes makes the reverse procedure of "unmixing" or *separation processes* one of the most challenging categories of engineering problems. We can now define separation processes as

> **Those Operations that Transform a Mixture of Substances into Two or More Products that Differ from Each Other in Composition**

The separation of mixtures, including enrichment, concentration, purification, refining and isolation are of extreme importance to chemists and chemical engineers. Separation technology has been practiced for millennia in the food, material processing, chemical and petrochemical industry. In the food, material processing and petrochemical industry, most processes do not involve a reaction step but are merely used for the recovery and purification of products from natural resources. Examples are oil refining (Figure 1.1), metal recovery from ores and the isolation and purification of sugar from sugar beets or sugar cane schematically shown in Figure 1.2. In this process a combination of separation techniques (washing, extraction, pressing, drying, clarification, evaporation, crystallization and filtration) are used to isolate the desired product, sugar, in its right form and purity.

Within the chemical industry, separation technology is mostly used in conjunction with chemical reactors. One of the main characteristics of all these chemical processes is that many more separation steps are used compared to the amount of chemical reaction steps. An illustrative example is the oxidation of ethylene to ethylene oxide shown in Figure 1.3. After the reaction, several absorption, desorption and distillation steps are required to obtain the ethylene oxide as a pure product.

The above examples serve to illustrate the importance of separation operations in the majority of industrial chemical processes. Looking at the flow sheets in Figures 1.2 and 1.3 it should not be surprising that separation processes *account for 40–90% of capital and operating costs* in industry and their proper application can significantly reduce costs and increase profits. As illustrated by Figure 1.4, separation operations are employed to serve a variety of functions:

https://doi.org/10.1515/9783110654806-001

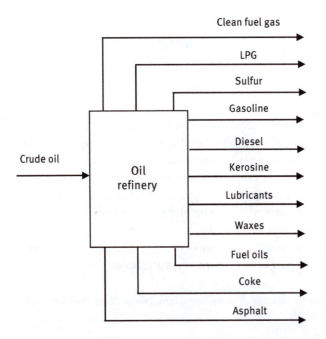

Figure 1.1: Very simple schematic of a refinery for converting crude oil into products.

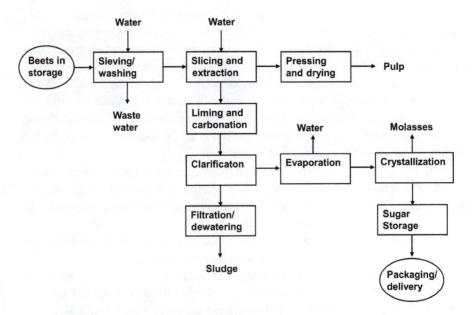

Figure 1.2: Processing sequence for producing sugar from sugar beets.

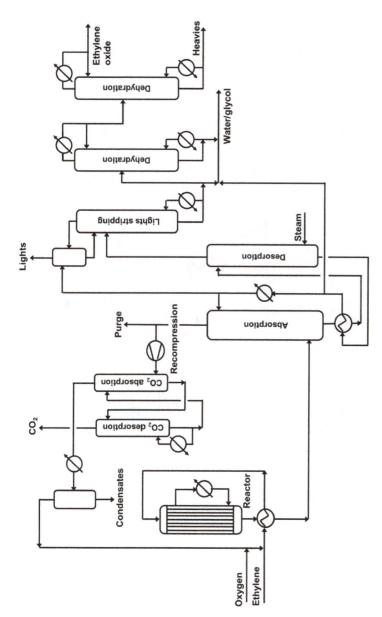

Figure 1.3: Ethylene oxide production by oxygen oxidation.

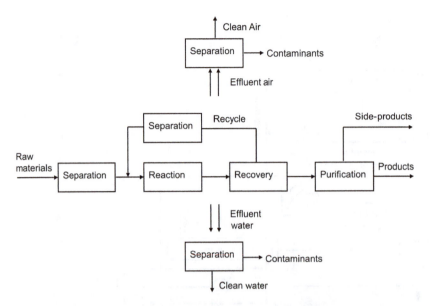

Figure 1.4: Schematic of the functions of separations in manufacturing.

- Removal of impurities from raw materials and feed mixtures
- Recycling of solvents and unconverted reactants
- Isolation of products for subsequent purification or processing
- Purification of products, product classes and recycle streams
- Recovery and purification of by-products
- Removal of contaminants from air and water effluents

1.2 Characteristics of separation processes

1.2.1 Categorization

There are many examples where a separation of a *heterogeneous feed* consisting of more than one phase is desired. In such cases it is often beneficial to first use some mechanical means based on gravity, centrifugal force, pressure reduction or an electric and/or magnetic field to separate the phases. Such processes are called *mechanical separation processes*. For example, a filter of a centrifuge serves to separate solid and liquid phases from slurry. Vapor–liquid separators segregate vapor from liquid. Then, appropriate separation techniques can be applied to each phase.

Most other separation processes deal with single, *homogeneous mixtures* (solid, liquid or gas) and involve a diffusion transfer of material from the feed stream to one of the product streams. Often a mechanical separation is employed to separate the product phases in one of these processes. These operations are referred to as *molecular separation processes*. Most molecular separation processes operate through equilibration of two immiscible phases, which have different compositions at equilibrium. Examples are evaporation, absorption, distillation and extraction processes. We shall call these *equilibrium-based processes*.

In another class of molecular separation processes the efficiency is mainly controlled by the transport rate through media resulting from a gradient in partial pressure, temperature, composition, electric potential or the like. They are called *rate-controlled processes* such as adsorption, ion exchange, crystallization and drying. Important in their design is the factor time that determines in combination with the transport rates their optimal, usually cyclic way of operation.

1.2.2 Separating agents

A simplified scheme of a separation process is shown in Figure 1.5. The feed mixture may consist of one or several streams and can be vapor, liquid or solid. From the fundamental nature of separation, there must be at least two product streams, which differ from each other in composition. If the feed mixture is a homogeneous solution a second phase must be developed before separation of chemical species can be achieved. This second phase can be created by an *energy-separating agent* (ESA), *mass-separating agent* (MSA) and **barrier *or external fields*** as shown in Figure 1.6.

The most common industrial technique, Figure 1.6a, is the application of an energy-separating agent to create a second phase (vapor, liquid or solid) that is immiscible with the feed phase. Such applications of an energy-separating agent involve energy (heat) transfer to or from the mixture. For example, in evaporation the separating agent is the heat (energy) supplied that causes the formation of a second (vapor) phase. A vapor phase may also be created by pressure reduction. A second technique is to create a second phase in the system by the introduction of a

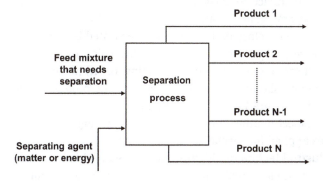

Figure 1.5: General separation process.

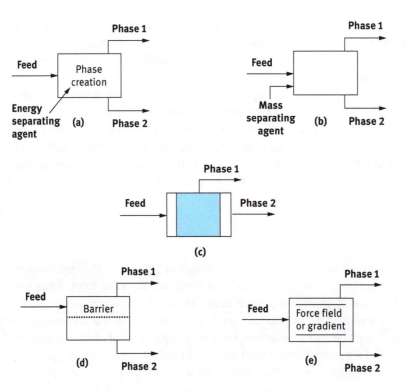

Figure 1.6: General separation techniques (a) by phase creation, (b) solvent addition, (c) solid mass-separating agent, (d) by barrier and (e) by force of field gradient.

mass-separating agent. Mass-separating agents may be used in the form of a liquid (absorption, extractive distillation and extraction) or a solid (adsorption and ion exchange). In extraction processes, Figure 1.6b, the separating agent is a solvent that selectively dissolves some of the species in the feed mixture. Of growing importance are techniques that involve the addition of solid particles, Figure 1.6c, which selectively adsorb solutes on their internal surface to create separation. Some separation processes utilize more than one separating agent. An example is *extractive distillation*, where a mixture of components with close boiling points is separated by adding a solvent (mass-separating agent), which serves to enhance the volatility of some components stronger than others, and heat (energy separating agent) in a distillation scheme to generate a more volatile product and a less volatile product. Less common is the use of a barrier, Figure 1.6d, which restricts the movement of certain chemical species with respect to other species. External fields, Figure 1.6e, of various types are sometimes applied for specialized separations.

The most common industrial technique, Figure 1.6a, is the application of an energy-separating agent to create a second phase (vapor, liquid or solid) that is immiscible with the feed phase. Such applications of an energy-separating agent

involve energy (heat) transfer to or from the mixture. For example, in evaporation the separating agent is the heat (energy) supplied that causes the formation
of a second (vapor) phase. A vapor phase may also be created by pressure reduction. A second technique is to create a second phase in the system by the introduction of a mass-separating agent. Mass-separating agents may be used in the
form of a liquid (absorption, extractive distillation and extraction) or a solid (adsorption and ion exchange). In extraction processes, Figure 1.6b, the separating
agent is a solvent that selectively dissolves some of the species in the feed mixture. Of growing importance are techniques that involve the addition of solid
particles, Figure 1.6c, which selectively adsorb solutes on their internal surface to
create separation. Some separation processes utilize more than one separating agent.
An example is *extractive distillation*, where a mixture of components with close
boiling points is separated by adding a solvent (mass-separating agent), which
serves to enhance the volatility of some components stronger than others, and heat
(energy separating agent) in a distillation scheme to generate a more volatile product
and a less volatile product. Less common is the use of a barrier, Figure 1.6d, which
restricts the movement of certain chemical species with respect to other species.
External fields, Figure 1.6e, of various types are sometimes applied for specialized
separations.

For all of the general techniques of Figure 1.6, the separations are achieved by enhancing the rate of mass transfer of certain species relative to all species. The driving
force and direction of mass transfer by diffusion is governed by thermodynamics, with
the usual limitations of equilibrium. Thus both transport and thermodynamic considerations are crucial in the design of separation operations. *Mass transfer governs the
rate of separation, while the extent of separation is limited by thermodynamic
equilibrium.*

1.2.3 Separation factors

An important consideration in the selection of feasible separation methods is the *degree of separation* that can be obtained between two key components of the feed.
Since the object of a separation device is to produce products of differing compositions,
it is logical to define this degree of separation in terms of product compositions: This is
commonly done through the *separation factor* for the separation of component A
from component B between phases 1 and 2, defined for a single stage of contacting as

$$SF_{A,B} = \frac{C_{A,1}/C_{A,2}}{C_{B,1}/C_{B,2}} = \frac{x_{A,1}/x_{A,2}}{x_{B,1}/x_{B,2}} \qquad (1.1)$$

where C and x are composition variables, such as mole fraction, mass fraction or
concentration. The value of the separation factor is limited by thermodynamic
equilibrium, except in the case of membrane separations that are controlled by

relative rates of mass transfer through the membrane. An effective separation is accomplished when the separation factor is significantly different from unity. If *SF* > 1 component A tends to concentrate in the product 1 while component B accumulates in product 2. On the other hand, if *SF* < 1, component B tends to concentrate preferentially in product 1. In general components A and B are designated in such a manner that the separation factor is larger than unity. Consequently, *the larger the value of the separation factor is, the more feasible the particular operation.*

In real processes the separation factor can reflect the differences in equilibrium compositions and transport rates as well as the construction and flow configuration of the separation device. For this reason it is convenient to define in Section 1.4 an inherent separation factor, the *selectivity*, which would be obtained under idealized conditions. For equilibrium separation processes the selectivity corresponds to those product compositions, which will be obtained when simple equilibrium is attained between the product phases. For rate-controlled separation processes the selectivity represents those product compositions, which will occur in the presence of the underlying physical transport mechanism alone. Complications from competing transport phenomena, flow configurations or other extraneous effects are excluded.

1.3 Industrial separation methods

1.3.1 Exploitable properties

The objective of the design of a separation process is to *exploit property differences* in the most economical manner to obtain the required separation factors. Mechanical separations exploit the property differences between the coexisting phases by applying a certain force to the heterogeneous mixture. Table 1.1 presents an overview of the different mechanical separations classified according to the principle involved. The achieved extend of separation in molecular separations depends on the exploitation of differences in molecular, thermodynamic and transport properties of the different chemical species present in the feed. Some important bulk thermodynamic and transport properties are vapor pressure, adsorptivity, solubility and diffusivity. These differences in bulk properties result from differences in properties of the molecules themselves such as

Molecular weight	Polarizability
Van der Waals volume	Dielectric constant
Van der Waals area	Electric charge
Molecular shape	Radius of gyration
Dipole moment	

Table 1.2 indicates the importance of the main molecular properties in determining the value of the separation factor for various separation processes. Classification of

Table 1.1: Separation principles of mechanical separations.

Technique	Applied mechanical force	Technique	Applicable for particles (micron)	Separation principle
Sedimentation Ch 9	Gravity	Settlers Classifiers	>100	Density difference
	Centrifugal	Centrifuges Cyclones Decanters	1–1,000	Density difference
	Electrostatic	Electrostatic precipitators	0.01–10	Charge on fine solid particles
	Magnetic	Liquid + Solid		Attraction by magnetism
Filtration Ch 10 & Ch 11	Gravity	Sieves Filtration	>100	Particle size larger than pore size of filter medium
	Pressure	Filtration Presses Sieves Membranes	0.001–1,000	Particle size larger than pore size of filter medium
	Centrifugal	Centrifuges	1–1,000	Particle size larger than pore size of filter medium
	Impingement	Filters Scrubbers Impact separators	0.1–1,000	Size difference

separation processes in terms of the molecular properties that primarily govern the separation factor can be quite useful for the selection of candidate processes for separating a given mixture. *Processes that emphasize molecular properties in which the components differ to the greatest should be given special attention.* Although this categorization only provides general guidelines, the basic differences in the importance of different molecular properties in determining the separation factors for different separation processes are apparent from Table 1.2. For example, the separation factor in distillation reflects vapor pressures, which in turn reflect primarily the strength of intermolecular forces. The separation factor in crystallization, on the other hand, reflects primarily the ability of molecules of different kinds to fit together, and simple geometric factors of size and shape become much more important.

Values of these properties for many substances are available in handbooks, specialized reference books and journals. Many of them can also be estimated using computer-aided process simulation programs. When not available, these properties must be estimated or determined experimentally if a successful application of the appropriate separation operation(s) is to be achieved.

Table 1.2: Relation of the separation factor to difference in molecular properties.

	Molecular weight	Molecular size	Molecular shape	Dipole moment and polarizability	Molecular charge	Chemical reaction
Distillation	++	+	–	++	–	–
Crystallization	+	++	++	+	++	–
Extraction	+	+	+	++	–	++
Absorption	+	+	++	+++	–	
Adsorption	+	++	++	++	–	++
Membrane filtration	–	++	+++	–	–	–
Ion exchange	–	–	–	–	–	+++
Drying	++	+-	–	+	–	–

1.3.2 Important molecular separations

An overview of the most commonly used industrial molecular separations divided into equilibrium and rate-based separations is given in Table 1.3. The table shows the phases involved, the separating agent and the physical or chemical principle on which the separation is based. It should be noted that the feed to a separation unit usually consists of a single vapor, liquid or solid phase. If the feed comprises two or more coexisting phases, it should be considered to separate the feed stream first into two phases by some mechanical means and then sending the separated phases to different separation units. Table 1.3 is not intended to be complete but serves as an overview of the most widely applied industrial separations.

Some separation operations are well understood and can be readily designed for a mathematical model and/or scaled up to a commercial size from laboratory data. This *technological and use maturity* has been analyzed and is graphically illustrated in Figure 1.7 for the molecular separation technologies from Table 1.3. At the "technological asymptote" it is assumed that everything is known about the process and no further improvements are possible. If a process is used to the fullest extent possible, it has approached its "use asymptote." As expected, the degree to which a separation operation is technologically mature correlates well with its commercial use. Those operations ranked near the top are frequently designed without the need for any laboratory of pilot-plant test. Operations near the middle usually require laboratory data, while operations near the bottom require extensive research before they can be designed. Distillation, the most frequently used process, is closer to its technological and use asymptotes than any other process.

Table 1.3: Commonly used molecular separation methods.

Separation method	Chapter	Phase of the feed	Separation agent	Products	Separation principle
Equilibrium-based processes					
Flash	2	Liquid and/or vapor	Heat transfer	Vapor + liquid	Difference in volatility
Distillation	2	Liquid and/or vapor	Heat transfer	Vapor + liquid	Difference in volatility
Absorption	3	Vapor	Liquid absorbent	Liquid + vapor	Difference in volatility
Stripping	3	Liquid	Vapor stripping agent	Vapor + liquid	Difference in volatility
Extractive distillation		Liquid and/or vapor	Liquid solvent and heat transfer	Vapor + liquid	Difference in volatility
Azetropic distillation		Liquid and/or vapor	Liquid entrainer and heat transfer	Vapor + liquid	Difference in volatility
Extraction	5	Liquid	Liquid solvent	Liquid + liquid	Difference in solubility
Rate-controlled processes					
Gas adsorption	6	Vapor	Solid adsorbent	Gas + solid	Difference in adsorbability
Liquid adsorption	6	Liquid	Solid adsorbent	Liquid + solid	Difference in adsorbability
Ion exchange	6	Liquid	Ion exchanger	Liquid + solid	Difference in chemical reaction
Leaching		Solid	Liquid solvent	Liquid + solid	Difference in solubility
Drying	7	Solid	Heat transfer	Vapor + solid	Difference in volatility
Crystallization	8	Liquid	Heat transfer	Liquid + solid	Difference in solubility or melting point

1.4 Inherent selectivities

1.4.1 Equilibrium-based processes

For separation processes based upon the equilibration of immiscible phases the selectivity is always based on the *equilibrium ratio* of the components, defined as

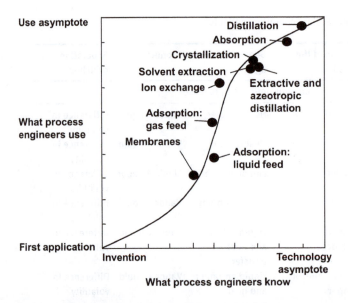

Figure 1.7: Technological and use maturities of separation processes.

the ratio of the mole fraction (concentration) in phase 1 to the mole fraction in phase 2, *at equilibrium*:

$$K_A \equiv \frac{x_{A,1}}{x_{A,2}} \tag{1.2}$$

Introduction in the previous equation for the separation factor *SF*, defined in eq. (1.1), provides the relation that defines the *inherent selectivity* S_{AB} in terms of the mole fractions as well as equilibrium ratios:

$$S_{AB} = \frac{K_A}{K_B} = \frac{x_{A,1}/x_{A,2}}{x_{B,1}/x_{B,2}} \tag{1.3}$$

For processes based on equilibration between gas and liquid phases the selectivity can be related to vapor pressures P^0 and activity coefficients y:

$$p_A = y_A P_{\text{tot}} = \gamma_A x_A P_A^0 \quad \text{or} \quad K_A \equiv \frac{y_A}{x_A} = \frac{\gamma_A P_A^0}{P_{\text{tot}}} \tag{1.4}$$

For a mixture that forms nearly ideal liquid solutions Raoult's law applies ($y = 1$). Then, the inherent selectivity for vapor–liquid separation operations employing energy as the separating agent such as partial evaporation, partial condensation or distillation is given by the following:

$$S_{AB} = \alpha_{AB} = \frac{P_A^0}{P_B^0} \tag{1.5}$$

The inherent selectivity α_{AB} in a vapor–liquid system is commonly called the *relative volatility*. In case of nonideal solutions in vapor–liquid separation processes, the expressions for the K values of the key components must be corrected by the liquid-phase activity coefficients according to eq. (1.4) and the relative volatility becomes

$$\alpha_{AB} = \frac{\gamma_A\, P_A^0}{\gamma_B\, P_B^0} \tag{1.6}$$

Noting that thermodynamic equilibrium requires equal activities of the components in both phases, the distribution coefficient K_A for equilibrium between two nonideal immiscible liquid phases 1 and 2 follows from the introduction of the activity $a_A = \gamma_A\, x_A$ in eq. (1.2):

$$K_A \equiv \frac{x_{A,1}}{x_{A,2}} = \frac{a_{A,1}/\gamma_{A,1}}{a_{A,2}/\gamma_{A,2}} = \frac{\gamma_{A,2}}{\gamma_{A,1}} \tag{1.7}$$

Hence now the inherent selectivity, for liquid–liquid extraction commonly denoted as β_{AB}, becomes

$$S_{AB} = \beta_{AB} = \frac{\gamma_{A,2}/\gamma_{A,1}}{\gamma_{B,2}/\gamma_{B,1}} \tag{1.8}$$

1.4.2 Rate-controlled processes

In adsorption processes a dynamic equilibrium is established for the distribution of the solute between the fluid and the solid surface. In Chapter 6, It is shown that at low amounts adsorbed, the isotherm should approach a linear form and the following form of Henry's law:

$$q_A = H_A\, p_A \qquad or \qquad q_A = H'_A\, c_A \tag{1.9}$$

where q is the equilibrium loading or amount adsorbed per unit mass of adsorbent, p is the partial pressure of a gas and c is the concentration of the solute. The *inherent selectivity of the adsorbent* is determined similarly to the inherent selectivity defined in the previous section. For small amounts of adsorbed material, the inherent adsorption selectivity simply corresponds to the ratio of the equilibrium Henry constants H_A/H_B (or H'_A/H'_B) which equals the ratio of adsorption constants b_A/b_B (see eq. (6.4)):

$$S_{AB} = \frac{q_A/p_A}{q_B/p_B} = \frac{H_A}{H_B} = \frac{q_{max}\cdot b_A}{q_{max}\cdot b_B} = \frac{b_A}{b_B} \tag{1.10}$$

The inherent selectivity for permeation through a barrier is defined in a similar way. By definition, the flux of a component through a membrane is proportional to its permeability, P_M (see Chapter 11). Under ideal circumstances the *inherent selectivity of permeation through a barrier* is then estimated by the ratios of the permeabilities $P_{M,A}$ and $P_{M,B}$:

$$S_{AB} = \frac{P_{M,A}}{P_{M,B}} \tag{1.11}$$

Nomenclature

a	Activity	–
b	Adsorption constant	see Ch. 6
c_A	Concentration of component A	mol m^{-3}
$c_{A,i}$	Concentration of component A in phase i	mol m^{-3}
K_A	Equilibrium ratio (eq. (1.2))	–
p	Partial pressure	–
p^o	Vapor (saturation) pressure	N m^{-2}
P_{tot}	Total pressure	N m^{-2}
P_M	Permeability	see Ch 11
Q	Equilibrium amount adsorbed	see Ch 6
S_{AB}	Inherent selectivity (eq. 1.3)	–
SF	Separation factor (eq. (1.1))	–
$x_{A,i}$	Mole fraction of component A in liquid phase I	–
x, y	Mole fraction of A in liquid and vapor phases, respectively	–
α_{AB}	Relative volatility (eq. (1.5))	–
	(=S_{AB} for ideal vapor–liquid systems)	
β_{AB}	Inherent selectivity in liquid–liquid extraction (eq. (1.8))	–
γ	Activity coefficient	–

Exercises

1 Compare and discuss the advantages and disadvantages of making separations using an energy separating agent (ESA) versus using a mass-separating agent (MSA).

2 The system benzene–toluene adheres closely to Raoult's law. The vapor pressures of benzene and toluene at 121 °C are 300 and 133 kPa.
   Calculate the relative volatility.

3 As a part of the life support system for spacecraft it is necessary to provide a means of continuously removing carbon dioxide from air. It is not possible to rely upon gravity in any way to devise a CO_2–air separation process.

Suggest at least two separation schemes, which could be suitable for continuous CO_2 removal from air under zero-gravity conditions.

4 Gold is present in seawater to a concentration level between 10^{-12} and 10^{-8} weight fraction, depending upon the location. Briefly evaluate the potential for recovering gold economically from seawater.

5 Assuming that the membrane characteristics are not changed, will the product–water purity in a reverse-osmosis seawater desalination process increase, decrease or remain constant as the upstream pressure increases.

6 Propylene and propane are among the light hydrocarbons produced by thermal and catalytic cracking of heavy petroleum fractions. Although propylene and propane have close boiling points, they are traditionally separated by distillation. Because distillation requires a large number of stages and considerable reflux and boil up flow rates compared to the feed flow, considerable attention has been given to the possible replacement of distillation with a more economical and less energy-intensive option.

Based on the given properties of both species, propose some alternative properties that can be exploited to enhance the selectivity of propylene and propane separation. What kind of separation processes are based on these alternative properties?

Property	Propylene	Propane
Molecular weight	42.08	44.10
Vd Waals volume (m^3/kmol)	0.0341	0.0376
Vd Waals area (m^2/kmol)	5.06	5.59
Acentric factor	0.142	0.152
Dipole moment (debye)	0.4	0.0
Radius of gyration ($m*10^{10}$)	2.25	2.43
Melting point (K)	87.9	85.5
Boiling point (K)	225.4	231.1
Critical temperature (K)	364.8	369.8
Critical pressure (MPa)	4.61	4.25

Chapter 2
Evaporation and Distillation

2.1 Separation by evaporation

2.1.1 Introduction

A large part of the separations of individual substances in a homogeneous liquid mixture or complete fractionation of such mixtures into their individual pure components is achieved through evaporative separations. Evaporative separations are based on the difference in composition between a liquid mixture and the vapor formed from it. This composition difference arises from differing effective vapor pressures, or *volatilities*, of the components in the liquid mixture. The required vapor phase is created by partial evaporation of the liquid feed through adding heat, followed by total condensation of the vapor. Due to the difference in volatility of the components, the feed mixture is separated into two or more products whose compositions differ from that of the feed. The resulting condensate is enriched in the more volatile components, in accordance with the vapor–liquid equilibrium (VLE) for the system at hand. When a difference in volatility does not exist, separation by simple evaporation is not possible.

The basis for planning evaporative separations is knowledge of the VLE. Technically, evaporative separations are the most mature separation operations. Design and operating procedures are well established. Only when VLE or other data are uncertain, a laboratory and/or pilot plant study is necessary prior to the design of a commercial unit. The most elementary form is *simple distillation* in which the liquid mixture is brought to boiling, partially evaporated and the vapor formed is separated and condensed to form a product. This technique is commonly used in the laboratory for the recovery and/or purification of products after synthesis in an experimental setup as illustrated in Figure 2.1.

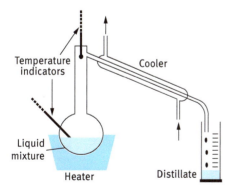

Figure 2.1: Laboratory distillation setup.

https://doi.org/10.1515/9783110654806-002

2.1.2 Vapor–liquid equilibria

It is clear that equilibrium distributions of mixture components in the vapor and liquid phases must be different if separation is to be made by evaporation. At thermodynamic equilibrium, the compositions are called VLE that may be depicted as in Figure 2.2. For binary mixtures, the effect of distribution of mixture components between the vapor and liquid phases on the thermodynamic properties is illustrated in Figure 2.3. Figure 2.3a shows a representative boiling point diagram with equilibrium compositions as functions of temperature at a constant pressure. Commonly, the more volatile (low boiling) component is used to plot the liquid and vapor compositions of the mixture. The lower line is the *liquid bubble point line*, the locus of points at which a liquid on heating forms the first bubble of vapor. The upper line is the *vapor dew point line*, representing points at which a vapor on cooling forms the first drop of condensed liquid. The region between the bubble and dew point lines is the *two-phase region*, where vapor and liquid coexist in equilibrium. At the equilibrium temperature T_{eq}, the liquid with composition x_{eq} is in equilibrium with vapor composition y_{eq}. Figure 2.3b displays a typical y–x diagram that is obtained by plotting the vapor composition that is in equilibrium with the liquid composition at a fixed pressure or a fixed temperature. At equilibrium, the concentration of any component i present in the liquid mixture is related to its concentration in the vapor phase by the equilibrium ratio as defined in eq. (1.2), also called the *distribution coefficient K_i*:

$$K_i \equiv \frac{y_i}{x_i} \tag{2.1}$$

where y_i is the mole fraction of component i in the vapor phase and x_i the mole fraction of component i in the liquid phase. The distribution coefficient is a function of

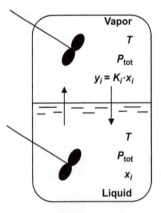

Figure 2.2: Schematic vapor–liquid equilibrium.

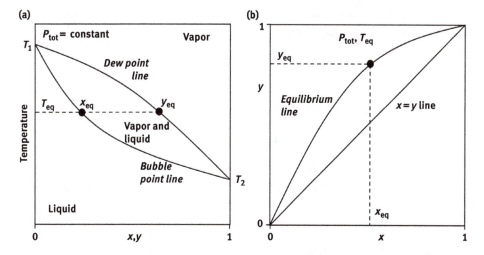

Figure 2.3: Isobaric vapor–liquid equilibrium diagrams: (a) dew and bubble point and (b) y–x diagram.

temperature and pressure only, not of compositions. The more volatile components in a mixture will have the higher values of K_i, whereas less volatile components will have lower values of K_i. The key separation factor in distillation is the selectivity of a component relative to a reference component. This *selectivity* in VLE usually referred to as *relative volatility* is defined in eq. (1.3) as follows:

$$\alpha_{ij} = \frac{K_i}{K_j} = \frac{y_i/x_i}{y_j/x_j} \qquad (2.2)$$

where component i is the more volatile one. The higher the value of the relative volatility, the more easily components may be separated by distillation. In ideal systems, the behavior of vapor and liquid mixtures obeys Dalton's and Raoult's laws. *Dalton's law* relates the concentration of a component present in an ideal gas or vapor mixture to its partial pressure:

$$p_i = y_i P_{tot} \qquad (2.3)$$

where p_i is the partial pressure of component i in the vapor mixture and P_{tot} is the total pressure of the system given by the sum of the partial pressures of all components in the system:

$$P_{tot} = \sum_{i=1}^{N} p_i \qquad (2.4)$$

Raoult's law relates the partial pressure of a component in the vapor phase to its concentration in the liquid phase, x_i:

$$p_i = x_i\, P_i^o \tag{2.5}$$

where P_i^o is the vapor pressure of pure component i at the system temperature (saturation pressure). Combining eqs. (2.3) and (2.5) yields:

$$y_i\, P_{tot} = x_i\, P_i^o \tag{2.6}$$

This results in the following relations between the pure component vapor pressure, P_i^o, its distribution coefficient, K_i, and the relative volatility of two components in an ideal mixture, α_{ij}:

$$K_i = \frac{y_i}{x_i} = \frac{P_i^o}{P_{tot}} \qquad \text{and} \qquad \alpha_{ij} = \frac{K_i}{K_j} = \frac{P_i^o}{P_j^o} \tag{2.7}$$

The latter equation shows that for ideal systems, the relative volatility is independent of pressure and composition. For a binary system with components A and B, where $y_A = 1 - y_B$ and $x_A = 1 - x_B$, the relative volatility equation, eq. (2.2), can be rearranged to give

$$\frac{y}{1-y} = \alpha \cdot \frac{x}{1-x} \qquad \text{or} \qquad y = \frac{\alpha x}{1 + (\alpha - 1)\, x} \tag{2.8}$$

where x and y represent the mole fractions of the more volatile component A. This equation is used to express the concentration of a component in the vapor as a function of its concentration in the liquid and relative volatility. This function is plotted in Figure 2.4 for various values of relative volatility. When relative volatility increases, the concentration of the most volatile component in the vapor increases. When the relative volatility is equal to 1, the concentrations of the most volatile component in the liquid and vapor phases are equal and a vapor/liquid separation is not feasible. At very low concentrations, eq. (2.8) reduces to a linear relation,[1] represented as tangents in Figure 2.4:

$$y = a \cdot x \tag{2.9}$$

Since vapor pressures of components depend on temperature, equilibrium ratios are a function of temperature. However, the relative volatility considerably less sensitive to temperature changes because it is proportional to the *ratio* of the vapor pressures. In general, the vapor pressure of the more volatile component tends to increase at a

---

[1] See also page 25.

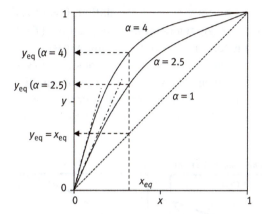

Figure 2.4: Vapor–liquid equilibrium compositions as a function of relative volatility.

slower rate with increasing temperature than the less volatile component. Therefore, the relative volatility generally decreases with increasing temperature and increases with decreasing temperature.

Most liquid mixtures are nonideal and require Raoult's law to be modified by including a correction factor called the liquid phase activity coefficient:

$$p_i = \gamma_i x_i P_i^o \tag{2.10}$$

Although at high pressures, the vapor phase may also depart ideal vapor mixture, the common approach in distillation is to assume ideal vapor behavior and to correct nonideal liquid behavior with the liquid phase activity coefficient. The standard state for reference for the liquid phase activity coefficient is commonly chosen as $\gamma_I \to 1$ for the pure component. The liquid phase activity coefficient is strongly dependent upon the composition of the mixture. Positive deviations from ideality ($\gamma_I > 1$) are more common when the molecules of different compounds are dissimilar and exhibit repulsive forces. Negative deviations ($\gamma_I < 1$) occur when there are attractive forces between different compounds that do not occur for either component alone. For nonideal systems the relations for the distribution coefficient and relative volatility as given in eq. (2.7) now become:

$$K_i = \frac{\gamma_i P_i^o}{P_{tot}}, \qquad \alpha_{ij} = \frac{K_i}{K_j} = \frac{\gamma_i P_i^o}{\gamma_j P_j^o} \tag{2.11}$$

In nonideal systems, the distribution coefficients and relative volatility are dependent on composition because of the composition dependence of the activity coefficients. When the activity coefficient of a specific component becomes high enough, an *azeotrope* may be encountered, meaning that the vapor and liquid compositions are equal and the components cannot be separated by conventional distillation. Figure 2.5 shows binary vapor–liquid composition (x–y), temperature–composition (T–x) and pressure–composition (P–x) diagrams for nonazeotrope, minimum azeotrope and maximum azeotrope systems. A *minimum boiling azeotrope* boils at a lower temperature than each of the components in their pure states. When separating the components of this type of system by distillation, such as ethanol–water, the overhead product is the azeotrope. A *maximum boiling azeotrope* boils higher than

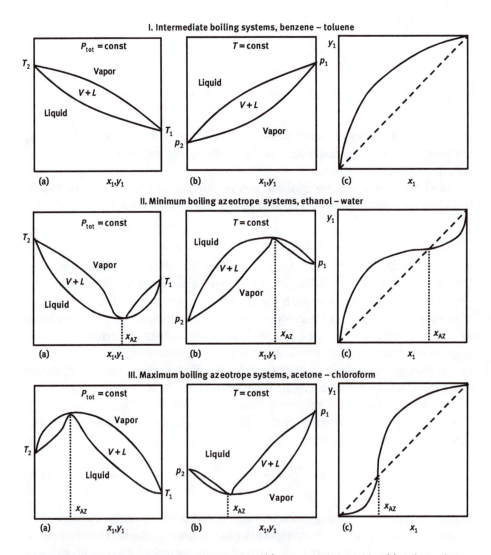

Figure 2.5: Types of binary temperature–composition (a), pressure–composition (b) and x–y phase diagrams for vapor–liquid equilibrium (c).

either component in their pure states and is the bottom product of distillation. An example of this type of system is acetone–chloroform.

Vapor pressures for many compounds are published in literature (see page 50) and often correlated as a function of temperature by the *Antoine equation*:

$$\ln P_i^o = A_i - \frac{B_i}{T + C_i} \tag{2.12}$$

where A_i, B_i and C_i are Antoine constants of component i, and T represents the temperature. Unfortunately, in some reference books, these data are given in non-SI units such as pressure in mmHg and, sometimes, temperature in °F. Then appropriate conversions are in order.

2.2 Separation by single-stage partial evaporation

2.2.1 Differential distillation

Two main modes are utilized for single-stage separation by partial evaporation. The most elementary form is *differential distillation*, in which a liquid is charged to the still pot and heated to boiling. As illustrated in Figure 2.6, the method has a strong resemblance with laboratory distillation, shown previously in Figure 2.1. The vapor formed is continuously removed and condensed to produce a distillate.

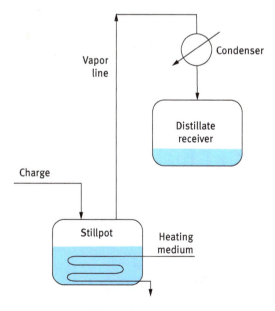

Figure 2.6: Simple differential distillation.

Usually the vapor leaving the still pot with composition y_D is assumed to be in equilibrium with perfectly mixed liquid in the still at any instant. The distillate is richer in the more volatile components and the residual unvaporized bottoms are richer in the less volatile components. As the distillation proceeds, the relative amount of volatile components composition of the initial charge and distillate decrease with time. Because the produced vapor is totally condensed, $y_D = x_D$, there is only one single equilibrium stage, the still pot. Simple differential distillation is not widely used in industry, except for the processing of high-valued chemicals in small production quantities or for distillations requiring regular sanitation.

2.2.2 Flash distillation

Flash distillation is the continuous form of simple single-stage equilibrium distillation. In a flash, a continuous feed is partially vaporized to give a vapor richer in the more volatile components than the remaining liquid. Because of the single stage, usually only a limited degree of separation is achieved. However, in some cases, such as the seawater desalination, the large volatility difference between the components results in complete separation. A schematic diagram of the equipment for flash distillation is shown in Figure 2.7. The pressurized liquid feed is heated and either flashed adiabatically across a nozzle to a lower pressure or, without the valve, partially vaporized isothermally. Either way, the created vapor phase is separated from the remaining liquid in a flash drum. A demister or

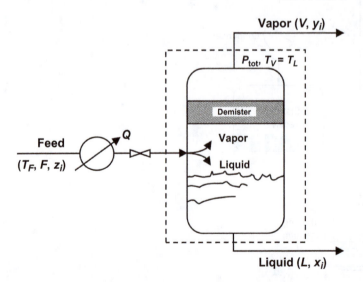

Figure 2.7: Flash drum: adiabatic flash distillation with valve or isothermal flash without valve.

entrainment eliminator is often employed to prevent liquid droplets from being entrained in the vapor. When properly designed, the system in the flash chamber behaves very close to an equilibrium stage ($T_V = T_L$) due to the intimate contact between liquid and vapor.

The designer of a flash system needs to know the liquid and vapor compositions that correspond to the operating pressure and temperature in the flash drum. The design calculations[2] are based on the mass balances for the balance envelope shown as a dashed line in Figure 2.7 and on VLE data; x, y and z represent the compositions (mole fractions) of the liquid, vapor and feed, respectively.

In a binary system, the two independent mass balances are the overall mass balance and the component balance for the more volatile component:

$$F = V + L \quad \text{and} \quad Fz = Vy + Lx \tag{2.13}$$

By introducing the liquid vapor ratio L/V, the latter two equations reduce to the dimensionless operating line equation:

$$y = -\frac{L}{V}x + \left(1 + \frac{L}{V}\right)z \tag{2.14}$$

Equilibrium data may be represented by relations such as given in eq. (2.1), which take the form in case of a binary system with components A and B:

$$y = K_A x \quad \text{and} \quad 1 - y = K_B (1 - x) \tag{2.15}$$

By combining these two equilibrium line may be expressed equally well in terms of relative volatility:

$$y = \frac{\alpha x}{1 + (\alpha - 1)x} \quad \text{with} \quad \alpha = K_A / K_B \tag{2.16}$$

In case of ideal gas and liquid phases, the relative volatility α is the VL-equilibrium constant, also referred to as K (see, e.g., eq. (3.21) and (4.5)).

In case of an *isothermal flash* no valve is used while the specified temperature is set through the heat exchanger. Gibbs' phase rule[3] states that in a binary two-phase system, *two* degrees of freedom exist; thus, one more parameter can be chosen freely. Given the feed composition, z, one option to define the system would be to specify temperature and liquid–vapor ratio, another option being the specification of temperature and total pressure. In the following two sections, we will show the choice of the relevant set of equations and how to solve these.

---

2 Estimation of the drum size related to the liquid and vapor fluxes is outside the scope of this textbook.

3 Degrees of freedom = number of components − number of phases + 2.

2.2.2.1 Specifying T and L/V

In this case, the total pressure is not known beforehand and, noting that K_A and K_B are inversely proportional to P_{tot}, eq. (2.16) should be used as equilibrium line. Vapor and liquid compositions follow from solving eqs. (2.8) and (2.14) with known values of α and L/V. The partial pressures are calculated from eq. (2.5) and the total pressure from eq. (2.4).

A graphical solution is very convenient since the simultaneous solution of eqs. (2.16) and (2.14) is represented by the intersection of the equilibrium curve and the operating line. Instead of the vapor–liquid ratio, often the *fraction feed remaining liquid q* and *fraction feed vaporized 1 – q* are utilized:

$$q \equiv \frac{L}{F} \quad \text{and} \quad (1-q) = \frac{V}{F} \tag{2.17}$$

The operating line equation now becomes

$$y = -\frac{q}{1-q} x + \frac{1}{1-q} z \tag{2.18}$$

The x–y diagram in Figure 2.8 shows an example for a benzene–water separation, three straight operating lines and the equilibrium line[4] according to eq. (2.16). Each operating line represents a fraction of feed evaporated. The intersection points give the vapor and liquid compositions, y_{eq} and x_{eq}, leaving the flash drum as the fraction of feed evaporated $(1 - q)$ varies from 0.001 to 0.002 to 0.005.

2.2.2.2 Specifying T and P_{tot}

Setting temperature and total pressure defines the liquid–vapor system in the flash drum. This is shown in Figure 2.3a, where at a given T and P_{tot} the compositions of the coexisting phases, x_{eq} and y_{eq}, are fixed. In this case, q (hence the slope of the operating line) is unknown beforehand; hence, the solution procedure now differs from that outlined in the previous section. However, three boundary conditions can be defined:

- the feed composition, z, must fall between x_{eq} and y_{eq} to assure that a VLE exists at the given temperature and pressure,
- in a vapor–liquid system q is be bounded between zero and unity, and
- the condition that the sum of mole fractions in each phase equals unity, but in the form $\Sigma x_i = \Sigma y_i$.

Now, the usual numerical procedure is to first find q by elimination of x and y from the operating line (eq. (2.14)) and equilibrium line equation (eq. (2.15)), noting that at given T and P_{tot} the distribution coefficients K_i are known constants.

---

4 Almost a straight line because of the very low values of the benzene mole fraction in the mixture.

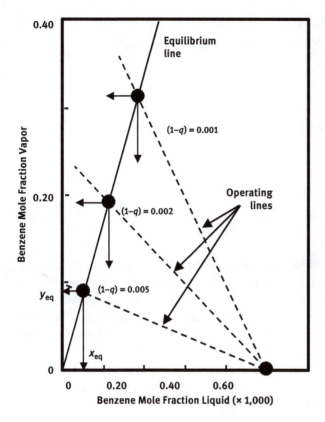

Figure 2.8: Graphical determination of the equilibrium benzene mole fraction in the vapor and liquid from a flash drum for several $(1 - q)$ values.

In a multicomponent mixture $y_i = K_i \cdot x_i$ and the operating line equation becomes

$$x_i K_i = -\frac{q}{1-q}x_i + \frac{z_i}{1-q} \qquad \text{or} \qquad x_i = \frac{z_i}{K_i + (1 - K_i)q} \qquad (2.19)$$

But

$$y_i = K_i x_i = \frac{K_i z_i}{K_i + (1 - K_i)q} \qquad (2.20)$$

and combination of eqs. (2.19) and (2.20) with the boundary condition $\Sigma x_i = \Sigma y_i$ leads to

$$\sum_i \frac{z_i(1 - K_i)}{K_i + (1 - K_i)q} = 0 \qquad (2.21)$$

This procedure for a multicomponent mixture thus leads to a single equation, which can be solved for the one unknown q. Once q is found between zero and unity, all x_i and y_i are calculated from eqs. (2.19) and (2.20).

Solving eq. (2.21) for q in a *binary* system ($z_A = z$ and $z_B = 1 - z$) and substitution of the result in eq. (2.19) gives

$$q = -z\frac{K_B}{1 - K_B} - (1 - z)\frac{K_A}{1 - K_A} \quad \text{and} \quad x = \frac{1 - K_B}{K_A - K_B} \tag{2.22}$$

Note that K_A and K_B are functions of temperature and total pressure.

This expression for x can also be obtained by elimination of y and α from eqs. (2.6), (2.7) and (2.8).

2.3 Multistage distillation

2.3.1 Distillation cascades

Flash distillation is a very simple unit operation that in most cases produces only a limited amount of separation. Increased separation is possible in a cascade of flash separators that produce one pure vapor and one pure liquid product. Within the cascade, the intermediate product streams are used as additional feeds. The liquid streams are returned to the previous flash drum, while the produced vapor streams are forwarded to the next flash chamber. Figure 2.9 shows the resulting *countercurrent cascade*, because vapor and liquid streams go in opposite directions. The advantages of this cascade are that there are no intermediate products and the two end products can both be pure and obtained in high yield.

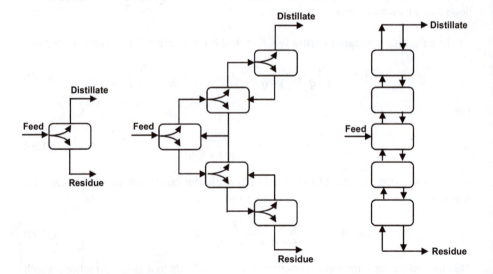

Figure 2.9: Flash drum cascades.

Although a significant advance, this multiple flash drum system is seldom used indus-trially. Operation and design is easier if part of the top vapor stream is condensed and returned to the first stage, *reflux*, and if part of the bottom liquid stream is evaporated and returned to the bottom stage, *boil up*. This allows control of the internal liquid and vapor flow rates at any desired level by applying all of the heat required for the distillation to the bottom reboiler and do all the required cooling in the top condenser. Partial condensation of intermediate vapor streams and partial vaporization of liquid streams is achieved by heat exchange between all pairs of passing streams. This is most effectively achieved by building the entire system in a column instead of the se-ries of individual stages shown in Figure 2.9. Intermediate heat exchange is the most efficient with the liquid and vapor in direct contact. The final result is a much simpler and cheaper device, the *distillation column* shown in Figure 2.10.

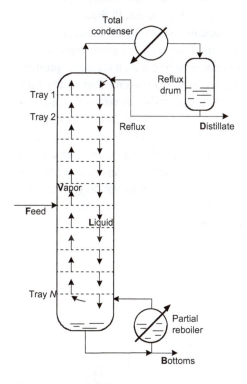

Figure 2.10: General flowchart of a plate distillation column.

2.3.2 Column distillation

Most commercial distillations involve some form of multiple staging in order to obtain better separation than is possible by single vaporization and condensation. It is the most widely used industrial method of separating liquid mixtures in the chemical process industry. As shown by Figure 2.10, most multistage distillations are continuously operated column-type processes separating components of a liquid mixture according to their different boiling points in a more volatile distillate and a less volatile bottoms or residue. The feed enters the column at the equilibrium feed stage. Vapor and liquid phases flow countercurrently within the mass transfer zone of the column where trays or packings are used to maximize interfacial contact between the phases. The section of column above the feed is called the *rectification section* and the section below the feed is referred to as the *stripping section*. Although Figure 2.10 shows a distillation column equipped with only 12 sieve trays, industrial columns may contain more than 100. The liquid from a tray flows through a *downcomer* to the tray below and vapor flows upward through the holes in the sieve tray to the tray above. Intimate contact between the vapor and liquid phases is created as the vapor passes through the holes in the sieve tray and bubbles through the pool of liquid residing on the tray. The vapors moving up the column from equilibrium stage to equilibrium stage are increasingly enriched in the more volatile components. Similarly, the concentration of the least volatile components increases in the liquid from each tray going downward.

The overhead vapor from the column is condensed to obtain a *distillate* product. The liquid distillate from the condenser is divided into two streams. Part of it is withdrawn as overhead product (D) while the remaining distillate is refluxed (L_o) to the top tray to enrich the vapors. The *reflux ratio* is defined as the ratio of the reflux rate to the rate of product removal. The required heat for evaporation is added at the base of the column in a *reboiler*, where the bottom tray liquid is heated and partially vaporized to provide the vapor for the stripping section. Plant-size distillation columns usually employ external steam-powered kettle or vertical thermosyphon-type heat exchangers (Figure 2.11). Both can provide the amount of heat transfer surface required for large installations. Thermosyphon reboilers are favored when the bottom product

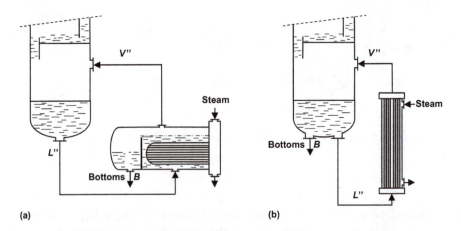

(a) (b)

Figure 2.11: Industrial reboilers: (a) kettle type and (b) vertical thermosyphon type.

contains thermally sensitive compounds, only a small temperature difference is available for heat transfer and heavy fouling occurs. The vapor from the reboiler is sent back to the bottom tray and the remaining liquid is removed as the bottom product.

2.3.3 Feasible distillation conditions

Industrial distillation processes are restricted by operability of the units, economic conditions and environmental constraints. An upper limit exists for feasible operating temperatures. One reason is the *thermal stability* of the species in the mixtures to be separated. Many substances decompose at higher temperatures and some species are not even stable at their normal boiling points. A second reason for a maximum temperature limitation is the means of heat supply. In most cases, the required energy is supplied by condensing steam. The pressure of the available steam places an upper limit on temperature levels that can be achieved. Only in special cases (crude oil, sulfuric acid), higher temperatures may be realized by heating with hot oil or natural gas burners. With high-pressure steam, the maximum attainable temperatures are limited to 300 °C, which can be raised to temperatures as high as 400 °C when other media are used. The temperature of the coolant for the overhead condenser dictates the lower limit of feasible temperatures in distillation columns. In most cases, water is used, resulting in a minimum temperature in the column of 40–50 °C.

In most cases, temperature constraints can be met by selecting a proper *operating pressure*. Operating temperatures can be decreased by the use of a vacuum. It is technically feasible to operate distillation columns at pressures up to 2 mbar. Lower pressures are seldom used because of the high operating and capital costs of the vacuum-producing equipment. A column can be operated at higher pressure to increase the boiling point of low boiling mixtures. The upper limit for the operating pressure lies in the range of the critical pressures of the constituents. In addition to temperature and pressure, the feasible number of theoretical stages of industrial columns is also limited. Only in exceptional cases, commercial distillation columns are constructed that contain more than 100 stages.

2.3.4 External column balances

Once the separation problem has been specified, the task of the engineer is to calculate the design variables. The first step is the development of mass balances around the entire column using the *balance envelope* shown by the dashed outline in Figure 2.12. The aim of these balances is to calculate flow rates and compositions for the distillate and bottom products, D and B, for the given problem specification. Analogously to the flash calculations in Section 2.2.2, the overall mass balance becomes

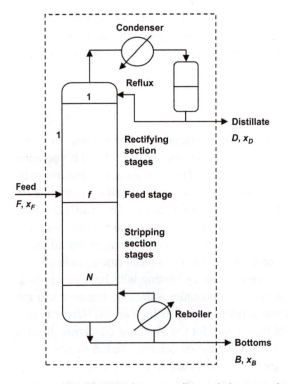

Figure 2.12: Distillation column overall mass balance envelope.

$$F = D + B \tag{2.23}$$

In binary distillation, the specification of x_D and x_B for the more volatile component fixes the distillate and bottoms flow rates D and B through the additional overall component material balance:

$$x_F F = x_D D + x_B B \tag{2.24}$$

2.4 McCabe–Thiele analysis

2.4.1 Internal balances

If the molar heat of vaporization for any component has approximately the same value, condensation of 1 mol of vapor will vaporize 1 mol of liquid. Thus, liquid and vapor flow rates tend to remain approximately constant as long as heat losses from the column are negligible and the pressure is uniform throughout the column. This simplified situation is closely approximated for many distillations; *constant molar vapor and liquid flow* may be assumed in each section of the distillation column, given a value for the distillate flow D. For constant molar flow conditions, it is not necessary to consider energy balances in either the rectifying or stripping sections. Only material balances and a VLE curve are required. Based on these assumptions, the *McCabe–Thiele* method provides a graphical solution to the material balances and equilibrium relationships inside the column. In this chapter, the graphical McCabe–Thiele method is used to visualize the essential aspects of multistage distillation calculations required to understand the fundamentals behind modern, computer-aided design methods. In addition to the number of equilibrium stages N, also the required minimum number of stages N_{min}, minimal reflux ratio R_{min} and optimal stage for feed entry are derived. Finally, the overall energy balances are derived to determine condenser and reboiler duties.

2.4.1.1 Rectifying section

As illustrated in Figure 2.13, the rectifying section extends from the top stage 1 to just above the feed stage f. For the indicated balance envelope, the material balance for the more volatile component over the total condenser and a top portion of rectifying stages 1 to n is written as

$$V'_{n+1} y_{n+1} = L'_n x_n + D x_D \qquad (2.25)$$

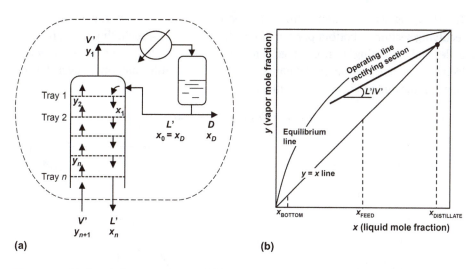

(a)   **(b)**

Figure 2.13: (a) Mass balance envelope and (b) operating line for the rectifying section.

Since V' and L' are constant throughout the rectifying section, eq. (2.25) can be rewritten as:

$$y_{n+1} = \left(\frac{L'}{V'}\right) x_n + \left(\frac{D}{V'}\right) x_D \qquad (2.26)$$

Equation (2.22) is the *operating line* in the rectifying section. It relates the concentrations y_{n+1} and x_n of the two passing streams V' and L' in the column, and thus represents the mass balances in the rectifying section. The slope of the operating line is L'/V', which is constant and <1 because in the rectifying section $V' > L'$. The liquid entering the top stage is the external reflux rate, L', and its ratio to the distillate rate, L'/D, is defined as the reflux ratio R. Since $V' = L' + D$, the slope of the operating line is readily related to the reflux ratio:

$$\frac{L'}{V'} = \frac{L'}{L' + D} = \frac{R}{R + 1} \quad \text{and} \quad \frac{D}{V'} = \frac{D}{D + L'} = \frac{1}{R + 1} \tag{2.27}$$

Substituting into eq. (2.26) produces the most useful form of the operating line for the rectifying section:

$$y_{n+1} = \left(\frac{R}{R+1}\right) x_n + \left(\frac{1}{R+1}\right) x_D \tag{2.28}$$

As shown in Figure 2.13, the operating line plots as a straight line in the yx-diagram with an intersection at $y = x_D$ on the $y = x$ line, for specified values of R and x_D.

2.4.1.2 Stripping section

The stripping section extends from the feed to the bottom stage. Analogous to the rectifying section, a bottom portion of the stripping stages, including the partial reboiler and extending up from stage N to stage $m + 1$, located somewhere below the feed is considered. A material balance for the more volatile component over the envelope indicated in Figure 2.14 results in

$$L''_m x_m = V''_{m+1} y_{m+1} + B x_B \tag{2.29}$$

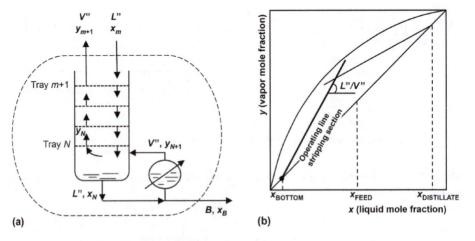

Figure 2.14: (a) Mass balance envelope and (b) operating line for the stripping section.

where the constant-molar-flow conditions rearrange into the *operating line* representing the mass balance of the passing streams V'' and L'' in the stripping section:

$$y_{m+1} = \left(\frac{L''}{V''}\right) x_m - \left(\frac{B}{V''}\right) x_B \tag{2.30}$$

The slope of this operating line for the stripping section is seen to be >1 because $L'' > V''$. This is the reverse of the conditions in the rectifying section. For known values of L'', V'' and x_B, eq. (2.30) can be plotted as a straight line with an intersection at $y = x_B$ on the $y = x$ line and a slope of L''/V'', as shown in Figure 2.14.

2.4.1.3 Feed stage considerations

Thus far we have not considered the feed to the column. In the rectifying and stripping sections of the column, the operating line represents the mass balances. When material is added or withdrawn from the column, the mass balances will change and the operating lines will have different slopes and intercepts. The addition of the feed increases the liquid flow in the stripping section, increases the vapor flow in the rectifying section, or does both. In the rectifying section, the vapor flow is greater than the liquid flow rate. In the stripping section, the liquid flow is greater than the vapor flow rate. Obviously, the phase and temperature of the feed affect the vapor and liquid flow rates in the column. For instance, if the feed is liquid, the liquid flow rate below the feed stage must be greater than the liquid flow above the feed stage, $L'' > L'$. Similarly V' should be larger than V'' if the feed is a vapor. These effects can be quantified by writing mass and energy balances around the feed stage shown schematically in Figure 2.15a. The overall mass balance for the balance envelope is

$$F + V'' + L' = L'' + V' \tag{2.31}$$

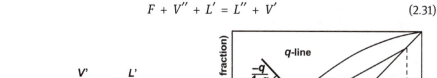

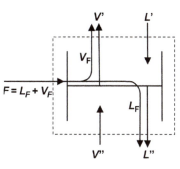

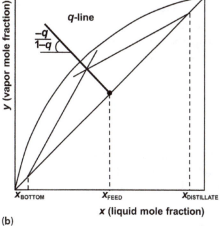

(a) (b)

Figure 2.15: Mass balance envelope (a) and q-line (b) for the feed stage.

and the related balance over just the liquid phase reads:

$$L_F = L'' - L' \equiv q \cdot F \tag{2.32}$$

where q is defined as the liquid fraction in the feed at feed stage temperature, similar to the q-factor defined in Section 2.2.2. Under the condition that the feed is at feed stage temperature, q represents the moles of the liquid flow in the stripping section that result from the introduction of each mole of feed. Assuming that the molar vapor enthalpies h_V nor the molar liquid enthalpies h_L vary much with composition, the energy balance across the feed stage equals

$$F h_F + V'' h_V + L' h_L = L'' h_L + V' h_V \tag{2.33}$$

Rewriting the mass balance equation (2.31) in terms of $V''-V'$ and substituting into the energy balance gives

$$F h_F + (L'' - L') h_V - F h_V = (L'' - L') h_L \tag{2.34}$$

which is easily rearranged to yield

$$q \equiv \frac{L'' - L'}{F} = \frac{h_V - h_F}{h_V - h_L} \tag{2.35}$$

Hence, the liquid feed fraction equals the ratio of the heat required to completely evaporate a mole of feed $h_V - h_F$ and the molar heat of evaporation $h_V - h_L$. Since the liquid and vapor enthalpies can be estimated, we can easily calculate q from eq. (2.35) for the condition that the feed consists of a partially evaporated liquid and vapor mixture ($1 > q > 0$). The feed may also be a saturated liquid at its bubble point ($q = 1$), or a saturated vapor at its dew point ($q = 0$).

The energy balance in eq. (2.33) does not comprise a term to bring the feed to the feed stage temperature. Thus, if the feed is introduced as cold liquid ($q > 1$) or as a superheated vapor ($q < 0$), the value of q calculated from eq. (2.35) should be corrected for the heat involved.[5] The required correction is outside the scope of this textbook but can be found in the literature references.

2.4.1.4 Feed line
After quantifying the value for q, the changes in liquid and vapor flow rates due to the introduction of the feed on the feed stage are calculated. The contribution of the feed stream to the internal flow of liquid is $q \cdot F$ (eq. (2.32)), so the total flow rate of liquid and vapor in the stripping section becomes

---

5 We will denote q-values subject to this temperature correction as q'. For small temperature differences, $q' \approx q$.

$$L'' = L' + qF \quad \text{and} \quad V'' = V' - (1-q)F \tag{2.36}$$

Although with these values for L'' and V'' in eq. (2.30) can be used to locate the stripping operating line, it is more common to use an alternative method that involves the *feed-line* or *q-line*. The intersection point of the rectifying and stripping operating lines is derived by writing the mass balance equations for both sections for constant molar flow:

$$V'y = L'x + Dx_D \quad \text{and} \quad V''y = L''x - Bx_B \tag{2.37}$$

Subtracting both equations gives

$$y(V' - V'') = x(L' - L'') + Dx_D + Bx_B = x(L' - L'') + Fx_F \tag{2.38}$$

which can be rearranged into

$$y = \frac{L'' - L'}{V'' - V'}x - \frac{F}{V'' - V'}x_F \tag{2.39}$$

Substituting $L''-L'$ and $V''-V'$ from eq. (2.36) and some rearrangements results in the equation for the *feed-line* or *q-line*:

$$y = -\frac{q}{1-q}x + \frac{1}{1-q}x_F \tag{2.40}$$

This equation represents a straight line of slope $-q/(1-q)$ passing through the point (y_F,x_F) on which every possible intersection point of the two operating lines for a given feed must fall. The position of the line depends only on x_F and q'. Thus, by changing the reflux ratio, the point of intersection is changed but remains on the q-line. From the definition of q', it follows that the slope of the q-line is governed by the nature of the feed. Examples of the various types of feeds and the slopes of the q-line are illustrated in Figure 2.16. Note that all the feed lines intersect at one point, which is at $y = x = x_F$. Following the placement of the rectifying section operating line and the q-line, the stripping section operating line is located by drawing a straight line from the point $(y = x_B, x = x_B)$ on the $y = x$ line through the point of intersection with the q-line and the rectifying section operating line as shown in Figure 2.15.

2.4.2 Required number of equilibrium stages

2.4.2.1 Graphical determination of stages and location of feed stage
Once the equilibrium line, operating lines and q-line have been plotted, we can continue with the determination of the number of equilibrium stages required for the entire column and the location of the feed stage. The operating lines relate the solute concentration in the vapor passing upward between two stages to the solute

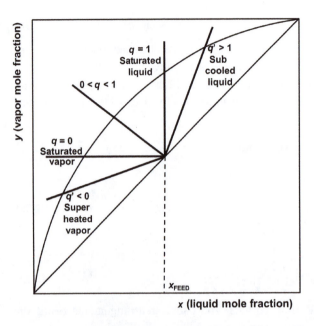

Figure 2.16: Dependence of q-line on feed condition (for q' see footnote 5 on page 36).

concentration in the liquid passing downward between the same two stages. The equilibrium curve relates the solute concentration in the vapor leaving an equilibrium stage to the solute concentration in the liquid leaving the same stage. This makes it possible to determine the required number of stages and the location of the feed stage by *constructing a staircase* between the operating line and the equilibrium curve, as shown in Figure 2.17.

Although the construction can start either from the top, the bottom or from the feed stage, it is common that the staircase is stepped of from the top and continued all the way to the bottom. Starting from the point ($y_1 = x_D$, $x_0 = x_D$) on the rectifying section operating line and the $y = x$ line, a horizontal line is drawn to the left until it intersects the equilibrium curve at (y_1, x_1), the compositions of the *equilibrium phases* leaving the top equilibrium stage. A vertical line is now dropped until it intersects the operating line at the point (y_2, x_1), the compositions of the *two phases passing each other* between stages 1 and 2. The horizontal and vertical line construction is continued down the rectifying section until the feed stage is reached. At the feed stage, we switch to the stripping section operating line because of the different operating line in the stripping section due to the addition of feed. Although the separation shown in Figure 2.17 would require exactly six equilibrium contacts (five equilibrium stages plus an equilibrium partial reboiler), in practice hardly ever an integer number of stages result, but rather a fractional stage will appear at the top or at the bottom. Starting the staircase construction from the feed stage to both the top and the bottom stages, at either side generally a fraction of a theoretical

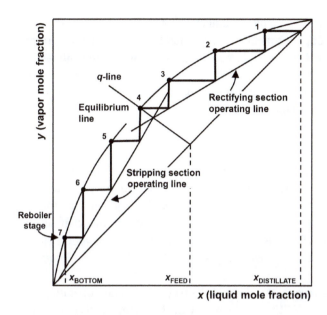

Figure 2.17: McCabe–Thiele graphical determination of the number of equilibrium stages.

stage will be found (see Figure 2.18a). Either fraction of a theoretical stage can be determined from

$$\text{fraction} = \frac{\text{distance from operating line to } x_B}{\text{distance from operating line to equilibrium line}} \qquad (2.41)$$

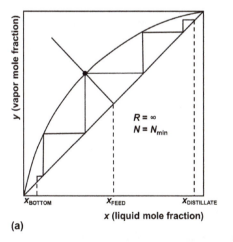

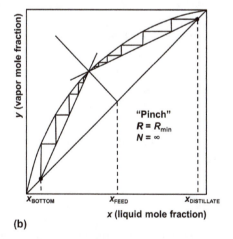

(a)

(b)

Figure 2.18: Determination of (a) minimum number of theoretical stages and (b) minimum reflux ratio in a McCabe–Thiele diagram.

In Figure 2.17, it is seen that the smallest number of total stages occurs when the transfer is made at the first opportunity after a horizontal line of the staircase crosses the q-line. This is the *optimal feed stage location* because the separation will require the fewest total number of stages when feed stage 3 is used. For binary distillation, the optimum feed stage will always be the stage where the step in the staircase includes the point of intersection of the two operating lines.

2.4.2.2 Limiting conditions

For any distillation operation, there are infinite combinations of reflux ratios and numbers of theoretical stages possible. The larger the reflux ratio, the fewer theoretical stages are required but the more energy is consumed. For a given combination of feed, distillate and bottom compositions, there are two constraints that set the boundary conditions within which the reflux ratio and number of theoretical stages must be. The *minimum number of theoretical stages*, N_{min}, and the *minimum reflux ratio*, R_{min}.

As the reflux ratio is increased, the slope of the rectifying section operating line increases to a limiting value of $L'/V' = 1$, meaning that the system is at total reflux. Correspondingly, the slope of the stripping section operating line decreases to a limiting value of $L''/V'' = 1$. The minimum number of theoretical stages occurs when both operating lines coincide with the $y = x$ line and neither the feed composition nor the q-line influences the staircase construction. An example of the McCabe–Thiele construction for this limiting condition is shown in Figure 2.18a. Because the operating lines are located as far away as possible from the equilibrium curve, a minimum number of stages is required.

Column operation with minimum internal gas and liquid flow (i.e., minimum reflux) separates a mixture with the lowest energy input. As the reflux ratio decreases, the intersection of the two operating lines and the q-line moves from the $y = x$ line toward the equilibrium curve. The number of equilibrium stages required increases because the operating lines move closer to the equilibrium curve. Finally, a limiting condition is reached when the point of intersection is on the equilibrium curve, as shown in Figure 2.18b. To reach that stage from either the rectifying section or the stripping section, an infinite number of stages would be required to achieve the desired separation. The corresponding *minimum reflux ratio* is easily determined from the slope of the limiting operating line $(L'/V')_{min}$ for the rectifying section:

$$R_{min} = \left(\frac{L'}{D}\right)_{min} = \left(\frac{L'}{V'-L'}\right)_{min} = \frac{(L'/V')_{min}}{1 - (L'/V')_{min}} \tag{2.42}$$

Both of these limits, the minimum number of stages and the minimum reflux ratio serve as valuable guidelines within which the practical distillation conditions must lie. The operating, fixed and total cost of a distillation system are a strong function of the relation of the operating reflux ratio to the minimum reflux ratio. As shown by Figure 2.19, first, the fixed cost decreases by increasing the reflux ratio because fewer stages are required but then rises again as the diameter of the column increases at higher vapor and liquid loads. Similarly, the operating cost for energy increases almost linearly as the operating reflux ratio increases. For most commercial operations, the *optimal operating reflux ratios* are in the range of 1.1–1.5 times the minimum reflux ratio.

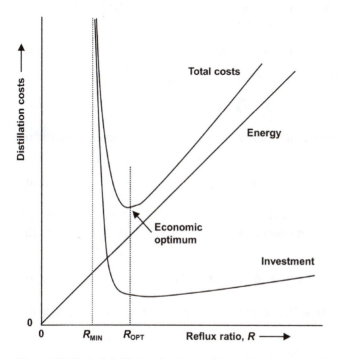

Figure 2.19: Typical distillation fixed, operating and total costs as a function of reflux ratio.

2.4.2.3 Fenske and Underwood equations

Alternative to the graphical determination, approximate values for the minimum number of stages and minimum reflux can also be obtained from the Fenske and Underwood equations, respectively. Both equations are based on the special situation that *the relative volatility remains approximately constant throughout the column*. The case of the minimum number of theoretical stages, N_{min}, is achieved at total reflux, as discussed in the previous section. Total reflux exists when no product is withdrawn:

$$D = B = 0 \tag{2.43}$$

hence

$$V' - L' = L'' - V'' = 0 \tag{2.44}$$

and the slope of both operating lines is unity; see eqs. (2.28) and (2.30). This means that the composition of the streams passing between two subsequent stages is equal:

$$y_2 = x_1, \; y_3 = x_2, \; \text{or, generally,} \; y_{n+1} = x_n \tag{2.45}$$

For a binary equilibrium between component A and the less volatile component B on stage 1 (see Figure 2.13) eq. (2.1) applies:

$$y_1 = K_A \cdot x_1 = K_A \cdot y_2 \tag{2.46}$$

For stage 2, it follows that

$$y_1 = K_A^2 \cdot y_3 \tag{2.47}$$

Similarly, repeating this procedure for a succession of N theoretical stages, the result becomes

$$y_1 = K_A^N \cdot y_{N+1} \tag{2.48}$$

At the top of the column $y_1 = x_0 = x_D$ for a total condenser while for the total reboiler $y_{N+1} = x_N = x_B$. Hence, the connection between the mole fraction of A and B in the top and bottom products is

$$x_D = K_A^N \cdot x_B \quad \text{and} \quad 1 - x_D = K_B^N \cdot (1 - x_B) \tag{2.49}$$

Finally, note that $K_A/K_B = \alpha_{AB}$, N_{min} stages are needed to obtain the preset top and bottom compositions:

$$\frac{x_D}{1 - x_D} = (\alpha_{AB})^{N_{min}} \frac{x_B}{1 - x_B} \tag{2.50}$$

Solving the equation for N_{min} by logarithms gives the *Fenske equation* that can be used to estimate the minimum number N_{min} of required theoretical stages ($N_{min} - 1$ plates plus the reboiler) in the column:

$$N_{min} = \frac{\ln\left[\left(\frac{x_D}{1 - x_D}\right) \Big/ \left(\frac{x_B}{1 - x_B}\right)\right]}{\ln \alpha_{AB}} \tag{2.51}$$

The Fenske equation is also applicable for multicomponent mixtures when the relative volatility is based on the *light key* relative to the *heavy key*. If the change in relative volatility from the bottom of the column to the top is moderate, a mean value of the relative volatility is generally calculated by taking the average of the relative volatility at the top, α_D, and the bottom of the column, α_B as follows:

$$\alpha_{av} = \sqrt{\alpha_B \cdot \alpha_D} \tag{2.52}$$

In a similar way, the minimum reflux ratio R_{min} can be obtained analytically from the physical properties of a binary system. Starting again from a material balance (Figure 2.13), this time over the top portion of the column comprising n theoretical stages, gives:

$$V' y_{n+1} = L' x_n + D x_D \tag{2.53}$$

and

$$V' (1 - y_{n+1}) = L' (1 - x_n) + D (1 - x_D) \tag{2.54}$$

Under conditions of minimum reflux, a column has to have an infinite number of plates. For large values of n, approaching the pinch (see Figure 2.18b), the composition on plate n is equal to that on plate $n + 1$, so near the pinch

$$y_{n+1} = y_\infty \text{ and } x_n = x_\infty \tag{2.55}$$

Application of the equilibrium relation equation (2.16), dividing eqs. (2.53) and (2.54) and introduction of the relative volatility α_{AB} provide:

$$\frac{y_\infty}{1 - y_\infty} = \frac{\alpha_{AB} x_\infty}{1 - x_\infty} = \frac{L' x_\infty + D x_D}{L' (1 - x_\infty) + D (1 - x_D)}. \tag{2.56}$$

Solving for L'/D, this rearranges into the *Underwood expression* for estimating the minimum reflux ratio required to separate the liquid composition on a plate near the pinch into the desired distillate composition:

$$R_{min} = \left(\frac{L'}{D}\right)_{min} = \frac{\frac{x_D}{x_\infty} - \alpha_{AB} \frac{1 - x_D}{1 - x_\infty}}{\alpha_{AB} - 1} \tag{2.57}$$

This expression applies to any feed condition q because the relation between the pinch and the feed composition, x_F, is not defined yet. For $q = 1$, saturated liquid, $x_\infty = x_F$ and the minimum reflux ratio for the desired separation of the feed composition becomes:

$$R_{min} = \frac{\frac{x_D}{x_F} - \alpha_{AB} \frac{1 - x_D}{1 - x_F}}{\alpha_{AB} - 1} \tag{2.58}$$

If the distillate product is required as a pure substance ($x_D = 1$), this equation reduces to

$$R_{min} = \frac{1}{(\alpha_{AB} - 1) x_F} \tag{2.59}$$

2.4.2.4 Use of Murphree efficiency

The McCabe–Thiele method assumes that the two phases leaving each stage are in thermodynamic equilibrium. In industrial equipment, the equilibrium is usually not fully approached. The most commonly used stage efficiencies used to describe individual tray performance for individual components are the Murphree vapor and liquid efficiencies. The *Murphree vapor efficiency* E_{MV} for stage n is defined as the ratio of the actual change in vapor composition over the change in vapor composition for an equilibrium stage:

$$E_{MV} = \frac{y_n - y_{n+1}}{y_n^* - y_{n+1}} \tag{2.60}$$

where y_n^* is the composition of the hypothetical vapor phase in equilibrium with the liquid composition leaving stage n. Once the Murphree efficiency is known for every stage, it can easily be used on a McCabe–Thiele diagram. The denominator represents the vertical distance from the operating line to the equilibrium line while the numerator is the vertical distance from the operating line to the actual outlet concentration. In stepping of stage, the Murphree vapor efficiency represents the fraction of the total vertical distance to move from the operating line to the equilibrium line. As only E_{MV} of the total vertical path is traveled, we get the result shown in Figure 2.20.

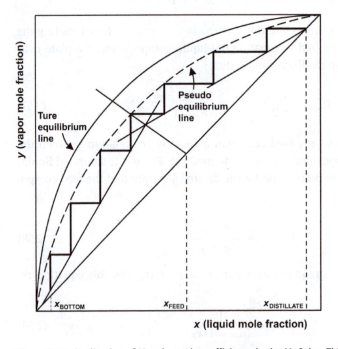

Figure 2.20: Application of Murphree plate efficiency in the McCabe–Thiele construction.

2.4.3 Energy requirements

Following the determination of the feed condition, reflux ratio and number of theoretical stages, estimates of the *heat duties* of the condenser and reboiler can be
made. When the column is well insulated, all of the heat transfer takes place in the
condenser and reboiler, column pressure is constant, *the feed is at the bubble point*
and a total condenser is used, the energy balance over the entire column gives

$$\text{Reboiler in} = \text{Condenser out} \tag{2.61}$$

or

$$Q_R = Q_C \tag{2.62}$$

The energy balance can by approximated by applying the assumptions of the
McCabe–Thiele method, yielding for the reboiler and condenser duty:

$$Q_R = Q_C = D\,(R+1)\,\Delta H_{\text{vap}} \tag{2.63}$$

If saturated steam is the heating medium for the reboiler, the *required steam rate*
becomes

$$\Phi_{m,\,\text{steam}} = \frac{Q_R \cdot M_{\text{steam}}}{\Delta H_{\text{vap, steam}}} \tag{2.64}$$

where $\Delta H_{\text{vap, steam}}/M_{\text{steam}}$ is the specific enthalpy of vaporization of steam
($\approx 2{,}100$ kJ kg^{-1}). The *cooling water rate* for the condenser is

$$\Phi_{m,\,C} = \frac{Q_C}{C_{P,\,\text{water}}\,(T_{\text{out}} - T_{\text{in}})} \tag{2.65}$$

where $C_{P,\text{water}}$ is the specific heat capacity of water (≈ 4.2 kJ kg^{-1} K^{-1}). In general, the
cost of cooling water can be neglected during a first evaluation because the annual
cost of reboiler steam is an order of magnitude higher.

2.5 Advanced distillation techniques

2.5.1 Batch distillation

In most of the large chemical plants, distillations are run continuously. For small
production units, where most chemical processes are carried out in batches it is
more convenient to distil each batch separately. In batch distillation, a liquid mixture is charged to a vessel where it is heated to the boiling point. When boiling begins, the vapor is passed through a fractionation column and condensed to obtain a
distillate product, as indicated in Figure 2.21. As with continuous distillation, the

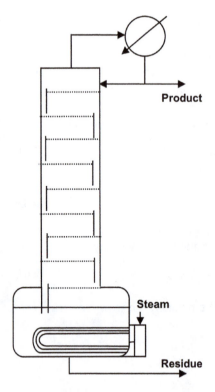

Product

Steam

Residue

Figure 2.21: Schematic of a batch distillation unit.

purity of the top product depends on the still composition, the number of plates of the column and on the reflux ratio used. In contrast to continuous distillation, batch distillation is usually operated with a variable amount of reflux. Because the lower boiling components concentrate in the vapor and the remaining liquid gradually becomes richer in the heavier components, the purity of the top product will steadily drop. This is generally compensated by a gradual increase in reflux ratio during the distillation process to maintain a constant quality of the top product. To obtain the maximum recovery of a valuable component, the charge remaining in the still after the first distillation may be added to the next batch.

The main advantage of batch distillation is that *multiple liquid mixtures can be processed in a single unit*. Different product requirements are easily taken into account by changing the reflux ratio. Even multicomponent mixtures can be separated into the different components by a single column when the fractions are collected separately. An additional advantage is that batch distillation can also handle sludge and solids. The main disadvantages of batch distillation are that for a given product rate, the equipment is larger and the mixture is exposed to higher temperature for a longer time. This increases the risk of thermal degradation or decomposition. Furthermore, it requires more operator attention, energy requirements are higher and its dynamic nature makes it more difficult to control and model.

2.5.2 Continuous separation of multiple product mixtures

Very often a separation unit incorporated in the process separates the stream into several products by means of several columns. Where the columns are located in the flow diagram will have a very great influence on the economics of the separation unit. Case studies have shown that operating costs alone may vary by a factor of 2 depending on the flow sheet chosen. If each component of a multicomponent distillation is to be essentially pure when recovered, the number of required columns is equal to the number of components minus one. Thus, a three-component mixture requires a two- and a four-component mixture requires three separate columns. Those columns can be arranged in many different ways as illustrated in Figure 2.22 for the two possible separation configurations of a ternary mixture.

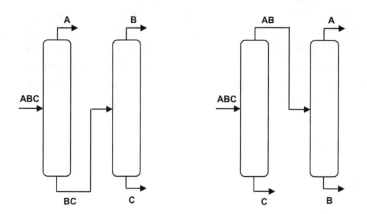

Figure 2.22: Possible distillation configurations for separating ternary mixtures.

Note that it takes a sequence of two ordinary distillation columns to separate a mixture in three products. Because of the many possible flow sheet configurations for mixtures with more than three components it would be far too complex to find the best flow diagram by systematic study. Fortunately, some *heuristic rules* are available to provide guidelines for defining flow diagrams to accelerate the search for the optimal separation sequence of multicomponent mixtures:
1. Remove corrosive and hazardous materials first
2. First, eliminate majority components
3. Start out with easy separations
4. Sequences that remove the components one by one in column overheads should be favored
5. Give priority to separations that give a more nearly equimolal division of the feed between the distillate and bottoms product
6. End up with difficult separations such as azeotropic mixtures

7. Separations involving very strict product specifications should be reserved until late in a sequence
8. Favor sequences that yield the minimum necessary number of products
9. Avoid vacuum distillation and refrigeration if possible

2.5.3 Separation of azeotropes

When due to minimal difference in boiling point and/or highly nonideal liquid behavior, the relative volatility becomes lower than 1.1, ordinary distillation may be uneconomic and in case an azeotrope forms even impossible. In that event, enhanced distillation techniques should be explored. For these circumstances, the most often used technique is *extractive distillation* where a large amount of a relatively high boiling solvent is added to increase the relative volatility of the key components in the feed mixture. In order to maintain a high concentration throughout the column, the solvent is generally introduced above the feed entry and a few trays below the top. It leaves the bottom of the column with the less volatile product and is recovered in a second distillation column as shown in Figure 2.23. Because the high boiling solvent is easily recovered by distillation when selected in such a

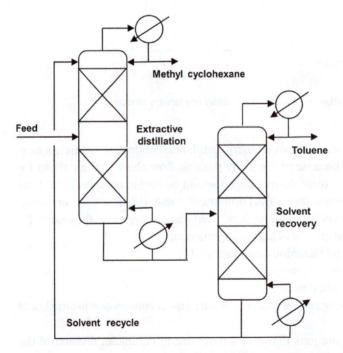

Figure 2.23: Extractive distillation separation of toluene from methylcyclohexane (MCH) using high boiling polar solvents such as NMP, sulfolane, phenol and glycols.

way that no new azeotropes are formed, extractive distillation is less complex and more widely used than azeotropic distillation.

In *homogeneous azeotropic distillation*, an entrainer is added to the mixture that forms a homogeneous minimum or maximum boiling azeotrope with one or more feed components. The entrainer can be added everywhere in the column. If an entrainer is added to form an azeotrope, the azeotrope will exit the column as the overhead or bottom product leaving behind component(s) that may be recovered in the pure state. A classical example of azeotropic distillation is the recovery of anhydrous ethanol from aqueous solutions. Organic solvents such as benzene or cyclohexane are used to form desirable azeotropes, allowing the separation to be made. As shown in Figure 2.24, the ethanol azeotrope from the crude column overhead is fed to the azeotropic distillation column where cyclohexane is used to form an azeotrope with water. Ethanol is recovered as the bottom product. The overhead azeotropic cyclohexane/water mixture is condensed, where water-rich and cyclohexane-rich liquid phases are formed. Residual cyclohexane is removed from the water-rich phase in a stripper.

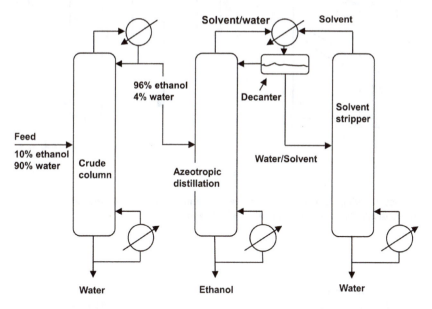

Figure 2.24: Dehydration of alcohol by azeotropic distillation with solvents such as benzene or cyclohexane as entrainer.

It is well known that many minimum boiling azeotrope-containing mixtures allow the position of the azeotrope to be shifted a change in system pressure. This effect can be exploited to separate a binary azeotrope containing mixture when appreciably changes (>5 mol%) in azeotropic composition can be achieved over a moderate pressure range. *Pressure swing distillation* uses a sequence of two columns operated at different pressures for the separation of pressure-sensitive azeotropes. This is illustrated in Figure 2.25, where the effect of pressure on the temperature and composition of a minimum boiling azeotrope is given. The binary azeotrope can be crossed by first separating the component boiling higher than the azeotrope at low pressure. The composition of the overhead should be as close as possible to that of the azeotrope at this pressure. As the pressure is increased, the azeotropic composition moves toward a higher percentage of A and component B can be separated from the azeotrope as the bottom product in the second column. The overhead of the second column is returned to the first column.

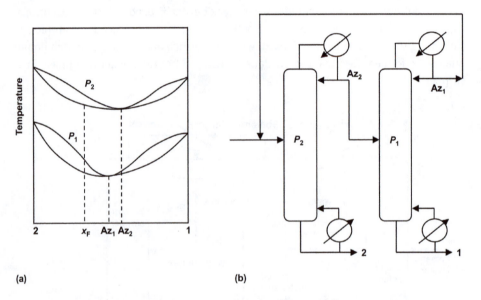

(a)   (b)

Figure 2.25: Pressure swing distillation: T–y–x curves for minimum boiling azeotrope at pressures P_1 and P_2 (a) and distillation sequence for minimum boiling azeotrope (b).

Nomenclature

B	Bottom product flow	mol s^{-1}
C_P	Specific heat capacity	J kg^{-1} K^{-1}
D	Top product flow	mol s^{-1}
D_{AB}	Coefficient of diffusion of A in a mixture of A and B	m^2 s^{-1}
E_{MV}	Murphree or plate efficiency, eq. (2.60)	–
F	Feed flow	mol s^{-1}
ΔH_{vap}	Molar heat of vaporization	J mol^{-1}
K_i	Distribution coefficient of component i	–
L	Liquid flow (in rectifying section L', in stripping section L'')	mol s^{-1}
N_{OV}	Overall number of gas-phase transfer units	–

B	Bottom product flow	mol s^{-1}
N	Number of (theoretical) stages	–
P	Partial pressure	N m^{-2}
p^0	Saturation pressure	N m^{-2}
P_{tot}	Total pressure	N m^{-2}
Q	Fraction-saturated liquid feed, eq. (2.17)	–
Q	Heat flow	J s^{-1}
R	Reflux ratio L'/D, eq. (2.27)	–
T	Temperature	K
x, y, z	Mole fraction (liquid, vapor, feed)	–
V	Vapor flow (in rectifying section V', in stripping section V'')	mol s^{-1}
Φ_m	Mass flow rate	kg s^{-1}
α_{ij}	Selectivity, relative volatility or equilibrium constant, eqs. (1.5), (2.2), (2.7), (2.8)	–
Γ	Activity coefficient	–

Indices

A, B, i, j	Components
B	Bottom
C	Condenser
D	Distillate
F	Feed
L	Liquid
n, m	Stage number
R	Reboiler
V	Vapor

Exercises

1   In *Vapor-Liquid Equilibrium Data Collection*, the following form of Antoine equation is used:

$$\log P_i^0 = A' - \frac{B'}{T + C'}$$

with P_i^0 = saturation pressure in mmHg and temperature T in °C (760 mmHg = 1 atm).
For benzene in benzene–toluene mixtures, the following values are reported:

$$A' = 6.87987 \; [-], \; B' = 1,196.760 \; [-] \text{ and } C' = 219.161 \; °C$$

Calculate the constants A, B and C in the Antoine equation as defined in eq. (2.12):

$$\ln P_i^0 = A - \frac{B}{T + C}$$

with pressure units in atm and temperature in °C.

2 With temperature in degrees Fahrenheit and pressure in pounds per sq. inch, the following Antoine constants apply for benzene:

$$A' = 5.1606, \ B' = 2,154.2, \ C' = 362.49 \ (\text{psi}, \ °F, \ ^{10}\log P \, \text{expression})$$

Note that 1 atm = 14.696 psi = 1.01325 bar and $T \, (°F) = T \, (°C)·9/5 + 32$, and calculate the saturation pressure of benzene in bar at 80.1 °C.

3 VLE data for benzene–toluene are given at 1 atm and at 1.5 atm in the following T–x diagram. Expressing the saturated vapor pressure in atm as a function of temperature in K, the Antoine constants for benzene and toluene are, respectively

$$A = 9.2082, \ B = 2,755.64, \ C = -54.00 \ \text{and}$$

$$A = 9.3716, \ B = 3,090.78, \ C = -53.97 \ (\text{atm}, \text{K}, \text{eq.}(2.12))$$

Check the phase compositions at 100 °C and total pressures of 1.0 and 1.5 atm.

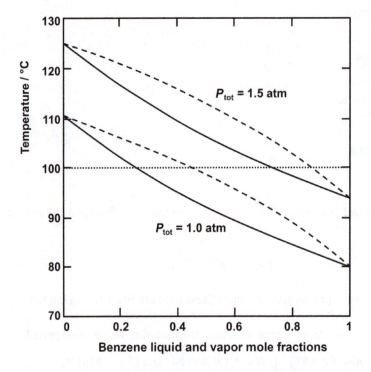

Benzene liquid and vapor mole fractions

4 A liquid benzene–toluene mixture with $z = 0.40$ should produce a liquid with $x = 0.35$ in a flash drum at 1.0 bar.
a. Calculate the required feed temperature.
b. Calculate the equilibrium vapor–liquid ratio in the flash drum.

5   We wish to flash distill isothermally a mixture containing 45 mol% of benzene
    and 55 mol% of toluene. Feed rate to the still is 700 mol/h. Equilibrium data for
    the benzene–toluene system can be approximated with a constant relative vola-
    tility of 2.5, where benzene is the more volatile component. Operation of the
    still is at 1 atm.
    a.   Plot the y–x diagram for benzene–toluene.
    b.   If 60% of the feed is evaporated, find the liquid and vapor compositions.
    c.   If we desire a vapor composition of 60 mol%, what is the corresponding
         liquid composition and what are the liquid and vapor flow rates?
    d.   Find the compositions and flow rates of all unknown streams for a two-
         stage flash cascade where 40% of the feed is flashed in the first stage and
         the liquid product is sent to a second flash chamber where 30% is flashed.

6   Distillation is used to separate pentane from hexane. The feed amounts 100 mol/s
    and has a mole ratio of pentane/hexane = 0.5. The bottom and top products have
    the compositions x_B= 0.05 and x_D= 0.98. The reflux ratio is 2.25. The column pres-
    sure is 1 bar. The feed, at the bubbling point, enters the column exactly on the feed
    tray. The tray temperature is equal to the feed temperature. The pentane vapor
    pressure is given by

    $$p_5^o = \exp\left(11^* \left(1 - \frac{310}{T(K)}\right)\right) \text{ (bar)} \qquad (1 \text{ atm} = 1.013 \text{ bar})$$

    The vapor pressure of hexane is 1/3 of the pentane vapor pressure over the
    whole temperature range. The average density of liquid pentane and hexane
    amounts to 8,170, respectively, 7,280 mol/m³. The heat of vaporization amounts
    to 30 kJ/mol. The distance between the trays amounts to 0.50 m.
    a.   Calculate the feed temperature.
    b.   Calculate the vapor stream from the reboiler.
    c.   Calculate the required energy in the reboiler.
    d.   Construct the y–x diagram.
    e.   Construct the operating lines and locate the feed line.
    f.   Determine the number of equilibrium stages.
    g.   Determine the height of the column.

7   Methanol (M) is to be separated from water (W) by atmospheric distillation. The
    feed contains 14.46 kg/h methanol and 10.44 kg/h water. The distillate is
    99 mol% pure, while the bottom product contains 5 mol% of methanol. The
    feed is subcooled such that q = 1.12.
    a.   Determine the minimum number of stages and minimum reflux.
    b.   Determine the feed stage location and number of theoretical stages required
         for a reflux ratio of R = 1.

VLE data (1 atm, mole fraction methanol)

x	0.0321	0.0523	0.075	0.154	0.225	0.349	0.813	0.918
y	0.1900	0.2940	0.352	0.516	0.593	0.703	0.918	0.963

8 A feed to a distillation unit consists of 50 mol% benzene in toluene. It is introduced to the column at its bubble point to the optimal plate. The column is to
produce a distillate containing 95 mol% benzene and bottoms of 95 mol% toluene. For an operating pressure of 1 atm, calculate:
 a. the minimum reflux ratio;
 b. the minimum number of equilibrium stages to carry out the desired separation;
 c. the number of actual stages needed, using a reflux ratio (L'/D) of 50% more
  than the minimum;
 d. the product and residue stream in kilograms per hour of product if the feed
  is 907.3 kg/h;
 e. the saturated steam required in kilograms per hour for heat to the reboiler
  using the following enthalpy data:
  Steam: $\Delta H_{vap} = 2{,}000$ kJ/kg
  Benzene: $\Delta H_{vap} = 380$ kJ/kg
  Toluene: $\Delta H_{vap} = 400$ kJ/kg

VLE data (1 atm, mole fraction benzene)

x	0.10	0.20	0.30	0.40	0.50	0.60	0.70	0.80	0.90
y	0.21	0.37	0.51	0.64	0.72	0.79	0.86	0.91	0.96

9 During the synthesis of nitrotoluene, *ortho*-nitrotoluene as well as *para*-
nitrotoluene are formed. The customer requires pure *para*-nitrotoluene, the
most volatile of the two isomers. The product requires a minimal purity of 95%.
Another customer is interested in *ortho*-nitrotoluene, also with a minimal purity
of 95%. The feed, of which 40% is in the vapor phase, amounts to 100 kmol/h
and consists of 55% *ortho*-nitrotoluene. At the chosen operation pressure and
temperature of the column, the relative volatility of *ortho*-nitrotoluene over
para-nitrotoluene is 2.0.
 a. Calculate the size of the top and bottom streams.
 b. Determine the minimum number of stages for this separation.
 c. Determine the minimum reflux ratio.
 d. The applied reflux ratio is 3.76, and determine the boilup ratio ($=V''/B$) in
  the reboiler.

10 About 550 kmol/h of a binary mixture of water/acetic acid (70 mol% water) is
separated by distillation in a bottom fraction with 98 mol% acetic acid and a

distillate containing 5 mol% acetic acid. The relative volatility of water over acetic acid is 1.85.

a. Calculate the amount of distillate and bottoms product this distillation yields?

b. Determine graphically the minimum number of stages.

The external reflux ratio R $(=L'/D)$ is set at 3. The boilup ratio in the stripping section $(=V''/B)$ is set to 10.

c. Determine graphically the required number of equilibrium stages under these conditions.

d. Determine graphically the optimal location for the feed stage.

e. Determine graphically the slope of the feed line and from that the vapor fraction in the feed.

11 Isopropyl alcohol (IPA, boiling point 82.3 °C) needs to be separated from a 30 mol% IPA solution in water. The total IPA production in the plant amounts to 50 kmol/h and the maximum residual concentration in the water should be only 1 mol%. The feed consists of 90% of liquid and the efficiency of the trays is 90%.

a. What is the maximum purity of the IPA that can be obtained by distillation under the current circumstances?

b. Calculate the size of the bottoms and distillate stream to obtain an IPA purity of 70 mol%.

c. Draw the yx-diagram including the pseudoequilibrium line.

d. Determine the minimal reflux ratio.

e. The real reflux ratio is 2.5, and determine the boilup ratio $(=V''/B)$ in the reboiler.

f. Determine graphically the number of real trays that is required for this distillation.

VLE data (1 atm, mole fraction IPA)

x	0.10	0.20	0.35	0.50	0.65	0.75	0.80	0.85	0.95
y	0.33	0.43	0.53	0.61	0.70	0.77	0.81	0.85	0.94

Chapter 3
Absorption and Stripping

3.1 Introduction

In absorption (also called gas absorption, gas scrubbing and gas washing) a gas mixture is contacted with a liquid (the *absorbent* or *solvent*) to selectively dissolve one or more components by transfer from the gas to the liquid. The components transferred to the liquid are referred to as solutes or *absorbates*. The operation of absorption can be categorized on the basis of the nature of the interaction between absorbent and absorbate into the following three general types:

1. *Physical solution*. In this case, the component being absorbed is more soluble in the liquid absorbent than the other gases with which it is mixed but does not react chemically with the absorbent. As a result, the equilibrium concentration in the liquid phase is primarily a function of partial pressure in the gas phase and temperature. Examples are the drying of natural gas with diethylene glycol or the recovery of ethylene oxide and acrylonitrile with water from the reactor product stream.

2. *Reversible reaction*. This type of absorption is characterized by the occurrence of a chemical reaction between the gaseous component being absorbed and a component in the liquid phase to form a compound that exerts a significant vapor pressure of the absorbed component. The most important industrial example is the removal of acid gases (CO_2 and $H_2 S$) with mono- or diethanolamine solutions.

3. *Irreversible reaction*. In this case, a reaction occurs between the component being absorbed and a component in the liquid phase, which is essentially irreversible. Sulfuric acid and nitric acid production by SO_3 and NO_2 absorption in water is the most widely used example of this application.

The use of physical absorption processes is usually preferred whenever feed gases are present in large amounts at high pressure and the amount of the component to be absorbed is relatively large. Chemical absorption usually has a much more favorable equilibrium relationship than physical absorption and is therefore often preferred when the components to be separated from feed gases are present in small concentrations and at low partial pressures.

Gas absorption is usually carried out in *vertical countercurrent columns* as shown in Figure 3.1. The solvent is fed at the top whereas the gas mixture enters from the bottom. The absorbed substance is washed out by the lean solvent and leaves the absorber at the bottom as a liquid solution. Usually, the *absorption column operates at a pressure higher than atmospheric pressure*, taking advantage of the fact that gas solubility increases with pressure. The loaded solvent is often recovered in a subsequent stripping or desorption operation. After preheating,

https://doi.org/10.1515/9783110654806-003

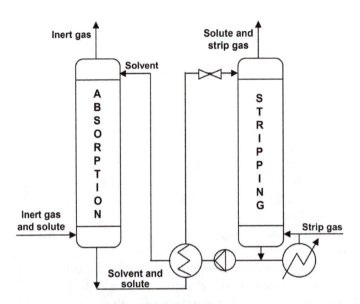

Figure 3.1: Basic scheme of an absorption installation with stripping for regeneration.

the rich solvent is transported to the top of a desorption column that usually operates under lower pressure than in the absorption column. This second step is essentially the reverse of absorption in which the absorbate is removed from the solvent. Desorption can be achieved through a combination of methods:

1. flashing the solvent to lower the partial pressure of the dissolved components
2. reboiling the solvent to generate stripping vapor by evaporation of part of the solvent
3. stripping with an inert gas or steam

Wide use is made of *desorption by pressure reduction* because the energy requirement are low. After depressurization, stripping with an inert gas is often more economic than thermal regeneration. However, because stripping is not perfect, the absorbent recycled to the absorber contains residual amounts of the absorbed solute. The desired purity of the absorbent determines the final costs of desorption. The necessary difference between the partial pressure of the absorbed key component over the regenerated solution and the purified gas serves as a criterion for determining the dimensions of the absorption and desorption equipment. The *lean solvent*, devoid of gas, flows through the heat exchanger, where part of the heat needed for heating the rich solvent is recovered, and then through a second heat exchanger, where it is cooled down to a desired temperature and flows into the absorption column. Usually a small amount of fresh solvent should be added to the column to replenish the solvent which was partly evaporated in the desorption column or underwent irreversible chemical reactions which take place in the whole system.

3.2 The aim of absorption

Generally the commercial purpose of absorption processes can be divided into *gas purification* or *product recovery*, depending on whether the absorbed or the unabsorbed portion of the feed gas has the greater value. Typical gas purification applications are listed in Table 3.1. The removal of CO_2 from hydrogen gas in ammonia production and the removal of acid gases (CO_2 and $H_2 S$) from natural gas are some of the most widespread applications that are being improved continuously by the development of new solvents, process configurations and design techniques. In both applications stripping is used for absorbent regeneration.

Table 3.1: Typical applications of absorption for gas purification.

Impurity	Process	Absorbent
Ammonia	Indirect process (Coke oven gas)	Water
Carbon dioxide and Hydrogen sulfide	Ethanolamine	Mono- or diethanolamine in water
	Benfield	Potassium carbonate and activator in water
	Selexol	Polyethylene glycol dimethyl ether
Carbon monoxide	Copper ammonium salt	Cuprous ammonium carbonate and formate in water
Hydrogen chloride	Water wash	Water
Toluene	Toluene scrubber	Toluene
Cyclohexane	Scrubber	Cyclohexane

Examples of absorption processes for product recovery are listed in Table 3.2. The absorption of SO_3 and NO_x in water to make concentrated sulfuric acid and nitric

Table 3.2: Typical applications of absorption for product recovery.

Product	Process	Absorbent
Acetylene	Steam cracking of hydrocarbons (Naphtha)	Dimethylformamide
Acrylonitrile	Ammoxidation of propylene	Water
Maleic anhydride	Butane oxidation	Water
Melamine	Urea decomposition	Water
Nitric acid	Ammonia oxidation (NO_x absorption)	Water
Sulfuric acid	Contact process (SO_3 absorption)	Water
Urea	Synthesis (CO_2 and NH_3 absorption)	Ammonium carbamate solution

acid, respectively, are probably the most widely used product recovery applications of absorption. Other frequently encountered examples are the recovery of various products from a gaseous product stream by inert absorbents such as water. In some cases the absorber is used as a reactor where the desired chemical compound is obtained by a liquid phase reaction of the absorbed gases. An illustration of such a process is the production of urea from CO_2 and ammonia.

3.3 General design approach

Both absorption and stripping can be operated as equilibrium stage operations with contact of liquid and vapor. The plate towers can be designed by following an adaptation of the *McCabe–Thiele method*. Packed towers can be designed by the use of HETP or preferably by mass transfer considerations. In both absorption and stripping, a separate phase is added as the separating agent. Thus the columns are simpler than those in distillation, in that reboilers and condensers are normally not used. Design or analysis of an absorber (or stripper) requires consideration of a number of factors, including the following:

1. Entering gas (liquid) flow rate, composition, temperature and pressure
2. Desired degree of recovery of one or more solutes
3. Choice of absorbent (stripping agent)
4. Operating pressure and temperature, and allowable gas pressure drop
5. Minimum absorbent (stripping agent) flow rate and actual absorbent (stripping agent) flow rate as a multiple of the minimum rate needed to make the separation
6. Number of equilibrium stages
7. Heat effects and need for cooling (heating)
8. Type of absorber (stripper) equipment
9. Height of absorber (stripper)
10. Diameter of absorber (stripper)

The initial step in the design of the absorption system is *selection of the absorbent* and overall process to be employed. There is no simple analytical method for accomplishing this step. In most cases, more than one solvent can meet the process requirements, and the only satisfactory approach is an economic evaluation, which may involve the complete but preliminary design and cost estimate for more than one alternative. The ideal absorbent should:

1. have a high solubility for the solute(s) to minimize the need for absorbent
2. have a low volatility to reduce the loss of absorbent and facilitate separation of solute(s)
3. be stable to maximize absorbent life and reduce absorbent makeup requirement
4. be noncorrosive to permit use of common materials of construction

5. have a low viscosity to provide low pressure drop and high mass and heat transfer rates
6. be nonfoaming when contacted with the gas
7. be nontoxic and nonflammable to facilitate its safe use
8. be available, if possible within the process, or be inexpensive.

The most widely used absorbents are water, hydrocarbon oils and aqueous solutions of acids and bases. For stripping the most common agents are water vapor, air, inert gases and hydrocarbon gases. Once an absorbent is selected, the design of the absorber requires the determination of basic physical property data such as density, viscosity, surface tension and heat capacity. The fundamental physical principles underlying the process of gas absorption are the solubility and heat of solution of the absorbed gas and the rate of mass transfer. Information on both must be available when sizing equipment for a given application. In addition to the fundamental design concepts based on solubility and mass transfer, many practical details have to be considered during actual plant design.

The second step is the selection of the operating conditions and the type of contactor. In general, operating pressure should be high and temperature low for an absorber to minimize stage requirements and/or absorbent flow rate. Operating pressure should be low and temperature high for a stripper to minimize stage requirements or stripping agent flow rate. However, because maintenance of a vacuum is expensive, strippers are commonly operated at a pressure just above ambient. A high temperature can be used, but it should not be so high as to cause undesirable chemical reactions. Choice of the contactor may be done on the basis of system requirements and experience factors such as those discussed in Section 3.6. Following these decisions it is necessary to calculate material and heat balance calculations around the contactor, define the mass transfer requirements, determine the height of packing or number of trays and calculate contactor size to accommodate the liquid and gas flow rates with the selected column internals.

3.4 Absorption and stripping equilibria

3.4.1 Gaseous solute solubilities

The most important physical property data required for the design of absorbers and strippers are *gas–liquid equilibria*. Since equilibrium represents the limiting condition for any gas–liquid contact, such data are needed to define the maximum gas purity and rich solution concentration attainable in absorbers, and the maximum lean solution purity attainable in strippers. Equilibrium data are also needed to establish the mass transfer driving force, which can be defined simply as the difference between the actual and equilibrium conditions at any point in a contactor. At equilibrium, a component of a gas in contact with a liquid has identical fugacities in both the gas and liquid phase. For ideal solutions *Raoult's law* applies, see also eq. (2.6):

$$y_A = \frac{P_A^0}{P_{tot}} x_A \qquad (3.1)$$

where y_A is the mole fraction of A in the gas phase, P_{tot} is the total pressure, P_A^0 is the saturation pressure, the vapor pressure of pure A and x_A is the mole fraction of A in the liquid. For nonideal mixtures Raoult's law modifies into:

$$y_A = \frac{y_A^\infty P_A^0}{P_{tot}} x_A \qquad (3.2)$$

where y_A^∞ is the activity coefficient of solute A in the absorbent at infinite dilution. A more general way of expressing solubilities is through the dimensionless *vapor–liquid distribution coefficient* K, defined in the previous chapter:

$$y_A \equiv K_A x_A \qquad (3.3)$$

Values of distribution coefficients are widely employed to represent hydrocarbon vapor–liquid equilibria in absorption and distillation calculations. Correlations and experimental data on the distribution coefficients of hydrocarbons are available from various sources. For moderately soluble gases with relatively little interaction between the gas and liquid molecules equilibrium data are usually represented by *Henry's law*:

$$p_A = P_{tot} y_A = H_A x_A \qquad (3.4)$$

where p_A is the partial pressure of A in the gas phase. H_A is a Henry constant,[1] which has the units of pressure. The Henry's law constants depend upon temperature and usually follow an *Arrhenius relationship*:

$$H_A = H_A^0 \exp\left(\frac{-\Delta H_{vap}}{RT}\right) \qquad (3.5)$$

A plot of $\ln H_A$ versus $1/T$ will then give a straight line. Usually a Henry constant increases with temperature, but is relatively independent of pressure at moderate levels. In general, for moderate temperatures, gas solubilities decrease with an increase in temperature. Henry's constants for many gases and solvents are tabulated in various literature sources. Examples of Henry's constants for a number of gases in pure water are given in Figure 3.2.

---

[1] Note that we use H_A = Henry coefficient of a component A, whereas ΔH_{vap} = H_V–H_L refers to enthalpies.

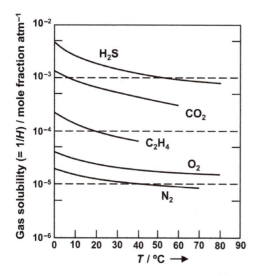

Figure 3.2: Solubilities of various gases in water expressed as the reciprocal of the Henry's Law constant (adapted from [15]).

3.4.2 Minimum absorbent flow

For each feed gas flow rate, absorbent composition, extent of solute absorption, operating pressure and operating temperature, a minimum absorbent flow rate exists that corresponds to an infinite number of countercurrent equilibrium contacts between the gas and liquid phases. As a result a tradeoff exists in every design problem between the number of equilibrium stages and the absorbent flow rates at rates greater than the minimum value. This *minimum absorbent flow rate* L_{min} is obtained from a mass balance over the whole absorber, assuming equilibrium is obtained between incoming gas and outgoing absorbent liquid in the bottom of the column. An overall mass balance over the column illustrated by Figure 3.3 gives the following result:

$$G y_{in} + L x_{in} = G y_{out} + L x_{out} \qquad (3.6)^2$$

The lower the absorbent flow, the higher the x_{out}. The theoretically highest possible concentration in the liquid x_{out} determines the minimum absorbent flow, such as an infinitely long column where equilibrium can be assumed between incoming gas and outgoing liquid ($x_{out} = y_{in}/K$):

---

2 Note that in absorption the gas phase usually is denoted by G instead of V(apor) as in the previous chapter.

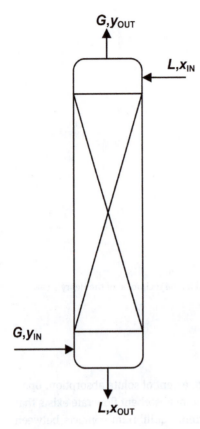

Figure 3.3: Continuous, steady-state operation in a counter-current column.

$$L_{min} = G \cdot \frac{y_{in} - y_{out}}{x_{max} - x_{in}} = G \cdot \frac{y_{in} - y_{out}}{\frac{y_{in}}{K} - x_{in}} \tag{3.7}$$

A similar derivation of the minimum stripping gas flow rate G_{min} for a stripper results in an analogous expression:

$$G_{min} = L \cdot \frac{x_{in} - x_{out}}{y_{max} - y_{in}} = L \cdot \frac{x_{in} - x_{out}}{K x_{in} - y_{in}} \tag{3.8}$$

3.5 Absorber and stripper design

3.5.1 Operating lines for absorption

The McCabe–Thiele diagram is most useful when the operating line is straight. This requires that the energy balances are automatically satisfied and the ratio of liquid to vapor flow rate is constant. In order to have the energy balances automatically satisfied, we must assume *that the heat of absorption is negligible and the oper-ation is isothermal*. These two assumptions will guarantee satisfaction of the en-thalpy balances. When the gas and liquid stream are both fairly dilute (say < 5%), the assumptions will probably be satisfied. We also desire a straight operating line at higher concentrations. This will automatically be true if we assume that the sol-vent is nonvolatile, the carrier gas is insoluble and define liquid and gas streams in terms of *solute free* solvent and carrier gas:

$$\frac{L'}{G'} = \frac{\text{moles nonvolatile solvent/s}}{\text{moles insoluble carrier gas/s}} \tag{3.9}$$

The results of these last two assumptions are that the mass balances for the solvent and carrier gas become

$$L'_N = L'_n = L'_0 = L' = \text{constant and } G'_{N+1} = G'_n = G'_1 = G' = \text{constant} \tag{3.10}$$

Now overall flow rates of gas and liquid are not used because in more concentrated mixtures a significant amount of solute may be absorbed which would change gas and liquid flow rates. This would result in a curved operating line. Using L' (=moles of nonvolatile solvent/s) and G' (=moles insoluble carrier gas/s), we must define compositions in such a way that we can write a mass balance for solute B. The cor-rect way to do this is to define the compositions as *mole ratios*:

$$Y = \frac{\text{moles solute B in gas}}{\text{moles pure carrier gas}} \quad \text{and} \quad X = \frac{\text{moles solute B in liquid}}{\text{moles pure solvent S}} \tag{3.11}$$

The mole ratios Y and X are related to the usual mole fractions by:

$$Y = \frac{y}{1-y} \quad \text{and} \quad X = \frac{x}{1-x} \tag{3.12}$$

Substitution of X and Y into the ideal equilibrium expression, eq. (2.8), gives the equilibrium line expressed in mole ratios:

$$Y = \alpha \cdot X \tag{3.13}$$

which *represents a straight line through the origin*. Note that both Y and X can be greater than unity. With mole ratio units we obtain for the gas and liquid stream leaving stage n:

$$Y_n G' = \frac{\text{moles B in gas stream}}{\text{moles carrier gas}} \cdot \frac{\text{moles carrier gas}}{s} = \frac{\text{moles B in gas stream}}{s} \tag{3.14}$$

and

$$X_n L' = \frac{\text{moles B liquid stream}}{\text{moles solvent}} \cdot \frac{\text{moles solvent}}{s} = \frac{\text{moles B liquid stream}}{s} \qquad (3.15)$$

Thus we can easily write the steady-state mass balance, moles solute B in s^{-1} = moles solute B out s^{-1}, in these units. The mass balance around the top of the column using the mass balance envelope shown in Figure 3.4 is

$$Y_{n+1} G' + X_0 L' = X_n L' + Y_1 G' \qquad (3.16)$$

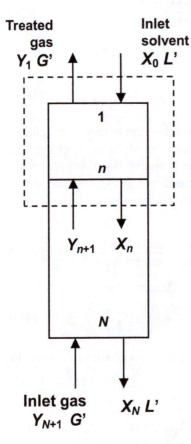

Figure 3.4: Top section mass balance for an absorber.

Solving for Y_{n+1} we obtain:

$$Y_{n+1} = \frac{L'}{G'} X_n + \left(Y_1 - \frac{L'}{G'} X_0 \right) \qquad (3.17)$$

This is a straight line with slope L'/G' and intercept $Y_1 - (L'/G') \cdot X_0$. It is the *operating line* for absorption. Thus if we plot ratios Y vs X we have a McCabe–Thiele type of graph as shown in Figure 3.5. The steps in the procedure are:

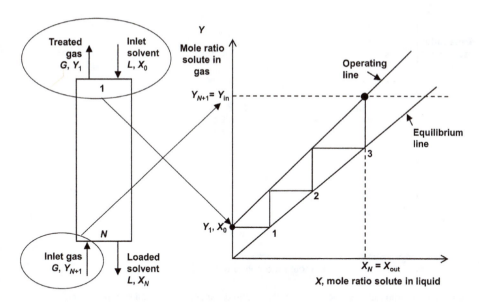

Figure 3.5: McCabe-Thiele diagram for absorption.

1. Plot Y vs X equilibrium data (convert from fractions to ratios)
2. Values of X_0, Y_{N+1}, Y_1 and L'/G' are known. Point (X_0, Y_1) is on the operating line since it represents passing streams; X_n follows from an overall balance.
3. Slope is L'/G'. Plot operating line.
4. Starting at stage 1, step off stages: equilibrium, operating line, equilibrium, etc.

Note that the *operating line in absorption is above the equilibrium line*, because solute is transferred from the gas to the liquid. In distillation we had material (the more volatile component) transferred from liquid to gas, and the operating line was below the equilibrium curve. The $Y = X$ line has no significance in absorption. As usual the stages are counted at the equilibrium curve.

The minimum L'/G' ratio corresponds to a value of X_N leaving the bottom of the tower in equilibrium with Y_{N+1}, the solute concentration in the feed gas. It takes an infinite number of stages for this equilibrium to be achieved. This minimal L'/G' ratio can be derived from the *McCabe–Thiele diagram* as shown in Figure 3.6. The selection of the actual operating absorbent flow rate is based on some multiple of L'_{min}, typically between 1.1 and 2. A value of 1.5 is usually close to the economically optimal conditions.

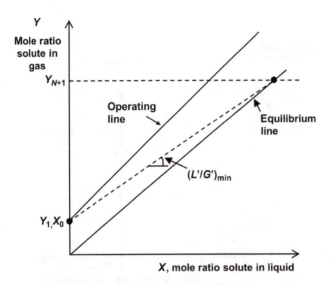

Figure 3.6: Determination of the minimum (L/G) ratio for absorption from a McCabe–Thiele diagram.

3.5.2 Stripping analysis

The number of equilibrium stages for the stripper is determined in a manner similar to that for absorption. Since stripping is very similar to absorption we expect a similar result. The mass balance for the column shown in Figure 3.7 is the same as for absorption and the operating line remains:

$$Y_{n+1} = \frac{L'}{G'} X_n + \left(Y_1 - \frac{L'}{G'} X_0 \right) \tag{3.18}$$

For stripping we know X_0, X_N, Y_{N+1} and L'/G', Y_1 follows from an overall balance. Since (X_N, Y_{N+1}) is a point on the operating line, we can plot the operating line and step off stages. This is illustrated in Figure 3.7. Note that the *operating line is below the equilibrium curve* because solute is transferred from liquid to gas. This is therefore similar to the stripping section of a distillation column. A maximum L'/G' ratio can be defined. This corresponds to the minimum amount of stripping gas. Start from the known point (Y_{N+1}, X_N) and draw a line to the intersection of $X = X_0$ and the equilibrium curve. For a stripper, $Y_1 > Y_{N+1}$ while the reverse is true in absorption. Thus the top of the column is on the right side in Figure 3.7 but on the left side in Figure 3.5.

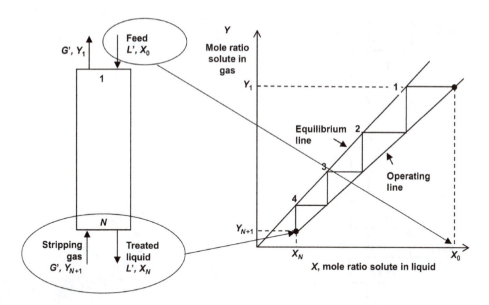

Figure 3.7: McCabe–Thiele diagram for stripping.

3.5.3 Analytical Kremser solution

McGabe–Thiele diagrams can be used to find graphically the number of equilibrium stages, no matter the operating and equilibrium expressions are linear or not. In case of linear and equilibrium lines, an analytical solution exists. When the solution is quite dilute (less than 1% in both gas and liquid), the total liquid and gas flow rates will not change significantly since only a small amount of solute is transferred. Now the entire analysis can be done with mole (or mass) fractions and constant total flow rates G and L. Alternatively, when the use of mole ratios and pure carrier gas and solvent flows result in linear operating and equilibrium lines, the following analysis applies equally well.[3] In this case the column shown in Figure 3.8 will look like the one in Figure 3.4 with $G' = G$ and $L' = L$. The mass balance around any stage n in terms of mole fractions and total flows:

$$y_{n+1} G + x_{n-1} L = x_n L + y_n G \qquad (3.19)$$

which can be rewritten as

$$\frac{L}{G} = \frac{y_{n+1} - y_n}{x_n - x_{n-1}} \qquad (3.20)$$

---

3 After conversion of mole fractions x, y into mole ratios X, Y and using L', G' instead of total flows L, G.

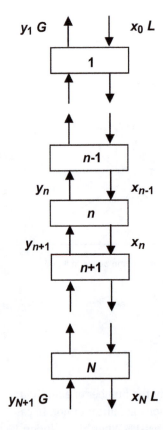

Figure 3.8: Coding of equilibrium stages, streams and fractions for the derivation of the Kremser equations.

It is essentially the same as eq. (3.16) except that the units are different. In case of the straight equilibrium line, that is, at low values of x_n and y_n, the equilibrium equation eq. (2.8) is reduced to

$$y_n = K x_n \qquad (3.21)^4$$

Equations (3.20) and (3.21) can be combined and, introducing the *absorption factor A*, rewritten as:

$$A \equiv \frac{L}{KG} = \frac{y_{n+1} - y_n}{K(x_n - x_{n-1})} = \frac{y_{n+1} - Kx_n}{y_n - Kx_{n-1}} \qquad (3.22)$$

Note that this absorption factor A is different from the separation factor as defined in eq. (1.1). For stage $n-1$ we find in an analogous way:

---

4 Here the equilibrium constant K replaces $\alpha = K_A/K_B$ in eq. (2.8).

$$\frac{y_n - Kx_{n-1}}{y_{n-1} - Kx_{n-2}} = \frac{L}{KG} = A \tag{3.23}$$

Elimination of x_{n-1} from eqs. (3.22) and (3.23) gives

$$\frac{y_{n+1} - Kx_n}{y_{n-1} - Kx_{n-2}} = A^2 \tag{3.24}$$

Continuing in a similar way we find for a cascade with N contacting equilibrium stages:

$$\frac{y_{N+1} - Kx_N}{y_1 - Kx_0} = \frac{y_{in} - Kx_{out}}{y_{out} - Kx_{in}} = A^N \tag{3.25}$$

Using the overall mass balance x_{out} can be eliminated from eq. (3.25):

$$x_{out} = x_{in} + \frac{G}{L} y_{in} - \frac{G}{L} y_{out} \tag{3.26}$$

resulting in

$$A^N = \left[\left(\frac{y_{in} - Kx_{in}}{y_{out} - Kx_{in}} \right) \left(1 - \frac{1}{A} \right) + \frac{1}{A} \right] \tag{3.27}$$

For specified values of y_{in}, y_{out} and x_{in} eq. (3.27) is rewritten to calculate the number of equilibrium stages N_{ts}:

$$N_{ts} = \frac{\ln \left[\frac{y_{in} - Kx_{in}}{y_{out} - Kx_{in}} \left(1 - \frac{1}{A} \right) + \frac{1}{A} \right]}{\ln A} \tag{3.28}$$

This equation is known as the *Kremser-equation for absorption*.

Equation (3.27) also allows calculating the fraction of the feed that is absorbed in a column with N equilibrium stages. Noting that $y_{in} - y_{out}$ is the actual change in gas composition and $y_{in} - K \cdot x_{in}$ is the maximum possible change in composition, that is, if gas leaving is in equilibrium with entering liquid, then from eq. (3.27) the *fraction absorbed* f_A can be derived:

$$\text{Fraction of a solute absorbed} = f_A = \frac{y_{in} - y_{out}}{y_{in} - Kx_{in}} = \frac{A^{N+1} - A}{A^{N+1} - 1} \tag{3.29}$$

Following an analogous derivation for stripping of a liquid by a gas, we are now interested in the change in liquid phase concentrations. Equation (3.25) is rewritten in terms of mole fractions solute in liquid:

$$\frac{x_{in} - y_{out}/K}{x_{out} - y_{in}/K} = \frac{1}{A^N} = S^N \tag{3.30}$$

where the solute *stripping factor* S is defined by $S \equiv \frac{K \cdot G}{L}$. Elimination of y_{out} with the overall mass balance provides

$$y_{out} = y_{in} + \frac{L}{G} x_{in} - \frac{L}{G} x_{out} \qquad (3.31)$$

resulting in

$$S^{N_{ts}} = \left[\left(\frac{x_{in} - y_{in}/K}{x_{out} - y_{in}/K} \right) \left(1 - \frac{1}{S} \right) + \frac{1}{S} \right] \qquad (3.32)$$

For specified values of x_{in}, x_{out} and y_{in}, the required number of equilibrium stages N_{ts} is then calculated from the *Kremser-equation for stripping*:

$$N_{ts} = \frac{\ln \left[\left(\frac{x_{in} - y_{in}/K}{x_{out} - y_{in}/K} \right) \left(1 - \frac{1}{S} \right) + \frac{1}{S} \right]}{\ln S} \qquad (3.33)$$

Analogously to the fraction absorbed, the *fraction of solute stripped*, f_S, is defined as the ration of the actual change in liquid composition and the maximum possible change. The degree of stripping obtained in a stripper with N_{ts} equilibrium stages:

$$\text{Fraction of a solute stripped} = f_S = \frac{x_{in} - x_{out}}{x_{in} - y_{in}/K} = \frac{S^{N_{ts}+1} - S}{S^{N_{ts}+1} - 1} \qquad (3.34)$$

Values of L and G in moles per unit time may be taken as the entering values. Values of K depend mainly on temperature, pressure and liquid phase composition.

3.6 Industrial absorbers

The main purpose of various industrial absorbers is to ensure large gas–liquid mass transfer area and to create such conditions that a high intensity of mass transfer is achieved. Although small-scale processes sometimes use batch-wise operation where the liquid is placed in the equipment and only gas is flowing, the continuous method of absorption is usually applied in large-scale industrial processes. There are various criteria for *classifying absorbers*. It seems that the best one is a widely used criterion that takes into account which of the phases (gas or liquid) is in a continuous or disperses form. Using this criterion, absorbers can be classified into the following groups:
1. Absorbers in which both phases are in a continuous form (packed columns)
2. Absorbers with a disperse gas phase and a continuous liquid phase (plate columns, bubble columns and packed bubbles columns)
3. Absorbers with a dispersed liquid phase and a continuous gas phase (spray columns)

The industrially most frequently used absorbers are packed columns, plate towers, spray columns and bubble columns.

3.6.1 Packed columns

Packed columns are the units most often used in absorption operations (Figure 3.9). Usually, they are cylindrical columns up to several meters in diameter and over 10 meters high. A large gas–liquid interface is achieved by introducing various packings into the column. The packing is placed on a support whose free cross section for gas flow should be at least equal to the packing porosity. Liquid is fed in at the top of the column and distributed over the packing through which it flows downwards. To guarantee a uniform liquid distribution over the cross section of the column, a liquid distributor is employed. Gas flows upward countercurrently to the falling liquid, which absorbs soluble species from the gas. The gas, which is not absorbed, flows away from the top of the column, usually through a mist eliminator. The mist eliminator separates liquid drops entrained by the gas from the packing. The separator may be a layer of the packing, mesh or it may be specially designed.

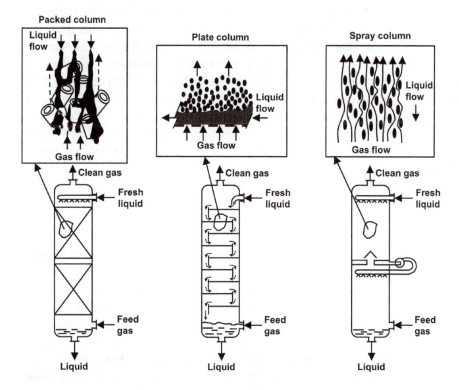

Figure 3.9: Schematic and operating principles of packed, tray and spray towers adapted from [18].

Packings may be divided into two main groups: random and structured. The most popular random packing is rings. *Raschig rings* have a large specific packing surface and high porosity. They are hollow cylinders with an external diameter equal to the ring height that can be made of ceramic, metal, graphite or plastic. Recently, the application of *structured packing* in absorption has increased rapidly owing to the fact that it has a larger mass transfer area than random packing. Another advantage is that by using this packing it is possible, in contrast to random packing, to obtain the same values of mass transfer coefficient in the entire column.

3.6.2 Plate columns

Another basic type of equipment widely applied in absorption processes is a plate column (Figure 3.9). The diameter and height of the column can reach 10 and 50 m, respectively, but usually they are much smaller. In plate columns various plate constructions guarantee good contact of the gas with the liquid. Taking into account the whole column, the flow of the gas with the liquid has a *countercurrent character*. Liquid is supplied to the highest plate, flows along it horizontally and, after reaching a weir, flows through a downcomer from plate to plate in cascade fashion. Gas is supplied below the lowest plate, and then it flows through perforations in the plate dispersers (e.g., holes in a sieve tray and slits in a bubble-cap tray) and bubbles through the flowing liquid at each plate. The application of such a flow pattern is aimed at ensuring the maximum mass transfer area and high turbulence of the gas and liquid phases, which results in obtaining high mass transfer coefficients in both phases. The distances between plates should be such that liquid droplets entrained by the gas are separated from the liquid and the gas is separated from the liquid in the downcomer. Usually, the plate-to-plate distance ranges from 0.2 to 0.6 m and depends mainly on the column diameter and the liquid load of the plate.

3.6.3 Spray and bubble columns

In spray towers liquid is sprayed as fine droplets, which make contact with a cocurrently or countercurrently flowing gas (Figure 3.9). The gas and liquid flows are similar to plug flow. Bubble columns (Figure 3.10) are finding increasing application in processes when absorption is accompanied by a chemical reaction. They are also widely used as chemical reactors in processes where gas, liquid and solid phases are involved. In bubble columns the liquid is a continuous phase while the gas flows through it in the form of bubbles, the character of the gas flow is well represented by plug flow, while that of the liquid is between plug and ideally mixed flow. These absorbers can operate in cocurrent and countercurrent phase flow.

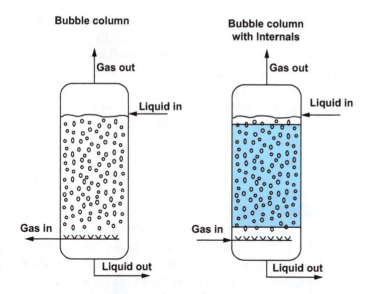

Figure 3.10: Schematic of a bubble column absorber, without and with internal packing.

3.6.4 Comparison of absorption columns

In each apparatus, due to different hydrodynamic conditions, various values of the mass transfer coefficients occur in both phases. Therefore, when choosing a given type of absorber, the following criteria should be taken into account:
1. the required method of absorption (continuous or semi-continuous)
2. the flow rate of the gas and the liquid entering the absorber (e.g. high gas flow rate and low liquid flow rate need different types of equipment than in the case where the two flow rates commensurate)
3. the required liquid hold-up (large or small liquid hold-up is needed)
4. which phase controls mass transfer in the absorber (gas phase or liquid phase)?
5. is it necessary to remove heat from the absorber?
6. the corrosiveness of the absorption systems
7. the required size of the interface (large or small mass transfer area)
8. the physicochemical properties of the gas and the liquid (particularly viscosity and surface tension).

Apart from these factors there are many others, which influence the selection of equipment. For instance, gas impurities or deposits formed during the absorption process require the application of a given absorber.

Packed columns are preferable to tray columns for small installations, corrosive service, liquids with a tendency to foam, very high liquid-to-gas ratios and low pressure drop applications. In the handling of corrosive gases, packing but not plates, can be made from ceramic or plastic materials. Packed columns are also advantageous in vacuum applications because the pressure drop, especially for regularly structured packings, is usually less than in plate columns. In addition packed columns offer greater flexibility because the packing can be changed with relative ease to modify column-operating characteristics.

Tray columns are particularly well suited for large installations and low-to-medium liquid flow rate applications. In general they offer a wider operating window for gas and liquid flow than a countercurrent packed column. That is, they can handle high gas flow rates and low liquid flow rates that would cause flooding in a packed column. Tray columns are also preferred in applications having large heat effects since cooling coils are more easily installed in plate towers and liquid can be withdrawn more easily from plates than from packings for external cooling. Furthermore they are advantageous for separations that require tall columns with a large number of transfer units because they are not subject to channeling of vapor and liquid streams which can cause problems in tall packed columns. The main disadvantages of tray columns are their high capital cost, especially when bubble-cap trays or special proprietary design are used, and their sensitivity to foaming.

Spray contactors are used almost entirely for applications where pressure drop is critical, such as flue gas scrubbing. They are also useful for slurries that might plug packings or trays. Other important applications include particulate removal and hot gas quenching. When used for absorption, spray devices are not applicable to difficult separations because they are limited to only a few equilibrium stages even with countercurrent spray column designs. The low efficiency of spray columns is believed to be due to entrainment of droplets in the gas and *back mixing* of the gas induced by the sprays. The high energy consumed for atomizing liquid and liquid entrainment in the gas outlet stream are two additional important disadvantages.

Bubble columns are particularly well suited for applications when significant liquid hold-up and long liquid residence time are required. An advantage of these columns is their relatively low investment cost, a large mass transfer area and high mass transfer coefficients in both phases. Disadvantages of bubble columns include a high-pressure drop of the gas and significant back mixing of the liquid phase. The latter disadvantage can be reduced by introducing an inert packing of high porosity to the bubble column. Such a packing eliminates to a large extent the effects of liquid phase mixing along the column height. In addition the packing may also cause an increase in the mass transfer surface area with relation to the bubble column at the same flows of both phases. The advantages of bubble columns appreciably exceed their disadvantages and therefore an increasing number of these columns are applied in industry.

Nomenclature

A	$L/K \cdot G$, absorption factor, eq. (3.22)	–
f	Fraction absorbed or stripped (eqs. (3.29) and (3.34))	–
G	Gas flow	mol s^{-1}
H	Henry coefficient	–
ΔH_{vap}	Molar heat of vaporization	J mol^{-1}
K	Equilibrium constant, eq. (3.21)	–
K_A	Distribution coefficient, eq. (3.3)	–
L	Liquid flow	mol s^{-1}
N	Number of (theoretical) stages	–
p	Partial pressure	N m^{-2}
p^0	Saturation pressure	N m^{-2}
P_{tot}	Total pressure	N m^{-2}
R	Gas constant	J mol^{-1} K^{-1}
S	$K \cdot G/L$, stripping factor, eq. (3.30)	–
T	Temperature	K
x, y	Mole fraction	–
X, Y	Mole ratio	–
α	Selectivity, relative volatility and equilibrium constant	–
γ	Activity coefficient	–

Indices

A	Components
n	Stage number
ts	Theoretical stages
vap	Evaporation

Exercises

1   A plate tower providing six equilibrium stages is employed for stripping ammonia from a waste water stream by means of countercurrent air at atmospheric pressure and 25 °C. Calculate the concentration of ammonia in the exit water if the inlet liquid concentration is 0.1 mole% ammonia in water, the inlet air is free of ammonia and 2,000 standard cubic meter (1 atm, 25 °C) of air are fed to the tower per m³ of waste water. The absorption equilibrium at 25 °C is given by the relation $y_{NH3} = 1.414\ x_{NH3}$.
$M_{water} = 0.018$ kg mol^{-1}; $\rho_{water} = 1,000$ kg m^{-3}; $R = 8.314$ J mol^{-1} K^{-1}; 1 atm = 1.01325 bar.

2   When molasses is fermented to produce a liquor containing ethanol, a CO_2-rich vapor containing a small amount of ethanol is evolved. The alcohol can be recovered by absorption with water in a sieve tray tower. For the following conditions, determine the number of equilibrium stages required for countercurrent flow of

liquid and gas if the liquid flow rate equals 1.5 times the minimum liquid flow rate, assuming isothermal, isobaric conditions in the tower and neglecting mass transfer of all components except ethanol.

Entering gas:	180 kmol/h, 98% CO_2, 2% ethyl alcohol, 30 °C, 1.1 bar
Entering absorbing liquid:	100% water, 30 °C, 1.1 bar
Required ethanol recovery:	95%

The vapor pressure of ethanol amounts 0.10 bar at 30 °C, and its liquid phase activity coefficient at infinite dilution in water can be taken as 7.5.

3   A gas stream consists of 90 mole% N_2 and 10 mole% CO_2. We wish to absorb the CO_2 into water. The inlet water is pure and is at 5 °C. Because of cooling coils the operation can be assumed to be isothermal. Operation is at 10 bar. If the liquid flow rate is 1.5 times the minimum liquid flow rate, how many equilibrium stages are required to absorb 92% of the CO_2? Choose a basis of 1 mole/h of entering gas. The Henry coefficient of CO_2 in water at 5 °C is 875 bar.

4   A vent gas stream in your chemical plant contains 15 wt% of a pollutant, the rest is air. The local authorities want to reduce the pollutant concentration to less than 1 wt%. You have decided to build an absorption tower using water as the absorbent. The inlet water is pure and at 30 °C. the operation is essentially isothermal. At 30 °C your laboratory has found that at low concentrations the equilibrium data can be approximated by $y = 0.5 \cdot x$ (where y and x are weight fractions of the pollutant in vapor and liquid). Assume that air is not soluble in water and that water is nonvolatile.

a.   Find the minimum ratio of water to air $(L'/G')_{MIN}$ on a solute-free basis

b.   With $(L'/G') = 1.22 \ (L'/G')_{MIN}$ find the total number of equilibrium stages and the outlet liquid concentration

5   A gas treatment plant often has both absorption and stripping columns as shown below. In this operation the solvent is continually recycled. The heat exchanger heats the saturated solvent, changing the equilibrium characteristics of the system so that the solvent can be stripped. A very common type of gas treatment plant is used for the drying of natural gas by physical absorption of water in a hygroscopic solvent such as diethylene glycol (DEG). In this case dry nitrogen is used as the stripping gas.

a)   At a temperature of 70 °C and a pressure of 40 bar the saturated vapor pressure of water is equal to 0.2 bar. It is known that water and DEG form a nearly ideal solution. Calculate the vapor–liquid equilibrium coefficient and draw the equilibrium line.

b) Construct the operating line for $x_{in} = 0.02$, $y_{out} = 0.0002$ and $L/G = 0.01$. Determine the number of stages required to reduce the water molefraction from $y_{in} = 0.001$ to $y_{out} = 0.0002$.

c) How many stages are required for $L/G = 0.005$. What happens for $L/G = 0.004$. Determine the minimal L/G ratio to obtain the desired separation.

d) Desorption takes place at 120 °C and 1 bar. The saturated vapor pressure of water is equal to 2 bar. Construct the equilibrium and operating lines for desorption using $y_{in} = 0$ and $L/G = (L/G)_{max}/1.5$. x_{out} and x_{in} are to be taken from the absorber operating at $L/G = 0.01$.

e) Calculate analytically the number of stages in both sections.

f) Comment on the chosen value of the liquid mole fraction at the outlet of the absorber.

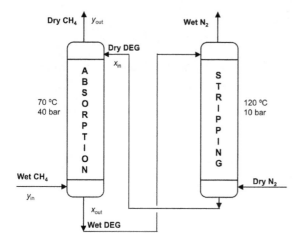

Schematic of natural gas absorptive drying operation

Chapter 4
General Design of Gas/Liquid Contactors

4.1 Introduction

In the previous chapters, it was shown how to calculate the number of theoretical stages to separate a given feed with two components of different volatility into two fractions of predefined composition. Distillation is usually conducted in vertical cylindrical vessels that provide intimate contact between the rising vapor and the descending liquid. A distillation column normally contains internal devices for such an effective vapor–liquid contact. *Their basic function is to provide efficient mass transfer between the two-phase vapor–liquid systems.* However, the short contact times between the two phases in distillation does not allow complete establishment of thermodynamic equilibrium.

The first aim of this chapter is to derive a tool to estimate the *column height* for a given separation. The number of real stages determines the height of the column. Therefore, we need to develop a model that translates the distance to the thermodynamic equilibrium to that number of real stages. The knowledge of the rate of interface mass transfer will appear to be very helpful in this matter. The second aim of this chapter is to estimate the *minimum diameter* of the column to be designed for a certain capacity.

Both goals are achieved by modeling the vapor–liquid contact, starting from a kinetic analysis of mass transfer between the two phases involved. To understand what factors influence the rate of interface mass transfer, we first derive rate expressions for such mass transfer in idealized hydrodynamic environments. Although parts of the models derived below are applicable to liquid–liquid extraction, the examples will focus on two important devices for distillation, absorption and stripping: the plate column and the packed column.

4.2 Modeling mass transfer

In gas–liquid contactors, two phases separated by an interface in between are present. Assume that a component A is diffusing from the gas phase to the liquid phase due to a difference in mole fraction as the driving force. Generally, diffusion as well as convection contributes to mass transport. Neglecting any other contribution, the rate of mass transfer of a component A at some distance z from the interface is given by

$$\Phi_A = -D^V_{AB} A_z \rho_V \frac{dy_A}{dz} + y_A \left(\Phi_A + \Phi_B\right) A_z \qquad (4.1a)$$

and

https://doi.org/10.1515/9783110654806-004

$$\Phi_A = -\, D^L_{AB}\, A_z\, \rho_L\, \frac{d\,x_A}{d\,z} + x_A\,(\Phi_A + \Phi_B)\, A_z \qquad (4.1b)$$

where ϕ_A is the molar transfer rate of A (mol s^{-1}); ρ_V, ρ_L are molar densities of vapor and liquid, respectively, (mol m^{-3}); D^V_{AB}, D^L_{AB} are coefficients of molecular diffusion in vapor and liquid, respectively (m^2 s^{-1}); A_z is the area at distance z from interface (m^2); y_A, x_A is the mole fraction of A in vapor and liquid, respectively (–); z is the distance to interface (m).

A mole fraction profile in case of spherical liquid or vapor droplets could be as shown in Figure 4.1a. In the stationary situation, the gradient shown is inversely proportional to the distance z squared.[1] It is very reasonable to assume that there would be no resistance to transfer across the interface[2]; hence, $y_{Ai} = K \cdot x_{Ai}$. Then, at low concentration ($x_A \ll 1$) or with *equimolecular diffusion* ($\phi_A = \phi_B$), the rate expressions given in eq. (4.1) reduce to

$$\Phi_A = -\, D^V_{AB}\, A_{LV}\, \rho_V\, \frac{d\,y_A}{d\,z}\bigg|_{z=0} = -\, D^L_{AB}\, A_{LV}\, \rho_L\, \frac{d\,x_A}{d\,z}\bigg|_{z=0} \qquad (4.2)$$

where A_{LV} is the interfacial area between liquid and vapor, comprising the total surface area of all droplets. This area depends on the droplet size, d_p. For n vapor

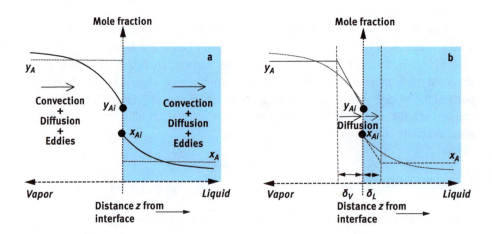

Figure 4.1: Mole fraction profiles around spherical liquid or vapor droplets: (a) actual profiles and (b) profiles according to film theory.

---

1 See Exercise 4.1.
2 Except for very fast reactions at the interface, see R.B. Bird, W.E. Stewart & E.N. Lightfoot, *Transport Phenomena*, 2nd Edition, John Wiley & Sons, **2007**.

droplets in a total volume V, the ratio of interfacial area A_{LV} and a vapor volume[3] $\varepsilon_V \cdot V$ reads (see also eq. (6.1))

$$\frac{A_{LV}}{\varepsilon_V \cdot V} = \frac{n \cdot \pi d_p^2}{n \cdot \frac{\pi}{6} d_p^3} = \frac{6}{d_p} \tag{4.3}$$

A useful simplification is achieved by adopting the *film mass transfer theory* to relate the driving force and the transfer rate. This theory is based on two assumptions (see Figure 4.1b):
  - a thin, laminar film exists at eitherside of the interface, and
  - in this film only molecular diffusion is taking place.

Hence, according to this theory, no gradients exist in the bulk of the two phases. Resistance to mass transfer is concentrated in the boundary layers where only *molecular diffusion* occurs. The stationary state rate expressions according to the film theory are

$$\Phi_A = k_V A_{LV} \rho_V (y_A - y_{Ai}) = k_L A_{LV} \rho_L (x_{Ai} - x_A) \tag{4.4}$$

where k_V, k_L is film or single-phase mass transfer coefficients (m s^{-1}).

When the boundary layers are very thin, the gradients in eq. (4.2) can be replaced by their difference quotients. Now, comparison with eq. (4.4) reveals that the mass transfer coefficient, k, equals the ratio of molecular diffusivity and film thickness, D_{AB}/δ, where $\delta = \Delta z$. Literature data of molecular diffusivities in the gas phase show values in the order of $0.1 \cdot 10^{-4} - 10^{-4}$ m^2 s^{-1}, whereas mass transfer coefficients appear to be in the order of 0.01 m s^{-1}. Thus, the order of magnitude of the film thickness in the gas phase, δ_V, is $10^{-3} - 10^{-2}$ m. Diffusivities in the liquid phase are much smaller in the order of 10^{-9} m^2 s^{-1} and experimental values of liquid mass transfer coefficients are around 10^{-4} m s^{-1}. Thus, the film thickness in the liquid phase, δ_L, is an order of magnitude smaller than that in the gas phase. In either case, the film thickness is sufficiently small to assume linear concentration gradients. The simplified picture according to the film theory is shown in Figure 4.1b. This gives only a rough, inaccurate model of transport through liquid vapor interfaces. Nevertheless, the film theory appears to be an adequate tool for many engineering applications, especially when it comes to understanding principles of mass transfer in absorption or distillation towers.

The interface compositions x_{Ai} and y_{Ai}, shown in Figure 4.1, are connected through eq. (4.4), which can be rearranged to show as a linear operating line in a y–x diagram:

---

3 ε_V = vapor volume fraction.

$$y_A - y_{Ai} = -\frac{k_L \, \rho_L}{k_V \, \rho_V}(x_A - x_{Ai}) \tag{4.5}$$

The slope of this straight line depends on the ratio of the two mass transfer coefficients involved. The absence of mass transport resistance in the interface suggests that *both interface compositions are in equilibrium*. This assumption gives the second equation required to calculate the interface compositions. A graphical method to do so is presented in Figure 4.2.

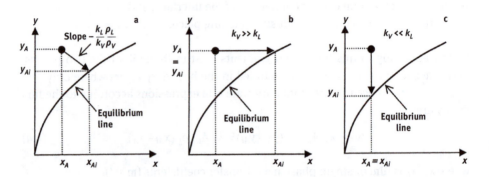

Figure 4.2: Driving force and interface compositions. (a) Transport in both liquid and vapor determines the overall transport rate; (b) transport in vapor is very fast; and (c) transport in liquid is very fast.

Equation (4.5) and the equilibrium curve are plotted in a y–x graph. The intersection of the straight line with the equilibrium curve gives the values of x_{Ai} and y_{Ai}. Two limiting situations can be considered. When the transport in the gas phase is very fast compared to that in the liquid phase, the gradient across the gas film can be neglected to that across the liquid film. All resistance to mass transport is said to be in the liquid phase; this situation is depicted in Figure 4.2b. Figure 4.2c shows the opposite situation where the resistance is entirely in the gas phase.

The application of the above method to model transfer rates in G/L contactors would be much easier to handle when the interface compositions could be eliminated from the driving force factor. The following mathematical treatment will do just that.[4] For the sake of simplicity, we assume a linear equilibrium expression:

$$y_A^* = K \cdot x_A \tag{4.6}$$

where y_A^* is a hypothetical gas phase composition in equilibrium with the actual mole fraction x_A in the liquid phase (see diagrams page 85). Combination with eq. (4.4) gives

---

4 See also Exercise 4.2.

$$y_A - y_{Ai} = \frac{\Phi_A}{k_V A_{LV} \rho_V} \quad \text{and} \quad K \cdot x_{Ai} - K \cdot x_A = y_{Ai} - y_A^* = \frac{K \cdot \Phi_A}{k_L A_{LV} \rho_L} \tag{4.7}$$

Summation of both expressions eliminates y_{Ai} giving (see eq. 4.8)

$$\Phi_A = k_{OV} A_{LV} \rho_V \left(y_A - y_A^* \right) \tag{4.8}$$

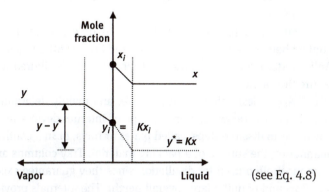

(see Eq. 4.8)

with

$$k_{OV} = \frac{k_V}{1 + \frac{k_V}{k_L} \cdot \frac{K \rho_V}{\rho_L}} = \frac{k_V}{1 + \frac{m k_V}{k_L}}$$

k_{OV} = overall mass transfer coefficient, based on the gas phase $\left(\text{ms}^{-1} \right)$

$m = \frac{K \rho_V}{\rho_L}$, volumetric distribution coefficient (see eq. 4.9)

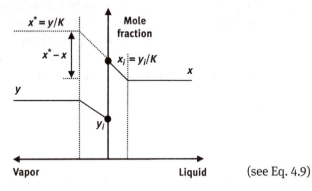

(see Eq. 4.9)

Similarly, elimination of x_{Ai} from eq. (4.4) gives a rate expression with an *overall mass transfer coefficient* k_{OL}, applicable when the main resistance against mass transfer is in the liquid phase:

$$\Phi_A = k_{OL} A_{LV} \rho_L \left(x_A^* - x_A \right) \tag{4.9}$$

with summarizing, the rate of mass transfer across a phase boundary depends on

- the driving force, $y_A - y_A^*$ or $x_A^* - x_A$, the asterisk $*$ referring to a hypothetical equilibrium;
- the contact area A_{LV} between the two phases; and
- an overall mass transfer coefficient, k_{OV} or k_{OL}.

The maximum mass transfer rate is obtained when these three factors on the right-hand side of eqs. (4.8) or (4.9) are as large as possible:

- The *largest driving force* is achieved when the overall flow pattern allows countercurrent contact between equal streams of gas and liquid without significant remixing. An important condition is that both phases are distributed uniformly over the entire flow area.
- A *large contact area* A_{LV} is desirable for mass transfer and mainly determined by the used column internals. Depending on the type of internal devices used, the contacting may occur in discrete steps, called plates or trays, or in a continuous differential manner on the surface of a packing material. Tray columns and packed columns are most often used for distillation since they guarantee excellent countercurrent flow and permit a large overall height. The internals provide a large mass transfer area. In tray columns, mass transfer areas range from 30 to 100 m^2 per unit tray area,[5] and in packed columns from 200 to 500 m^2 per unit column volume.
- The mass transfer coefficient increases proportionally with the relative velocities between the liquid and vapor phases and is also improved by constant regeneration of the contact area between the phases.

It is difficult to measure the overall mass transfer coefficients as such. The usual experimental setups designed to determine mass transport under well-defined conditions (temperature, pressure, concentration, hydrodynamic flow pattern) produce data to calculate the *product of mass transfer coefficient and contact area*. In the following sections, L/G mass transfer will be modeled in tray – or plate – columns as well as in packed columns.

4.3 Plate columns

Figure 4.3 shows the most important features of a tray column. The gas flows upward within the column through perforations in horizontal trays, and the condensed liquid

---

5 Specific interfacial area A_{LV} in gas–liquid systems may be estimated from gas volume fraction ε and bubble diameter d as $A_{LV} = 6\varepsilon/d$ in $m^2\ m^{-3}$ (see eq. (4.3), also eq. (6.2) for interfacial area in cylinders).

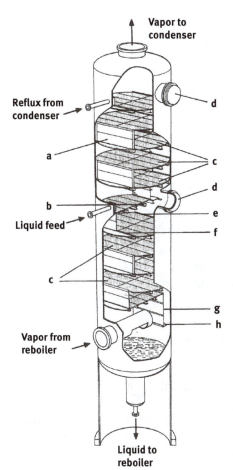

Vapor to condenser

Reflux from condenser → d

a

c

d

b

e

Liquid feed → f

c

g

h

Vapor from reboiler →

Liquid to reboiler

Figure 4.3: Cutaway section of a plate column: a, downcomer; b, tray support; c, sieve trays; d, man way; e, outlet weir; f, inlet weir; g, side wall of downcomer; and h, liquid seal. Reproduced with permission from [26].

flows countercurrently downward. However, as indicated in Figure 4.4, the two phases exhibit cross-flow to each other on the individual trays. The liquid enters the cross-flow plate from the bottom of the downcomer belonging to the plate above and flows across the perforated active or *bubbling area*. The ascending gas from the plate below passes through the perforations and aerates the liquid to form a *large interfacial area* between the two phases. It is in this zone where the main vapor–liquid mass transfer occurs. The vapor subsequently disengages from the aerated mass on the plate and rises to the tray above. The aerated liquid flows over the exit weir into a *downcomer*, where most of the trapped vapor escapes from the liquid and flows back to the interplate vapor space. Some of the liquid accumulates in each downcomer to compensate for the pressure drop caused by the gas as it passes through the tray. The liquid then leaves the plate by flowing through the downcomer outlet onto the tray below. In large diameter cross-flow plates, multiple liquid flow path

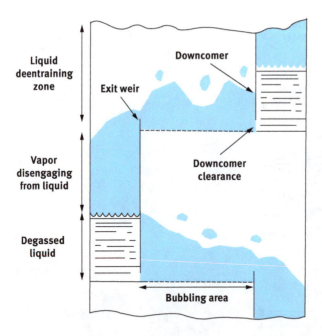

Figure 4.4: Schematic of flow pattern in a cross-flow plate distillation column.

plates with multiple downcomers are used to prevent the hydraulic gradient of the liquid flowing across the plate becomes excessive.

Three principal vapor–liquid contacting devices are used for cross-flow plate design: the sieve plate, the valve plate and the bubble cap plate (see Figure 4.5):

- *Bubble cap plates* have been used almost exclusively in the chemical industry until the early 1950s. As shown in Figure 4.5, their design prevents liquid from leaking downward through the tray. The vapor flows through a hole in the plate floor, through the riser, reverses direction in the dome of the cap, flows downward and exits through the slots in the cap. However, the complex bubble caps are relatively expensive and have a higher pressure drop than the former two designs. This limits their usage in newer installations to low liquid flow rate applications or to those cases where the widest possible operating range is desired.
- *Sieve plates* have become very important because they are simple, inexpensive, have high separation efficiency and produce a low pressure drop across the tray. Conventional sieve plates contain typically 1–12 mm holes and exhibit ratios of open area to active area ranging from 1:20 to 1:7. If the open area is too small, the pressure drop across the plate is excessive, if the open area is too large the liquid weeps or dumps through the holes.
- *Valve plates* are a relatively new development that represents a variation of the sieve plate with liftable valve units such as those shown in Figure 4.6 fitted in the holes. The liftable valves prevent the liquid from leaking at low gas loads

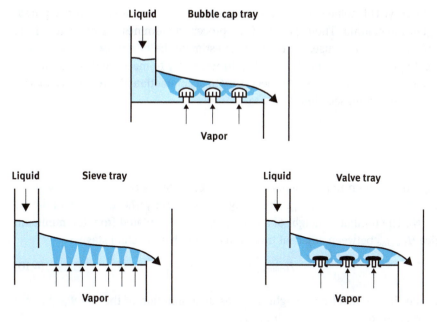

Figure 4.5: Schematic of a bubble cap, sieve and valve tray.

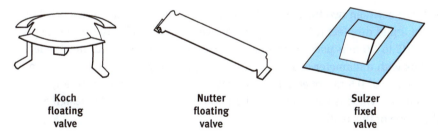

| Koch floating valve | Nutter floating valve | Sulzer fixed valve |

Figure 4.6: Examples of valves used in valve plates.

and to avoid excessive pressure increase at high gas loads. The main advantage is the ability to maintain efficient operation while being able to vary the gas load up to a factor of 4–5. This capability gives valve trays a much larger operational flexibility than any other tray design.

4.3.1 Dimensioning a tray column

Product specification of a distillation, absorption or stripping process includes at least product purity and capacity. In the previous chapters, methods are discussed to determine the number of theoretical stages required to achieve the desired product purity. This relates to the height of the column, the more trays are required, the higher the column. The capacity of the process determines the diameter of the column. This section focuses on methods to estimate the *minimum dimensions* of a tray column: the *minimum column height* necessary to meet the desired product purity and the *minimum diameter* necessary to accommodate the required capacity (at constant pressure, see Figure 2.3).

4.3.2 Height of a tray column

Although the column requirements are calculated in terms of *theoretical or equilibrium number of stages*, N_{ts}, the design must specify the *actual number of stages*, N_s. The minimum height of a column is then calculated from the minimum distance $H_{spacing}$ between adjacent trays and the number of actual stages:

$$H_{column} = H_{spacing} \cdot N_s \tag{4.10}$$

Thus, an estimate of column height involves determination of the *number of real stages*, N_s, and the *tray spacing*, $H_{spacing}$.

4.3.2.1 Tray spacing
The distance between two trays should be as small as possible but sufficiently large to prevent liquid droplets to reach the tray above. In industrial tray columns, spacing from 0.15 to 1.0 m is used. This value depends on vapor load and tray diameter. In columns with a diameter of 1 m or larger, a typical tray spacing of 0.3–0.6 m is found. A value of 0.5 m appears to be a reasonable initial estimate. In a detailed tray design, this value is subject to revision; however, that part of the design is outside the scope of this textbook.

4.3.2.2 Tray efficiency
Due to relatively high flow rates in gas–liquid towers, contact times on trays are not sufficient to establish thermodynamic equilibrium. A measure of the performance of a real tray in approaching equilibrium is the *efficiency* of a real tray. In the previous chapters, methods were discussed to determine the total number of theoretical *stages* for a given separation such as distillation, absorption or stripping. The *overall efficiency* E_o of a column is defined by the number of theoretical trays, N_{ts}, divided by the actual number of trays required to obtain the desired product purity.[6]

---

6 In distillation where the reboiler counts as one theoretical stage, this definition should be adjusted accordingly.

$$E_o \equiv \frac{N_{ts}}{N_s} \tag{4.11}$$

Most hydrocarbon distillation systems in commercial columns achieve overall tray efficiencies of 60–80%[7]; in absorption processes the range is 10–50%. These figures are just guidelines. They have no theoretical basis and are not suitable for design purposes. A more fundamental approach would be to derive an expression for the efficiency of a single plate in terms of local driving force for interphase mass transport and hydrodynamic conditions. First, we need a hydrodynamic model of a tray. A reasonable model would be to assume that the gas phase, passing at high velocity through the liquid layer on a tray, behaves as a plug flow. Further that its concentration changes only in the axial direction, and radial concentration gradients can be neglected. The gas flow causes extensive mixing in the liquid layer, so the liquid is assumed being ideally mixed. The composition y_n in gas phase leaving the nth tray is independent of the radial position. Now the tray, plate or *Murphree efficiency* E_{MV} compares the actual difference $y_{n+1} - y_n$, leaving two consecutive trays to the maximum achievable difference $y_{n+1} - y_n^*$:

$$E_{MV} = \frac{y_{n+1} - y_n}{y_{n+1} - y_n^*} \tag{4.12}$$

Figure 4.7 gives an amplification of the *tray efficiency* in an absorber. Here, the lighter component in the gas phase from tray $n + 1$ is absorbed by the liquid, which has the same composition x_n everywhere on the tray. At increasing height in the

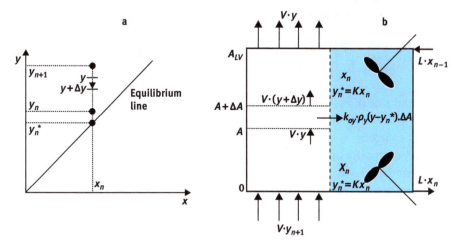

Figure 4.7: Mass transport on a model nonequilibrium tray in absorber: (a) distance to equilibrium and (b) elements in mass balance for interphase mass transport.

---

7 E_o sometimes is expressed as a percentage by multiplication with 100.

liquid layer, the mole fraction y decreases. On an equilibrium tray the composition of the gas phase leaving tray n would equal y_n^*, whereas on a real tray the exit value amounts to y_n. In the following paragraph, the concept of *transfer units* is used to estimate the number of real trays, N_s.

4.3.2.3 Transfer units

The next step is to relate the mole fractions in eq. (4.12) to kinetic parameters and flow rates. The hydrodynamic simplification for a plate defined above (liquid ideally mixed, plug flow in gas phase) allows setting up a mass balance in a thin liquid layer at a certain position h above the tray. Instead of height h, it is convenient to take the interphase A_{LV} present on the tray as independent parameter. A thin liquid layer with thickness Δh comprises ΔA unit surface area. The driving force for interphase transport at this position equals $y - y_n^*$. Note that y_n^*, the virtual gas phase composition in equilibrium with the liquid of composition x_n, is constant. In the stationary situation, the difference in molar flow of the absorbable component in the gas phase is balanced by interphase transport (see Figure 4.7). Applying the rate expression given in eq. (4.8), the mass balance becomes

$$V \cdot y = V \cdot (y + \Delta y) + k_{OV} \cdot \rho_V \cdot (y - y_n^*) \cdot \Delta A \tag{4.13}$$

For the sake of clarity, we restrict ourselves to the case that heat effects can be neglected and that the molar vapor flow rate, V, is constant, for example, *equal molal overflow* and/or the concentration is low (<10 mol%). Taking the differential and integrating from $y = y_{n+1}$ at the liquid–tray interphase where interface surface area $A = 0$ to $y = y_n$ at the top of the liquid layer, where total LV–surface area $A = A_{LV}$, it follows that

$$\int_{y_{n+1}}^{y_n} \frac{dy}{y_n^* - y} \equiv N_{OV} = \ln \frac{y_{n+1} - y_n^*}{y_n - y_n^*} = \frac{k_{OV} \cdot A_{LV}}{V / \rho_V} \tag{4.14}$$

The left-hand side of eq. (4.14) represents the ratio of actual change in composition to driving force and characterizes the difficulty of separation. This dimensionless quantity is called the *overall number of gas phase transfer units N_{OV}*. The larger the number of transfer units, the more efficient the tray.

Combination of eqs. (4.12) and (4.14) relates the number of transfer units to the *Murphree efficiency, E_{MV}*:

$$E_{MV} = 1 - e^{-N_{OV}} \tag{4.15}$$

This expression for the Murphree efficiency holds for mass transfer between a vapor phase plug flow and an ideally mixed liquid phase and is applicable to all three types of trays. The right-hand side of eq. (4.14) contains factors that depend on tray dimensions and operating conditions, such as liquid and vapor load. It can be interpreted as the ratio of interphase mass transport $k \cdot A$ ($m^3\ s^{-1}$) to convective vapor flow $Q = V/\rho_V$ ($m^3\ s^{-1}$). This ratio is connected to the *height of a transfer unit*, H_{tu}

$$N_{OV} = \frac{k_{OV} \cdot A_{LV}}{Q_V} = \frac{k_{OV} \cdot A_h}{Q_V} \cdot H \equiv \frac{H}{H_{tu}} \qquad (4.16)$$

where A_h is the interfacial area per unit length and H is the height of liquid layer. In a plate column, H_{tu} has little meaning for the determination of the real number of stages. In Section 4.4, the concept of the height of a transfer unit will be applied to estimate the height of a packed column.

4.3.2.4 Overall efficiency
Finally, the overall efficiency E_O in the column should be related to the tray efficiency E_{MV}. We have to find the number of real trays with efficiency E_{MV} that achieve the same change in composition as with N_{th} theoretical trays. A graphical method would be to replace the equilibrium line for theoretical stages in a McCabe–Thiele plot with a pseudoequilibrium line for real stages as shown in Figure 2.20. Real stages are stepped off between this pseudoequilibrium line and the operating line. Keep in mind that the reboiler in a distillation column is considered to count as one theoretical stage. If the operating and equilibrium lines are straight, an analytical expression is more convenient. In the x–y plot of Figure 4.8, step ABC represents an equilibrium

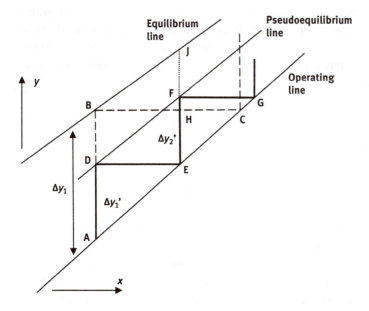

Figure 4.8: Construction of real, on-equilibrium trays subject to mass transport resistance. *ABC* represents an equilibrium plate; *ADE* and *EFG* represent real trays with constant tray efficiency E_{MV}.

tray with an arbitrary equilibrium change in vapor composition Δy_1. ADE and EFG illustrate the effect of two nonequilibrium trays on vapor composition, Δy_1^* and Δy_2^*, respectively. By definition, AD/AB and EF/EJ are the tray efficiencies E_{MV}. At constant values of E_{MV}, D and F are points of a straight *pseudo-equilibrium line*. In this case, the ratio of change in composition of two consecutive trays is constant:

$$\frac{\Delta y_2'}{\Delta y_1'} = \frac{EF}{AD} = \frac{EF}{EJ} \cdot \frac{EJ}{AD} = E_{MV} \cdot \frac{(JH + HE)}{AD} \tag{4.17}$$

but $JH = K \cdot HB = \frac{KV}{L} \cdot AD$ and $HE = AB - AD = \left(\frac{1}{E_{MV}} - 1\right) \cdot AD$; thus, the ratio $\frac{\Delta y_1^*}{\Delta y_2^*}$ is constant:

$$\frac{\Delta y_2'}{\Delta y_1'} = 1 + E_{MV}(S - 1) = s \tag{4.18}$$

with separation factor $S = S_S = KV/L$ for stripping (as in Figure 4.8) and $S = S_A = L/KV$ for absorption. The total change in composition achieved by N trays, each with efficiency $E_{MV} = \Delta y_1'/\Delta y_1$ equals:

$$\sum_{1}^{N} \Delta y_i' = \Delta y_1 \cdot E_{MV} \cdot (1 + s + s^2 + \cdots + s^{N-1}) = \Delta y_1 \cdot E_{MV} \cdot \frac{s^N - 1}{s - 1} \tag{4.19}$$

This result would be the first step to derive a general, *non-equilibrium Kremser equation* that transforms into the equilibrium Kremser equations (3.29) or (3.34) for $E_{MV} \to 1$. Here, we just need eq. (4.19), which relates the number of trays and the total composition change in one phase. N_s real trays should give the same total change in composition as N_{th} theoretical plates with $E_{MV} = 1$ and $s = S$. Application of eq. (4.19) to both types of trays gives

$$\Delta y_1 \cdot E_{MV} \cdot \frac{s^{N_s} - 1}{s - 1} = \Delta y_1 \cdot \frac{S^{N_{th}} - 1}{S - 1} \tag{4.20}$$

and, after rearranging

$$E_O = \frac{N_{th}}{N_s} = \frac{\ln[1 + E_{MV}(S - 1)]}{\ln(S)} \tag{4.21}$$

Now the overall efficiency E_O and the real number of stages N_s can be calculated from the separation factor and the tray efficiency. With the initial estimate of tray spacing and the number of real stages, the column height can be calculated from eq. (4.8). It should be noted that the hydrodynamic model used above is an oversimplification. Especially on larger trays, the assumption of complete mixing is far from realistic. In fact, cross-flow is more likely, as shown in Figure 4.4. Liquid flows from the downcomer clearance across the plate and leaves it via the exit weir. The liquid entering the plate is richer in the volatile component than is the liquid leaving the plate. Vapor enrichment is higher than with complete mixing, where the liquid composition anywhere on the tray equals the lower exit composition. Consequently, tray efficiency in cross-flow is higher than with complete mixing. Cross-flow models would be more appropriate to quantify this effect and are available in literature [5]. However, for a basic understanding of the phenomena determining the column size, the simple mixing-plug flow model is adequate.

4.3.3 Diameter of a tray column

In the previous section, methods are discussed to estimate column height for a given separation from first principles in thermodynamics and kinetics. Any approach involves calculation of the slope of an operating line, L/V. The molar flow rates of liquid, L and vapor, V, are also of great importance in estimating a column diameter. More detailed column dimensioning requires determination of the entire operating region bound by a range of liquid and vapor flow rates as shown in Figure 4.9. Here we limit ourselves to a discussion of the influence of flow rates on performance, only to illustrate the most important phenomena occurring on a tray. More attention will be paid to the maximum vapor load of the operating region

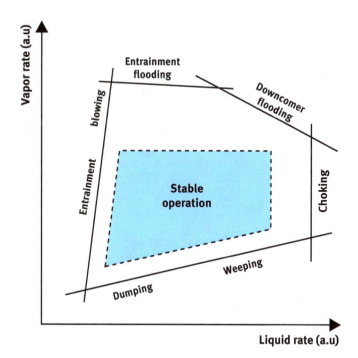

Figure 4.9: Schematic operating diagram for cross-flow plates.

because, at a given vapor volume flow rate Q_V in $(m^3 \, s^{-1})$, the minimum column diameter is determined by the maximum allowable vapor flow rate $u_{V,max}$ in $(m \, s^{-1})$.

Figure 4.9 shows several mechanisms that limit the operation range of a sieve plate. Boundaries to the region of stable operation exist at minimum and maximum vapor and liquid loads. The *weeping* (or *dumping*) line represents the minimum operable vapor flow rate at various liquid flow rates. Below the line, the vapor rate is too low to maintain the liquid on the plate and the liquid weeps, or even dumps, through the plate orifices. At low liquid loads, liquid droplets may be carried with the vapor to the tray above. This *entrainment* lowers the tray efficiency because liquid of lower volatility is recycled to a tray of higher volatility. *Coning* occurs, at higher vapor loads, when the vapor pushes aside the liquid from the orifices. This prevents a good interphase contact, also decreasing the plate efficiency. To minimize this effect, exit weirs (see Figure 4.3) should be high enough to ensure sufficient liquid on the tray. Flowing of the vapor upward through the downcomers bypassing the tray is called *blowing*, which may also occur at low liquid loads, especially if the downcomer slit is large. Extremely low liquid loads cause an uneven liquid distribution across the tray also decreasing mass transfer efficiency. These lower limits are not absolute, although below these limits separation efficiency becomes worse.

Unlike these lower limits, the upper limits discussed below are absolute. Above these upper limits, column operation stops. At a high liquid load, especially with small downcomer areas, *choking* may occur: the liquid in the downcomer drags vapor down to the lower tray. *Downcomer flooding*, also called *overflowing*, occurs when the liquid flow through the column becomes larger than the downcomer capacity. The liquid fills the downcomer and recycles back to the previous tray with the rising vapor and proper countercurrent column operation breaks down. High vapor rates increase pressure drop across the tray, encouraging flooding. At high vapor loads, liquid can be blown off the tray in the form of fine droplets. This *entrainment flooding* prevents the liquid from flowing countercurrently to the vapor. Again, proper column operation stops.

4.3.3.1 Flooding
Apart from geometrical ratios considering tray configuration, six factors can be identified as having a major importance considering these limits: superficial vapor and liquid velocity, u_V and u_L, vapor and liquid density[8] ρ_V and ρ_L, tray spacing $H_{spacing}$ and gravitational constant g. All six factors determine the pressure drop across a hole in a tray. From dimensional analysis, it appears that these six factors can be reduced to two dimensionless numbers: a *gas number* $\frac{u_V}{\sqrt{gH}}\sqrt{\frac{\rho_V}{\rho_L}}$ and a *flow*

---

8 Note that the dimensions of ρ_L and ρ_V are in $kg \, m^{-3}$.

parameter $\frac{u_L}{u_V}\sqrt{\frac{\rho_L}{\rho_V}}$. In literature, flooding properties are given in charts where gas numbers are usually plotted as a function of a flow parameter taking the form $\frac{Q_L}{Q_V}\sqrt{\frac{\rho_L}{\rho_V}}$. Q_V and Q_L are volume flow rates (m³ s⁻¹) obtained from the molar vapor and liquid flows emerging from the stage-wise calculations discussed in the previous two chapters.

The flow parameter represents the ratio of kinetic energies of the two phases and indicates the type of the two-phase mixture on a tray. A high value suggests a bubbly liquid with a dispersed gas phase, which is characteristic for absorption columns at high pressure. In such a system, the vapor rate can be increased until the point where the pressure drop across the tray balances the pressure drop in the down-comer. A higher liquid load starts downcomer flooding. A low value of the flow parameter indicates a system where the vapor phase is continuous with dispersed liquid, as found in vacuum distillation. In this region, too high a value of vapor rate causes entrainment flooding and liquid droplets are blown to the tray above.

In the following, a simple model for each of the two types of flooding, down-comer entrainment flooding, respectively, will be derived, predicting the maximum allowable vapor rate to avoid flooding. This maximum flow rate determines the *minimum tray diameter*. These derivations use two geometric ratios:

$$\varphi = \frac{A_{\text{holes}}}{A_{\text{bubble}}} \quad \text{and} \quad \sigma = \frac{A_{\text{slit}}}{A_{\text{bubble}}} \tag{4.22}$$

where φ is the fraction hole area relative to the active or bubbling area A_{bubble} of the tray. A_{bubble} is that part of the tray not blocked by the downcomers. A_{slit} is the vertical surface area of the slit at the exit of the downcomer, equal to the product of the width of the slit and the downcomer clearance (the height of the slit, see Figure 4.4).

4.3.3.2 Downcomer flooding

At high liquid loads, the vapor rate can be increased to a point where the pressure drop across the orifices or holes equals that across a completely filled downcomer. Any further increase would overflow the downcomer, backing up the liquid to the tray above. A fully loaded downcomer contains liquid as well as a turbulent two-phase mixture on top. In the pressure balance below, it is assumed that the effective liquid head is just half of the tray spacing. Reminding that $u_{L,\text{slit}} = \frac{u_L}{\sigma}$, $u_{V,\text{hole}} = \frac{u_V}{\varphi}$, and neglecting the contribution of the liquid layer on the tray to the pressure drop, a stationary state pressure balance gives

$$\Delta p = \Delta p_1 + \Delta p_2 \text{ or } c_1 \rho_V \left(\frac{u_{V,\text{max}}}{\varphi}\right)^2 = \rho_L g \frac{H_{\text{spacing}}}{2} - c_2 \rho_L \left(\frac{u_L}{\sigma}\right)^2 \tag{4.23}$$

The vapor pressure drop across the holes on the left-hand side of eq. (4.23) balances the downcomer pressure drop (static pressure by the liquid minus pressure drop across the slit) on the right-hand side. Rearrangement of eq. (4.23) and replacing the velocity ratio by the volume flow ratio leads to the following downcomer flooding line equation:

$$\frac{u_{V.\text{flood}}}{\sqrt{g \cdot H_{\text{spacing}}}} \sqrt{\frac{\rho_V}{\rho_L}} = \frac{\phi}{\sqrt{c_1}} \sqrt{\frac{0.5}{1 + \frac{c_2 \phi^2}{c_1 \sigma^2} \frac{\rho_L Q_L^2}{\rho_V Q_V^2}}} \tag{4.24}$$

At these higher liquid loads, smaller fractional hole areas are preferable. These values are plotted in Figure 4.10, for reasons of simplicity as a straight line ranging from ϕ = 6% to ϕ = 10%. In this plot, the two flooding lines, qualitatively depicted in Figure 4.9, are now substantiated.

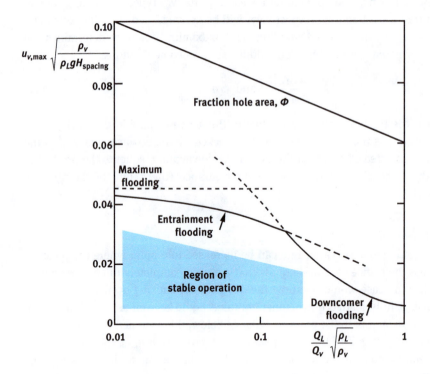

Figure 4.10: Flooding properties of a sieve tray.

The friction coefficients in eq. (4.24) (c_1 for holes and c_2 for slits) have to be determined by experiments. Estimates can be obtained from first principles. In the turbulent conditions on a sieve tray, reasonable values for c_1 and c_2, are 1.1 and 1.6, respectively. To obtain the *maximum allowable vapor rate*, one more unknown in eq. (4.24) has to be assigned a value to σ, the ratio of vertical downcomer slit area and active tray area (eq. (4.22)). In practice, a value between 0.01 and 0.05 is used. The gas number at flooding velocity is plotted in Figure 4.10 as a function of the flow parameter. In a more detailed design, the first estimate of σ need to be optimized, which requires recalculating the maximum vapor rate from the adjusted downcomer flooding line, which will not be discussed here.

4.3.3.3 Entrainment flooding

The second mechanism occurring on a sieve tray, causing flooding by entrainment, involves tearing apart the liquid by the high kinetic energy of the vapor. The line describing entrainment flooding is not available as a function explicit in the flow parameter. Moreover, it contains all geometrical ratios of a sieve tray and therefore would require a detailed design of the tray, which is outside the scope of this textbook. In Figure 4.10, we just give an entrainment flooding line for a nonspecified tray without deriving the implicit function describing it. In the region where either mechanism might dominate, the lower gas number prevails. The thick parts of both flooding lines indicate this.

The limiting value of the entrainment flooding line can be derived easily by noting that the *Weber number* for droplet formation relates the disruptive aerodynamic forces and the restoring surface tension forces:

$$We = \frac{\rho_L\, d\, (u_V/\varphi)^2}{\gamma_{LV}} \tag{4.25}$$

where d is the diameter of liquid droplet and γ_{LV} is surface tension (N m^{-1}). At a vapor rate, just sufficient to start droplet formation, the critical value of the Weber number is approximately 12; at lower values, surface tension restores the liquid surface. Once formed, the stationary falling rate of a droplet should at least equal the upward vapor velocity to avoid entrainment. In this limiting condition, the stationary state force balance states that apparent weight equals resistance force:

$$\frac{1}{6}\pi d^3 \cdot (\rho_L - \rho_V) \cdot g = C_w \cdot \frac{1}{4}\pi d^2 \cdot \frac{1}{2}\rho_V \left(\frac{u_{V,\,max}}{\varphi}\right)^2 \tag{4.26}$$

For a sphere in the turbulent region, the friction factor $C_w = 0.4$. Elimination of the droplet diameter d from the latter two equations, applying the critical value of the Weber number at $u_V = u_{V,\text{flood}}$, leads to the following expression for the *highest allowable gas number*:

$$\frac{u_{V,\text{flood}}}{\sqrt{gH_{\text{spacing}}}}\sqrt{\frac{\rho_V}{\rho_L}} = 2.51\left[\frac{\varphi^2\gamma_{LV}(\rho_L-\rho_V)}{\rho_L^2 gH_{\text{spacing}}^2}\right]^{0.25} \tag{4.27}$$

The fractional hole area ϕ at the lowest liquid loading is read from the upper line in Figure 4.10. The gas number thus calculated is shown as a horizontal square dotted line in Figure 4.20.

4.3.3.4 Tray diameter and pressure drop

At this stage, the maximum allowable vapor rate to avoid flooding can be calculated at any given value of the flow parameter. The flooding lines in Figure 4.10 are based on either eq. (4.24) (downcomer flooding) or on eq. (4.27) (entrainment flooding). The lowest value of the two is used for the diameter calculation. For safety reasons, a *value of 85% of the flooding velocity* should be used while the downcomer area will occupy a further 15% of the active tray (bubble) area. The active hole area and the minimum column diameter then follow from

$$A_{\text{bubble. min}} = \frac{Q_V}{0.85\cdot u_{V.\text{food}}} \quad \text{and} \quad D_{\text{min}} = \sqrt{\frac{4}{\pi}\cdot\frac{A_{\text{bubble. min}}}{0.85}} \tag{4.28}$$

Finally, the pressure drop across a sieve tray is calculated. The total pressure drop comprises two contributions, one due to the resistance in the holes (see eq. (4.23)) and another due to the static pressure of the liquid layer with height h_{liq} on the tray:

$$\Delta p_{\text{tray}} = \Delta p_{\text{dry}} + \Delta p_{\text{iq}} = c_1\cdot\rho_V\cdot\left(\frac{0.85\cdot u_{V.\text{flood}}}{\phi}\right)^2 + \rho_L\cdot g\cdot h_{\text{liq}} \tag{4.29}$$

4.3.3.5 Remark on flooding charts

Many authors use the flooding chart of Fair (as given in, e.g., *Perry's Chemical Engineers' Handbook* [16]), who plotted a nondimensionless capacity factor C_{sb} as a function of the flow parameter for various values of tray spacing. Vapor rate and capacity factor are related through

$$u_{V,\text{flood}} = C_{sb,\text{flood}}\left(\frac{\gamma_{LV}}{0.020}\right)^{0.2}\left(\frac{\rho_L-\rho_V}{\rho_V}\right)^{0.5} \tag{4.30}$$

In yet other published flooding charts, the superficial vapor rate is plotted as a function of superficial liquid rate, or as a gas number $\frac{u_V}{\sqrt{gH}}\sqrt{\frac{\rho_V}{\rho_L}}$ as a function of liquid number $\frac{u_L}{\sqrt{gH}}$. All these types of flooding diagrams are usually subject to some restrictions:

- the vapor–liquid system is nonfoaming;
- the weir height 15% of plate spacing or less;

- sieve plate perforations are 13 mm or less in diameter;
- if not specified in the chart, the fraction hole area is 0.1;
- if not specified in the chart, the surface tension is 0.020 N m^{-1}.

4.4 Packed columns

The most important features of a packed column are shown in Figure 4.11. Obviously, the most important element of packed column internals is the packing itself. To promote mass transfer, the packing should have a large surface area per unit of volume

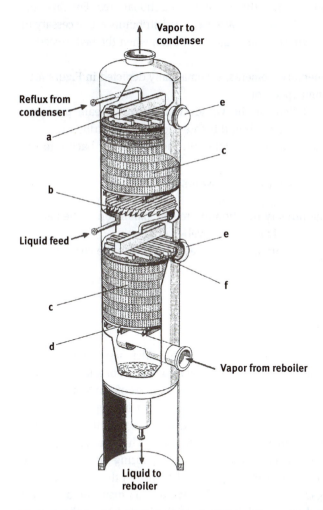

Figure 4.11: Cutaway section of a packed column with a structured packing: a, liquid distributor; b, liquid collector; c, structured packing; d, support grid; e, manway; and f, liquid redistributor. Reproduced with permission from [26].

and be wetted by the liquid as completely as possible. The internals of the column must offer minimum resistance to gas flow. Modern packings have a relatively free cross-sectional area of more than 90%. The vapor enters the bottom of the column and flows upward through the free cross-sectional area of the internals in countercurrent contact with the downflowing liquid. Because this countercurrent flow exists throughout the column, *packed columns are in principle more effective for mass transfer than a tray column*. However, the countercurrent flow of gas and liquid in a packed column is not perfect since the liquid flow is not uniform over the cross section. Well-known mechanisms causing *liquid maldistribution* are liquid channeling and wall flow. Good contacting efficiency between the two phases is only obtained with a uniform liquid distribution over the entire cross-sectional area. For this purpose, *liquid distributors* and *redistributors* are used. Redistributors are necessary to avoid the built up of a high degree of liquid maldistribution when the bed exceeds a height of 6 m.

Other important supplementary elements, schematically depicted in Figure 4.11, are necessary for proper column operation:

– *Liquid collectors* are installed for the withdrawal of side stream products, pump arounds and the collection of liquid before each liquid distributor.
– *Wall wipers* ensure that liquid accumulated on the wall is fed back into the packing.
– *Support grid* is installed to support the packing and the liquid holdup of the packing.
– *Hold down plate* has the primary function to prevent expansion of the packed bed as well as to maintain a horizontal bed level.
– *Gas distributor* is used to obtain a uniform gas flow across the column cross section.

4.4.1 Random packing

The type of packing used has a great influence on the efficiency of the column. Most industrial packed columns use random packings composed of a large number of specially formed particles. Presently, some 50 different types of random packings are offered on the market. Some examples of the more important types are shown in Figure 4.12. The oldest packing element is the *Raschig ring*, with its characteristic feature that the length of the ring is equal to its diameter. This feature makes the particles form quite a homogeneous bed structure during pouring into the column. Raschig rings have a rather high pressure drop since the walls of those rings in horizontal position block the gas flow. The *Pall ring* avoids this disadvantage since parts of the wall are punched out and deflected into the inner part of the ring. A Pall ring has the same porosity and the same volumetric area as a Raschig ring but a considerably lower pressure drop. The feasible particle size depends on column

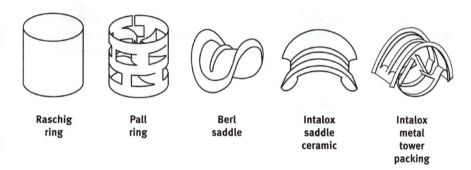

| Raschig ring | Pall ring | Berl saddle | Intalox saddle ceramic | Intalox metal tower packing |

Figure 4.12: Common random column packings.

diameter but should not exceed 1/10 to 1/30 of the column diameter. Most particles are made of metal or ceramics, but plastics are increasingly used.

4.4.2 Structured packing

A certain degree of inhomogeneity is unavoidable in any random packing. These inhomogeneities that cause liquid maldistribution are avoided by using ordered packing structures such as the corrugated sheet structure that was developed by Sulzer in the mid-1960s. Figure 4.13 shows that the corrugated sheets are assembled parallel in vertical direction with alternating inclinations of the corrugations of neighboring sheets. Since the packing does not fit perfectly into the cylindrical column shell, additional tightening strips have to be installed between packing and column wall. These structured packings provide a homogeneous bed structure, and a low pressure drop due to the vertical orientation of the sheets. It is therefore not surprising that their first applications were found in vacuum distillation. At present,

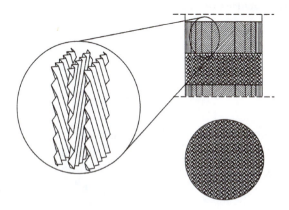

Figure 4.13: Schematic of corrugated structured packing (adapted from [26]).

structured packings are available in different types (gauze, sheet) and in a variety of materials (metals, plastics, ceramics, carbon). For most applications, structured sheet metal packings offer a more attractive performance to cost ratio than structured gauze packings because the cost of structured sheet is about one-third that of gauze while the efficiencies are about the same. Disadvantages of structured packings are high costs relative to trays, criticality of initial liquid and vapor distributions and the associated hardware required. Typically, the installed cost of structured sheet metal packing plus associated hardware is about three to four times that for conventional trays.

A good packing design should have high capacity, high separation efficiency and large flexibility to gas and liquid throughput. Figure 4.14 shows a rough comparison of capacity and separation efficiency of several metal packings. Structured packings are generally superior to random packings in both capacity and separation efficiency. Modern random packings are designed for low pressure drop and perform significantly better than the standard Raschig and Pall ring packings.

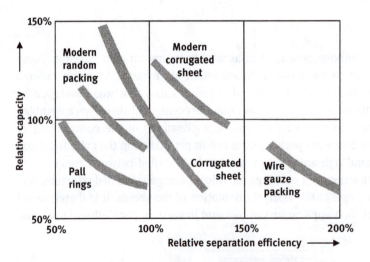

Figure 4.14: Comparison of relative capacity and relative separation efficiency of various random and structured packings (adapted from [26]).

4.4.3 Dimensioning a packed column

Like the design of tray columns, the design of packed columns starts with estimates of column height,[9] column diameter and pressure drop. The design method follows the same approach as that for tray columns. The optimal molar liquid and vapor

---

9 Column height actually means the height of the packed section of the column.

flows, L and V, respectively, and the number of theoretical stages, N_{ts}, are calculated at constant pressure following one of the methods discussed in the previous chapters. The height depends on the required product purity and is connected to the difficulty of separation. The diameter depends on the given capacity and the pressure drop follows from capacity, diameter and properties of the chosen packing.

4.4.4 Height of a packed column

Basically, two different but related approaches can be followed to obtain a first estimate of column height. In the first approach, the number of equilibrium stages determines the total height, H_{column}:

$$H_{column} = N_{ts} \cdot H_{ets} \tag{4.31}$$

using H_{ets} which is the *height equivalent to a theoretical stage*; vapor and liquid leaving a section of height H_{ets} of the column are in equilibrium. H_{ets} is also known as *HETP*, the *height equivalent of a theoretical plate*. Various forms of empirical relations exist to find the *HETP* of packed columns. The following equation gives the *HETP* for packed columns using 25 and 50 mm Raschig rings:

$$\text{HETP} = 18d_{ring} + 12K(V/L - 1) \tag{4.32}$$

where d_{ring} is the diameter of Raschig rings (m); K is the slope of the equilibrium line (–); V, L are molar vapor and liquid flow (mol m^{-2} s^{-1}).

HETP data on several common packing are given in Figure 4.15 as a function of vapor rate. It can be seen that smaller packings improve the *HETP*; however, this advantage is counteracted by a similar increase in pressure drop. Conventional types of packings, such as Raschig and saddle-type rings, exhibit a range of *HETP* values as a function of vapor rate, whereas newer types, such as Mini rings and Pall rings, give a fairly constant value over a wide range of vapor rates.

4.4.4.1 Transfer units

The other approach is based on a hydrodynamic model of mass transfer in a packed column. A packed G/L contactor can be seen at as a column where liquid and vapor flow countercurrently. Figure 4.16 shows the essentials to formulate the differential mass balance. The packed part of the column contains A_h unit surface area per unit length of column height. Applying the rate expression given in eq. (4.8), the mass balance over a thin slice of height Δh becomes

$$V \cdot y = V \cdot (y + \Delta y) + k_{OV} \cdot \rho_V \cdot (y - y^*) \cdot \Delta h \cdot A_h \tag{4.33}$$

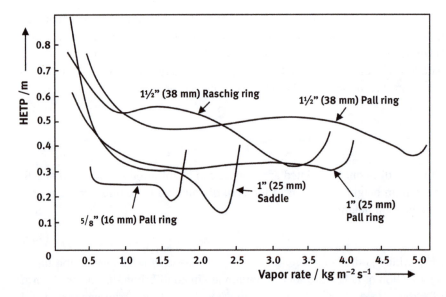

Figure 4.15: HETP, height equivalent of a theoretical plate for commercial available packings (adapted from [13]).

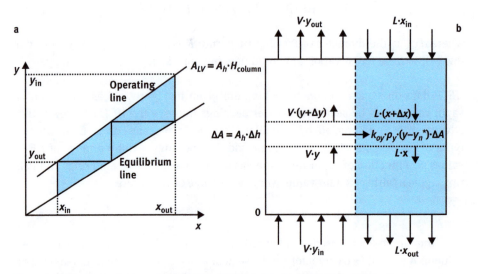

Figure 4.16: Mass transfer in an idealized packed bed absorber (a) McCabe-Thiele diagram, (b) elements in mass balance for interphase mass transport.

Integrating from $y = y_{in}$ at $h = 0$ to $y = y_{out}$ at $h = H_{column}$ and rearranging leads to

$$H_{column} = \frac{Q_v}{k_{OV} \cdot A_h} \int_{y_{in}}^{y_{out}} \frac{dy}{y^* - y} = H_{tu} \cdot N_{OV} \tag{4.34}$$

The first term at the right hand side of this equation is called the *height of a transfer unit*, H_{tu}, calculated from volume flow rate, overall mass transfer coefficient and surface area provided by the packing, similar to eq. (4.16). The integral is defined in eq. (4.14) as the overall number of transfer units, N_{OV}, where the *height* of a transfer unit reflects the ratio of convective transport and absorption rate, the *number of transfer units* determines the axial concentration gradient over the column:

$$H_{tu} \equiv \frac{Q_v}{k_{OV} \cdot A_h} \quad \text{and} \quad N_{OV} \equiv \int_{y_{in}}^{y_{out}} \frac{dy}{y^* - y} \tag{4.35}$$

In contrast with y^* in the analogous equation for a sieve tray (eq. (4.14)), y^* in the mass balance for a packed column is not a constant but a function of the local liquid composition x. The integral can be solved in case of linear operating (eq. (3.17) for absorption) and equilibrium lines (eq. (3.13)):

$$y - y_{out} = \frac{L}{V}(x - x_{in}) \tag{4.36}$$

and

$$y^* = K \cdot x \tag{4.37}$$

Elimination of x and y^* from eqs. (4.33)–(4.35) and putting the separation factor $S = S_{abs} = L/KV$ (in case of absorption S equals the absorption factor A) gives the following expression for N_{OV}:

$$N_{OV} = \frac{S}{S-1} \ln \frac{\frac{1}{S}(y_{out} - y_{in}) + y_{in} - Kx_{in}}{y_{out} - Kx_{in}} = \frac{S}{S-1} \cdot \ln \left[\frac{y_{in} - Kx_{in}}{y_{out} - Kx_{in}} \left(1 - \frac{1}{S} \right) + \frac{1}{S} \right] \tag{4.38}$$

With eqs. (4.34) and (4.38), the column height can be calculated. On the other hand, the number of transfer units, N_{OV}, is connected to the number of theoretical stages, N_{ts}, as calculated with the Kremser equation through the separation factor S. Comparing eqs. (3.28) (for the case of absorption $S = A$) and (4.36) gives:

$$N_{OV} = N_{ts} \cdot \frac{S \cdot \ln S}{S-1} \tag{4.39}$$

with separation factor $S = S_{abs} = A = L/KV$ in case of absorption or $S = S_{str} = KV/L$ in case of stripping.

Equation (4.39) defines the relation between the *number of transfer units* and the *number of theoretical stages*. As an alternative to eq. (4.31), the column height may also be calculated from:

$$H_{\text{column}} = H_{\text{tu}} \cdot N_{OV} \tag{4.40}$$

4.4.5 Minimum column diameter

In contrast to tray columns, Figure 4.17 illustrates that packed columns require a *minimum liquid load* to ensure sufficient mass transfer. Below this minimum value, only a very small part of the packing surface is wetted, and liquid and gas are no longer in intimate contact. This results in a considerable drop of separation efficiency. The lower capacity limit of liquid load depends on the type of the packing, the quality of the packing supplements and the physical properties of the liquid.

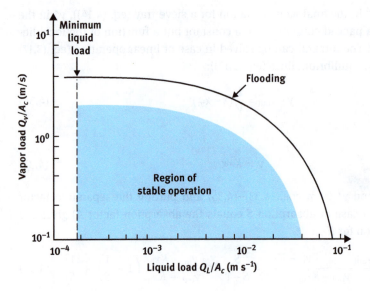

Figure 4.17: Operating region of a packed column.

The diameter of a packed bed column is determined by the required maximum capacity. Finding the maximum allowable vapor rate requires a basic understanding of how flooding and pressure drop are related to operating conditions and packing properties. The feasible operating region of packed columns differs considerably from tray columns and is limited by *flooding* and *wetting*. As illustrated by Figure 4.17, flooding sets the upper capacity limit. At the flood point the pressure drop of the gas flow through the bed increases so drastically that the liquid is no longer able to flow downward against the gas flow. Hence, the countercurrent flow of gas and liquid breaks down and the separation efficiency decreases dramatically.

Flooding correlations for packed beds are available in plots similar to those for tray columns. The abscissa scale term is the same *flow parameter* (a measure of the ratio of kinetic energies of liquid and vapor) used for trays. The dependent parameter should contain parameters characteristic for the actual flow rate, $v_{V,\text{flood}}$, and properties of the packing, such as ε/a, the ratio of volume available for flow and the wetted surface area of the packing, where ε is the fractional void volume of the dry packing and a the packing surface per unit column volume. Converting the actual flow rate, $v_{V,\text{flood}}$, to the superficial[10] (or empty tube) vapor rate $u_{V,\text{flood}}$ by $u_{V,\text{flood}} = \varepsilon \cdot V_{V,\text{flood}}$, then a *relevant dimensionless parameter appears* to be $u_{V,\text{flood}} \sqrt{\frac{a}{\varepsilon^3 g} \cdot \frac{\rho_V}{\rho_L}}$. The ratio a/ε^3 is a function of the packing properties only and is usually referred to as the *packing factor* F_p (m^{-1}). In Figure 4.18, the flooding properties of packed beds are given in a plot of $u_{V,\text{flood}} \sqrt{\frac{F_p}{g} \cdot \frac{\rho_V}{\rho_L}}$ as a function of the flow parameter $\frac{Q_L}{Q_V} \sqrt{\frac{\rho_L}{\rho_V}}$. The ordinate in this figure and in that of Figure 4.10 for sieve trays are similar, $1/F_p$ taking the place of H_{spacing}.

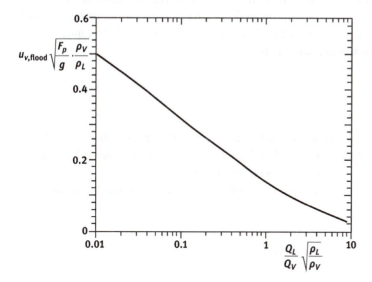

Figure 4.18: Flooding properties of a packed column.

The packing factor F_p for a certain type of random packing is either supplied by the manufacturer or estimated from ε and a. Like the flooding diagram of sieve trays, the

---

10 The total flow rate is divided by the cross-sectional area of the column.

latter diagram is also subject to some restrictions. The flooding line in Figure 4.18 is only applicable to

- nonfoaming systems,
- flooding defined as a pressure drop of 0.01 bar m^{-1} and
- systems with viscosity equal to that of aqueous solutions.

For systems with viscosity η different from water, the maximum vapor rate calculated from the ordinate value should be multiplied by $(\eta_{water}/\eta)^{0.1}$.

Once Q_V, the flow parameter and the packing factor are known, the flooding velocity of the vapor can be calculated from the value read at the ordinate. For design purposes, the *maximum allowable vapor rate* is usually taken at some *70% of the flooding velocity* $u_{V,flood}$. Once this value has been fixed, *minimum column cross-sectional area* and *minimum column diameter* can be calculated in a way similar to the dimensions of a tray column as in eq. (4.26):

$$A_{packing,\, min} = \frac{Q_V}{0.70 \cdot u_{V,flood}} \quad \text{and} \quad D_{min} = \sqrt{\frac{4}{\pi} \cdot A_{packing,\, min}} \tag{4.41}$$

Packing factors of Raschig rings range from 100 to several hundreds of reciprocal meters, depending on the size and material. Structured packings show much lower values; flooding velocities in structured packings can be two to three times higher, resulting in 40–70% smaller column diameters.

4.4.6 Pressure drop

The column diameter as estimated in the previous section is the minimum column diameter as far as flooding is considered. In this estimation, constant column pressure is assumed. This assumption should be checked.

The pressure drop ΔP over a packed column can be calculated from the well-known *Ergun equation:*

$$\frac{\Delta P}{L} = \frac{G_o^2}{D_p \cdot \bar{\rho}} \cdot \frac{1-\varepsilon}{\varepsilon^3} \cdot \left(1.75 + \frac{150}{Re}\right) \tag{4.42}$$

where L is the length of packed bed[11] (m); D_p is the effective diameter of packing material (m); ε is the bed porosity (–); $\bar{\rho}$ is the average gas phase density in column (kg m^{-3}); $Re = \frac{\rho v_o D_p}{(1-\varepsilon)\eta}$ (–); η is viscosity (Pa.s), (kg m^{-1} s^{-1}); v_o is superficial velocity (m s^{-1}); $G_o = \rho v_o$, mass flux (kg s^{-1} m^{-2}).

---

11 In case of distillation, L refers to the length of either the rectifying or stripping section.

It should be noted that while ρ and v_0 vary across the column, the product ρv_0 is constant. This Ergun equation has been applied successfully to gases by using the average density of the gas, provided the pressure drop over the column is sufficiently small. For Re > 1,000, the Ergun equation reduces to the Burke–Plummer equation for turbulent flow. For small values of the Reynolds number (Re < 10), the term 1.75 is small compared to 150/Re and eq. (4.42) transforms into the Blake–Kozeny equation for laminar flow (which is based on Darcy's law: $\Delta P/L = v_0\eta/$permeability).

Estimation of the total pressure drop starts with the assumption that at the conditions of the minimum column diameter, the pressure drop is reasonable small and that the ideal gas law is applicable. Then, the average density follows from

$$\bar{\rho} = \frac{\bar{M}}{RT}\left(P_{tot} - \frac{\Delta P}{2}\right) \tag{4.43}$$

$\bar{M}$ being the molecular weight of the gas phase mixture. The minimum column diameter defines the superficial velocity v_0 at the inlet of the column. With the known value of v_0 at that position G_0 and Re are calculated

$$G_0 = \frac{\bar{M}\cdot P_{tot}}{RT}\cdot v_0, \quad Re = \frac{G_0\cdot D_p}{(1-\varepsilon)\eta} \tag{4.44}$$

which are constants throughout the column. Elimination of $\bar{\rho}$, G_0 and Re by combining eqs. (4.42)–(4.44) results in a single equation with just one unknown: ΔP. A convenient way to solve the complete set of three equations is by iteration:

a. Assume a pressure drop of say 10% of the total pressure.
b. Calculate $\bar{\rho}$. (eq. (4.43)) and G_0 and Re (eq. (4.44)).
c. Calculate ΔP from eq. (4.42) and compare the result with the value assumed in a.
d. If ΔP from eq. (4.42) is larger than the value used in a, the pressure drop is to large, D_{min} should be increased to decrease v_0 and ΔP.
e. Otherwise, to refine the result, the calculation can be repeated starting at b with the smaller value of ΔP.

4.5 Criteria for column selection

Proper choice of equipment is very important for effective and economical distillation. Tray columns are generally employed in large diameter (>1 m) towers. The gas load must be kept within a relatively narrow range, only valve trays allow greater operational flexibility. The liquid load can be varied over a wide range even to very low liquid loads. For vacuum operation, their relatively high pressure drop (typically 7 mbar per equilibrium stage) is a disadvantage. Tray columns have a relatively high liquid holdup compensating for fluctuations in feed compositions. However, the resulting high liquid residence time in the column may lead to the decomposition of thermally unstable substances. Tray columns are also relatively insensitive to impurities in the liquid.

Packed columns are used almost exclusively in small diameter (<0.7 m) towers. Only structured packings permit the use of packed columns with very large diameters. They are extremely flexible with regard to gas load, but require a minimum liquid load for packing wetting. Structured packings with an extremely small pressure loss, typically under 0.5 mbar per equilibrium stage, are a tremendous advantage in vacuum operation. The danger of decomposition of thermally unstable substances is also less in packed columns because of very small liquid holdup.

Nomenclature

a	Surface area per unit column volume, page 28	$m^2\,m^{-3}$
A	Area (interfacial, column cross section)	m^2
A_h	Interfacial area per unit column height, eq. (4.14)	$m^2\,m^{-1}$
d, D	Diameter (of packing rings, column or particles)	m
D_{AB}	Coefficient of diffusion of A in a mixture of A and B	$m^2\,s^{-1}$
E_{MV}	Murphree or plate efficiency, eq. (4.10)	–
E_O	Overall efficiency, eq. (4.9)	–
F_p	Packing factor, page 29	m^{-1}
g	Acceleration constant due to gravity	$m\,s^{-2}$
G_o	Mass flux, eq. (4.40)	$kg\,m\,s^{-1}$
H	Height	m
$HETP, HETS, H_{ets}$	Height equivalent of a theoretical stage, eq. (4.29)	–
$H_{spacing}$	Distance between two adjacent plates	m
HTU, H_{tu}	Height of a transfer unit, eq. (4.14)	m
k	Mass transfer coefficient	$m\,s^{-1}$
K	Equilibrium constant	–
L	Liquid flow	$mol\,s^{-1}$
m	Volumetric distribution coefficient, eqs. (4.6), (4.7)	–
N_{OV}	Overall number of gas-phase transfer units, eq. (4.12)	–
N_s	Number of real stages	–
N_{ts}	Number of theoretical stages	–
Q	Volume flow	$m^3\,s^{-1}$
R	Gas constant	$J\,mol^{-1}\,K^{-1}$
S	Separation factor, $S = A = L/KV$ for absorption; $S = KV/L$ for stripping	–
u	Superficial velocity, page 28	$m\,s^{-1}$
v	Interstitial velocity, page 28	$m\,s^{-1}$
x, y	Mole fraction	–
V	Vapor flow	$mol\,s^{-1}$
V	Total system volume (liquid + vapor), in eq. (4.2)	m^3
z	Distance from interface	m
ϕ_A	Molar transfer rate of component A	$mol\,s^{-1}$
ϕ	Hole area over active or bubble area on a sieve tray	–
ε	Fractional void volume	–

δ	Film thickness		m
ρ	Density, Figures 4.10 and 4.18		kg m^{-3}
ρ	Molar density, eq. (4.1)		mol m^{-3}
η	Surface tension		N m^{-1}

Indices

A,B	Components	N	Stage number
abs	Absorption	OL, OV	Overall based on resistance
i	Interface	In liquid and vapor phase, respectively	
L	Liquid	str	Stripping
LV	Liquid–vapor interface	V	Vapor
		z	At distance z from interface

Exercises

1   An aqueous droplet with radius R is surrounded by a stagnant, inert vapor phase containing a low concentration of ammonia.

Show that $y|_r - y|_{r=R}$, the difference in ammonia concentration in the vapor phase and at the LV-interface is proportional to $1/R - 1/r$, where r is the distance to the center of the droplet.

Assume that the resistance against mass transfer from the vapor to the liquid is mainly situated in the vapor phase.

2   Derive the rate expression in eq. (4.6), starting from the stationary state balance given in eq. (4.3).

3   A tray column is to be designed to reduce the water content in natural gas (M_{methane} = 0.016 kg mol^{-1}) from 0.10 to 0.02 mol% by absorption of water at 70 °C and 40 bar in diethylene glycol (M_{DEG} = 0.106 kg mol^{-1}, ρ_{DEG} = 1,100 kg m^{-3}) containing 2.0 mol% H_2O. The equilibrium constant K = 5.0·10^{-3} at this temperature.

Preliminary calculations (see Exercise 3.5) resulted in a slope of the operating line L/V = 0.010 and N_{ts} = 2.25.

The column should have the capacity to treat Q_V = 3 m^3 gas per unit time at this condition (40 bar, 70 °C). Lab-scale experiments showed that the overall mass transfer coefficient k_{OV} = 3·10^{-3} m s^{-1} at this gas load.

The gas phase is assumed to obey the ideal gas law. H_{spacing} = 0.5 m, g = 9.81 m s^{-2}.

Calculate:

a.   The minimum column diameter

b.   The plate efficiency

c.   The height of the column

4 An aqueous waste stream of $0.015 \text{ m}^3 \text{ s}^{-1}$, saturated with benzene, should be purified by reducing the benzene content by at least 99.9%. This is possible by stripping with (pure) air. To this purpose, an existing tray column with five trays is available. The effective surface area of a single tray amounts to 1.77 m^2. Laboratory experiments at 1 bar and 294 K show that the overall mass transfer coefficient based on the gas phase, $k_{OV} = 0.0080 \text{ m s}^{-1}$.

The stripping process is carried isothermically out at the same conditions with an airflow[12] of $0.932 \text{ m}^3 \text{ STPs}^{-1}$. The distribution coefficient $K_{benzene} = 152$, and the saturation pressure of benzene is 0.104 bar. At this air flow, the interfacial surface area is 50 m^2 per m^2 tray area.

$M_{air} = 0.029 \text{ kg mol}^{-1}$, $M_{water} = 0.018 \text{ kg mol}^{-1}$, $M_{benzene} = 0.078 \text{ kg mol}^{-1}$, $g = 9.8 \text{ m s}^{-2}$, $R = 8.31 \text{ J mol}^{-1} \text{ K}^{-1}$, $\rho_{water} = 1{,}000 \text{ kg m}^{-3}$, $1 \text{ atm} = 1.01325 \text{ bar}$.

a. How many theoretical stages are required to reduce the benzene content by 99.9%?

b. Calculate the plate efficiency E_{MV}.

c. Show by calculation that at the given flow rate with this column the required reduction can be obtained.

d. Calculate the mole fraction of benzene in the effluent.

e. Comment on the chosen value of the airflow.

f. How should the exiting gas flow be treated to avoid dumping of benzene?

5 An air–ammonia mixture, containing 5 mol% NH_3 at a total flow rate of 5 mol s^{-1}, is scrubbed in a packed column by a countercurrent flow of $0.5 \text{ kg water s}^{-1}$. At 20 °C and 1 bar, 90% of the ammonia is absorbed.

$M_{air} = 0.029 \text{ kg mol}^{-1}$, $M_{ammonia} = 0.017 \text{ kg mol}^{-1}$, $M_{water} = 0.018 \text{ kg mol}^{-1}$, the density of water may be taken as $1{,}000 \text{ kg m}^3$.

Calculate the flooding velocity for two different packing materials:

a. 25 mm ceramic Raschig rings with a packing factor of 540 m^{-1}, and

b. 25 mm metal Hiflow rings with a packing factor of 125 m^{-1}.

6 To study the rate of absorption of SO_2 in water, a laboratory-scale column packed with plastic *Hiflow* rings is used. The diameter of the rings is 15 mm. The bed has a porosity $\varepsilon_{bed} = 0.90$ and a surface area per unit bed volume $a = 200 \text{ m}^2 \text{ m}^{-3}$. The height of the packing is 2.0 m, and the internal diameter of the column 0.33 m.

Air containing 2.0 mol% SO_2 is fed to the column at a rate of $V = 2.25 \text{ mol s}^{-1}$, and clean water is fed to the top of the column. The absorption of SO_2 in water is studied in countercurrent operation at 1 bar and 285 K. At this temperature, the distribution coefficient $K_{SO_2} = 32$.

---

12 1 m^3 STP means 1 m^3 at standard *temperature* (0 °C) and standard pressure (1 atm).

At a liquid flow rate $L = 65$ mol s^{-1}, the SO$_2$ content in the effluent is reduced to 0.50 mol%.

$M_{lucht} = 0.029$ kg mol^{-1}, $M_{water} = 0.018$ kg mol^{-1}, $g = 9.81$ ms^{-2}, $R = 8.31$ J mol^{-1} K^{-1}, $\rho_{water} = 1{,}000$ kg m^{-3}.

a. How much larger is the chosen value of L compared to the theoretical minimum value, L_{min}?

b. How many transfer units N_{OV} characterize this absorption process at the given process conditions?

c. Calculate the overall mass transfer coefficient based on the gas phase, k_{OV}.

d. At what (superficial) velocity of the gas feed flooding will start?

e. Calculate the concentration of SO$_2$ in the gaseous effluent in case the air feed is increased to 75% of the flooding velocity. Temperature, pressure and liquid flow rate remain unchanged.

7 Groundwater with a flow of 0.009 m^3 s^{-1}, consisting of water with a small amount (15 mg L^{-1}) dissolved MTBE, needs to be purified by reducing the MTBE content by 99.9%. This will be executed by stripping with (clean) air in an available existing column with 30 trays. The effective area of one tray is 1.10 m^2. The air flow amounts 0.900 m^3 s^{-1}. Laboratory experiments have shown that $k_{OV} = 0.012$ m s^{-1}. Stripping conditions are 1 bar and 293 K and the process can be considered isothermal. At this temperature, the equilibrium constant $K = 20$. For an air flow of 0.900 m^3 s^{-1}, the area of the bubbles is 50 m^2 per m^2 tray area.

Other data: $M_{air} = 0.029$ kg mol^{-1}, $M_{water} = 0.018$ kg mol^{-1}, MMTBE $= 0.088$ kg mol^{-1}

$g = 9.81$ m s^{-2}, $R = 8.31$ J mol^{-1} K^{-1}, $\rho_{water} = 1{,}000$ kg m^{-3}.

a. How many theoretical stages are required for the requested reduction of 99.9%?

b. Calculate the tray efficiency E_{MV}.

c. Show that for the given flows the column can achieve the requested reduction.

d. Calculate the residual amount (ppm) of MTBE in the water flow leaving the column.

e. Show that the air flow is sufficiently low to avoid flooding (use Figure 4.10).

8 The gaseous product stream of an ethylene oxide reactor contains 2 mol% ethylene oxide in ethylene, which can be considered as a gas. Water is used to absorb ethylene oxide at a temperature of 30 °C and a pressure of 20 bar. The gaseous feed stream amounts to 2,500 mol s^{-1} and the water stream 3,500 mol s^{-1}. At these operating conditions, the distribution coefficient K of ethylene oxide equals 0.85 (molal base).

Deabsorption takes place in a packed column with a packing height of 20 m and a diameter of 3 m that contains Pall rings with a specific area a of 115 m^2 m^{-3} and a packing factor F_p of 88 m^{-1}. $k_{OV} = 0.0025$ m s^{-1}.

Other data: $M_{ethylene} = 0.028$ kg mol^{-1}, $M_{water} = 0.018$ kg mol^{-1}, $M_{ethylene\ oxide} = 0.044$ kg mol^{-1}.

$g = 9.81$ m s^{-2}, $R = 8.31$ J mol^{-1} K^{-1}, $\rho_{water} = 1{,}000$ kg m^{-3}.

a. How many stages are required to absorb 99% of the ethylene oxide?
b. Calculate the average gas flow (m^3 s^{-1}) and average superficial gas velocity (m s^{-1}) in the column.
c. Calculate the interface area per unit column height A_h (m^2 m^{-1}) for this column assuming that the whole packing surface is wetted.
d. Show that for the given flows, the column can achieve the requested 99% ethylene oxide absorption.
e. How many equilibrium stages does the column contain and which fraction of the ethylene oxide can finally be absorbed?
f. At which % of the flooding capacity is the available column operated? (Use Figure 4.18.)

9 A CO$_2$-rich gas stream from fermentation contains small amounts of ethanol, which can be recovered by absorption in water using a packed column (metal cascade mini ring packing) with a height of 5 m and an effective diameter of van 2 m. Absorption conditions:

- constant temperature of 30 °C
- Gas feed flow (3,800 m^3 h^{-1})
- Gas composition: 98% CO$_2$ (molar mass 44 g mol^{-1}), 2% ethanol at 120 kPa
- $K = 0.57$
- $L/V = 2$
- $a = 300$ m^2 per unit bed volume

a. Determine the total number of gas transfer units (N_{OV}).
b. Determine the overall mass transport coëfficiënt (k_{OV}).
c. Calculate the residual amount of ethanol left in the gas stream after absorption.
d. Calculate the amount of water (m^3 h^{-1}) required to remove 95% of the ethanol.
e. Flooding occurred as a gas velocity of 1 m s^{-1}. Determine the bed porosity of this type of packing. (Use Figure 4.18.)

Chapter 5
Liquid–Liquid Extraction

5.1 Introduction

5.1.1 General Introduction

Since the introduction of industrial liquid–liquid extraction processes, a large number of applications have been proposed and developed. An overview of some important industrial example applications is presented in Table 5.1. The first and until now largest volume industrial application of solvent extraction is in the petrochemical industry, but also in the mining industry solvent extraction is very important. In the petroleum industry, extraction processes are well suited to separate liquid feeds according to the chemical type (e.g., aliphatic, aromatic, naphthenic) rather than by molecular weight or vapor pressure. With liquid extraction, it is thus possible to do separations of mixtures that cannot be separated by distillation because of the order in their volatility. In metals processing, the recovery of metals such as copper from acidic leach liquors and the refining of uranium, plutonium and other radioactive isotopes from spent fuel elements is done by liquid extraction. A recently added area is urban mining, a circular economy approach in which valuable metals are recovered from electronics waste. Other major applications include the purification of antibiotics and the recovery of vegetable oils from natural substrates. Recently, extraction is gaining increasing importance as a separation technique in biotechnology and biorefinery. In biorefineries, typically large aqueous feed streams are to be processed, and often heat-sensitive molecules are present, making liquid extraction preferred over distillation for many of the separations.

Liquid–liquid extraction is based on the partial miscibility of liquids and used to separate a dissolved component from its solvent by transfer to a second solvent. The principle of liquid–liquid extraction and some of the special terminology are illustrated in Figure 5.1. In the simplest case, the feed solution consists of the *carrier* solvent containing the desired *solute*. This liquid feed is contacted with a *solvent*, which is immiscible or only partly miscible with the liquid feed. In general, the solvent has a higher affinity for the solute than the carrier and the solute is extracted into the solvent phase. For efficient contact, a large interfacial area must be created across which the solute can transfer until equilibrium is closely approached. This is achieved by bringing the feed mixture and the solvent into intimate contact. The resulting solute-loaded solvent phase is called the *extract*, while the other liquid phase is designated the *raffinate*. When equilibrium is reached, the stage is defined as an ideal or theoretical stage, and the equilibrium conditions can be expressed in terms of the *extraction factor E* for the solute:

https://doi.org/10.1515/9783110654806-005

Table 5.1: Industrial liquid–liquid extraction processes.

Solute	Carrier	Solvents
Acetic acid	Water	Ethyl acetate, isopropyl acetate
Aromatics	Paraffins	Diethylene glycol, furfural, Sulfolane, NMP, DMSO
Caprolactam	Aqueous ammonium sulfate	Benzene, toluene, chloroform
Benzoic acid	Water	Benzene
Formaldehyde	Water	Isopropyl ether
Phenol	Water	Benzene
Penicillin	Broth	Butyl acetate
Vanilla	Oxidized liquors	Toluene
Vitamins A, D	Fish liver oils	Liquid propane
Vitamin E	Vegetable oils	Liquid propane
Copper	Acidic leach liquors	Chelating agents in kerosene
Uranium	Acidic leach liquors	Tertiary amines in kerosene

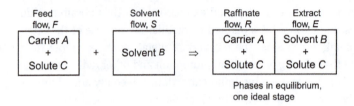

Figure 5.1: Principle of a single extraction contacting stage.

$$E = \frac{\text{amount of solute in extract phase}}{\text{amount of solute in raffinate phase}} = K \frac{\text{extract flow}}{\text{raffinate flow}} \qquad (5.1)$$

where K is the distribution coefficient of the solute between the extract and the raffinate. The larger the value of the extraction factor E, the greater the extent to which the solute it extracted. Large values of E result from large values of the distribution coefficient, K, or large solvent-to-carrier ratios.

After extraction, at least one distillation column (or other separation process) is required to separate the solvent from the extract and recycle the solvent. If the solvent is partially miscible with the feed, a second separation process (normally distillation) is required to recover solvent from raffinate. Figure 5.2 illustrates such an extraction system where two distillation columns are needed. This example is based on a *high boiling* solvent, which is recovered as the bottom product of the distillation column. A *low boiling* solvent would be recovered as the top product. *Solvent recovery is an important factor in the economics of industrial extraction processes*. Especially if the solvent and solute have close boiling points, the distillation column may require many trays and a high reflux ratio, resulting in a costly process. Sometimes azeotropic distillation is required to recover the solvent from the extract and/or raffinate, for example, in the industrial process to extract acetic acid with ethyl acetate.

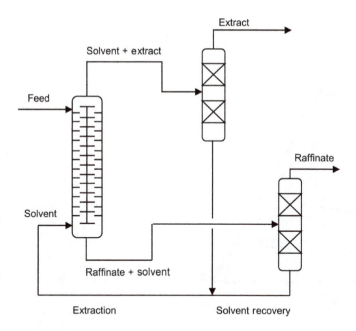

Figure 5.2: Schematic of a principle liquid–liquid extraction process.

The main disadvantage of extraction is that the necessity of a solvent increases the complexity and thereby costs of the process. Reasons to prefer extraction over distillation can be technical and economic in nature. Thus, either distillation is technically infeasible, or it is more expensive than liquid extraction. Applications where liquid extraction should be considered include the following examples:

Distillation is technically difficult or infeasible:
- Dissolved or complexed inorganic substances in organic or aqueous solutions.
- Recovery of heat-sensitive materials and low-to-moderate processing temperatures are needed.
- Mixtures that form azeotropes or exhibit low relative volatilities and distillation cannot be used. Making use of extraction instead can exploit chemical differences (molecular properties) other than the boiling point.

Distillation is not economic:
- Removal or recovery of components present in small concentrations
- When a high boiling component is present in relatively small quantities in a waste stream
- Separation of close melting or close boiling liquids, when solubility differences can be exploited due to differences in the chemical nature of the molecules

5.1.2 Liquid–liquid equilibria

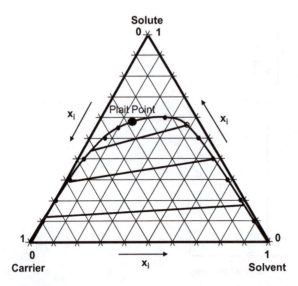

Figure 5.3: Triangular diagram for a type 1 system with the tie lines represented by the dashed lines.

Addition of a solvent to a binary mixture of the solute in a carrier can lead to the formation of several mixture types. The simplest ternary system is the type I system for one immiscible–liquid pair shown in Figure 5.3 (other ternary types show, for example, also a miscibility gap between the solute and the solvent, as in type II). In a type I system as displayed in Figure 5.3, the carrier and the solvent are essentially immiscible, while the solute is miscible with the carrier as well as the solvent. Such ternary systems are commonly represented in a *triangular diagram*, showing the two-phase envelope that encloses the region of overall compositions in which two phases exist in equilibrium. The equilibrium compositions of the extract and raffinate phases are connected by *tie lines* that can be used to determine distribution coefficients and selectivities. For each component, the *distribution coefficient* is given by the ratio of the concentrations in the two phases as follows:

$$K_i = \frac{x_{i,E}}{x_{i,R}} \tag{5.2}$$

where $x_{i,E}$ is the mole fraction of component i in the extract phase and $x_{i,R}$ is the mole fraction of component i in the raffinate phase. In contrary to the distillation literature, where always moles are used, in extraction literature it happens often that weight fractions are reported. Both are possible, and it is advisable to carefully check the units that are applied. The *selectivity* is defined as the ratio of K values, which is equivalent to the relative volatility in distillation:

$$\beta_{12} = \frac{K_1}{K_2} = \frac{(x_1/x_2)_{\text{EXTRACT}}}{(x_1/x_2)_{\text{RAFFINATE}}} \qquad (5.3)$$

At equilibrium, thermodynamics requires the activity of each of the components to be the same in the two liquid phases. However, due to the strong thermodynamic nonideal behavior of the two liquid phases (thermodynamic nonideality is a prerequisite to form two liquid phases!), the activity coefficients may differ largely. This can cause significant differences in equilibrium compositions:

$$\gamma_{i,E}\, x_{i,E} = \gamma_{i,R}\, x_{i,R} \qquad (5.4)$$

Replacing the composition ratio with the ratio of activity coefficients gives for the selectivity:

$$\beta_{12} = \frac{(\gamma_2/\gamma_1)_{\text{EXTRACT}}}{(\gamma_2/\gamma_1)_{\text{RAFFINATE}}} \qquad (5.5)$$

The required activity coefficients or distribution coefficients are generally calculated from thermodynamic models such as NRTL (nonrandom two liquid model) or UNIQUAC (universal quasichemical model). These models require experimental data to fit the parameters in the model. Alternatively, activity coefficients can be predicted with group contribution methods like UNIFAC (UNIQUAC Functional-group Activity Coefficients). For further reading on thermodynamic models, reference 31 is suggested. The use of UNIFAC is beneficial because it is fast and requires no experiments, but the results may be much less accurate. Therefore, for detailed process design, it is wise to validate the phase behavior with experiments. Figure 5.3 shows that as the concentration of the solute is increased, the tie lines become shorter because of the increased mutual miscibility of the two phases. At the plait point, the raffinate-phase and extract-phase boundary curves intersect, the tie line length becomes zero and the selectivity becomes equal to 1. An important additional use of the triangular diagram is the graphical solution of material balance problems, such as the calculation of the relative amounts of equilibrium phase obtained from a given overall mixture composition. Liquid–liquid equilibria having more than three components cannot be represented on a two-dimensional diagram.

5.1.3 Solvent selection

The key to an effective extraction process is the selection of the right solvent. A solvent may be a pure chemical compound or a mixture of chemical compounds, a composite solvent. When a solvent is a composite solvent, it often comprises a complexing agent also called the extractant, that is diluted in a physical (noncomplexing) solvent, which is called the diluent. Extractants are applied to improve the distribution coefficient and show affinity for the solute; for this reason they are also called affinity separating agents. Proposing a solvent requires the knowledge of the physical and chemical properties of the solvent, solute, carrier and other constituents of the extraction system. While pure component properties are easily found in the literature, physical properties of mixtures are available in the literature for very few systems. The following criteria are important to consider during the selection of a solvent:

- *Distribution coefficient:* A high value of distribution coefficient indicates high affinity of the solvent for the solute and permits lower solvent/feed ratios. Very high distribution ratios obtained through the use of extractants with high affinity for the solute might induce solvent regeneration problems and regenerability should be checked.
- *Selectivity:* A high value of the selectivity reduces the required number of equilibrium stages. If the feed is a complex mixture where multiple components need to be extracted from, selectivities between groups of components become important.
- *Density:* Higher density differences between extract and raffinate phases permit higher capacities in extraction devices using gravity for phase separation. The density difference must at least be large enough to ease the settling of the liquid phases.
- *Viscosity:* High viscosities lead to difficulties in pumping, dispersion and reduce the rate of mass transfer. Low viscosities benefit rapid settling, capacity and formation of smaller droplets, which have larger specific surface areas (see interfacial surface area A_{LV} in eq. (4.2)).
- *Solvent recoverability:* Recovery (regeneration) of the solvent should be easy. A solvent that boils much higher than the solute generally leads to better results, though solvents boiling lower than the solute are also used commercially.
- *Solubility of solvent:* Mutual solubilities of carrier and solvent should be low to avoid an additional separation step for recovering solvent from the raffinate.
- *Interfacial tension:* Low interfacial tension facilitates the phase dispersion but may require large volumes for phase separation due to slow coalescence. High interfacial tension permits a rapid settling due to an easier coalescence, allowing higher capacities. A too low interfacial tension leads to emulsification. Sometimes minor impurities in liquid–liquid systems can significantly enhance emulsification potential, which can be observed among others in biorefinery and biotechnology processes.
- *Availability and cost:* Solvent cost may represent a large initial expense for charging the system, as well as a heavy continuing expense for replacing solvent losses. Therefore, one should make sure that the solvent of interest is commercially available and relatively inexpensive.
- *Toxicity, compatibility and flammability:* These criteria are important occupational health and safety considerations a suitable solvent has to meet. Especially for food and pharmaceutical products, nontoxic solvents are highly

desired. In general, any hazard associated with the solvent will require extra safety measures and increases costs.

- *Thermal and chemical stability:* It is important that the solvent should be thermally and chemically stable because it is recycled. In particular, the solvent should resist breakdown during recovery in, for example, a distillation column.
- *Corrosivity:* Corrosive solvents can lead to increased equipment cost but might also require expensive pre- and posttreatment of streams.
- *Environmental impact:* The solvent should not only be compatible with the process, but also with the environment (minimal losses due to evaporation, solubility and entrainment).

In addition to being nontoxic, inexpensive and easily recoverable, a good solvent should be relatively immiscible with feed components other than the solute and have a different density from the feed to facilitate phase separation. Also, it must have a very high affinity for the solute, from which it should be easily separated by distillation, crystallization or other means. Obviously, no solvent will be best from all of these viewpoints, and the selection of a desirable solvent involves compromises between the various criteria. In this selection process, some of the criteria are essential for the separation while others are desirable properties that will improve the separation and/or make it more economical. The most important compromises include those between *solute solubility and selectivity that usually behave exactly opposite,* and *between high distribution and good regenerability of the solvent.* When a high selectivity is obtained, solubilities are usually low and the other way around. In most cases, the selectivity is the most important parameter, since this determines whether a certain separation can be accomplished.

5.2 Extraction schemes

In the simplest extraction scheme, feed and solvent are contacted in a single equilibrium stage as shown in Figure 5.1. The main disadvantage of single stage contacting is that residual solute is left behind in the carrier. Therefore, often a series of contacting stages is arranged in a cascade to accomplish a separation that cannot be achieved in a single stage and/or reduce the required amount of the mass-separating agent. Three frequently encountered extraction cascades are the cocurrent, crosscurrent and countercurrent arrangements (Figure 5.4). In this paragraph, we will explore their differences in extraction efficiency. All examples in this section refer to dilute systems where the distribution coefficient K can be considered constant.

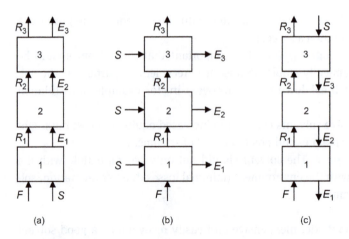

Figure 5.4: Schematic of cocurrent (a), crosscurrent (b) and countercurrent (c) extraction.

5.2.1 Single equilibrium stage

Considering the case that the solvent and the carrier are completely immiscible, the solvent contains initially no solute and solute concentrations in the extract and raffinate remain low (<1%). Under these conditions, the resulting extract flow will approximately equal the solvent flow, $E \approx S$, and the raffinate flow approximately equals the feed flow, $R \approx F$. Thus, the material balance over the solute in a single equilibrium stage becomes

$$x_F F = x_E S + x_R F \tag{5.6}$$

and the distribution of solute at equilibrium is given by

$$x_E = K x_R \tag{5.7}$$

where K is the distribution coefficient defined in terms of mass or mole ratios. Elimination of x_E gives

$$x_R = \frac{x_F F}{F + KS} = \frac{x_F}{1 + \frac{KS}{F}} \tag{5.8}$$

It is convenient now to introduce the previously defined extraction factor,[1] E, for the solute:

$$E = \frac{KE_1}{R_1} = \frac{KS}{F} \tag{5.1}$$

---

1 Note that E = extraction factor, not to be confused with E_1, E_2, . . ., the extract flow rate from stages 1, 2, . . . as in Figure 5.4.

The *fraction of solute* that is *not extracted* is now given by

$$\frac{x_R}{x_F} = \frac{1}{1 + E} \tag{5.9}$$

where it is clear that the larger the extraction factor, the greater the extent to which the solute is extracted. Large values of E result from large values of the distribution coefficient, K, or large ratios of solvent to carrier.

5.2.2 Cocurrent cascade

If additional stages are added in the cocurrent arrangement in Figure 5.4a, the equation for the first stage is that of a single stage:

$$\frac{x_1}{x_F} = \frac{1}{1 + E} \tag{5.10}$$

Since y_1 is in equilibrium with x_1 ($y_1 = K\,x_1$) this can be rewritten as

$$\frac{y_1}{x_F} = \frac{K}{1 + E} \tag{5.11}$$

For the second stage, a material balance for the solute gives

$$x_1 F + y_1 S = x_2 F + y_2 S \tag{5.12}$$

with

$$K = \frac{y_2}{x_2} \tag{5.13}$$

Combining eq. (5.11) with eqs. (5.12), (5.10) and $y_2 = K\,x_2$ to eliminate, respectively, y_1, x_1 and y_2 gives

$$\frac{x_2}{x_F} = \frac{1}{1 + E} = \frac{x_N}{x_F} \tag{5.14}$$

Comparison of eq. (5.14) with (5.10) shows that $x_N = x_1$, which means that no additional extraction takes place in the second stage. This is as expected because in the derivation we assumed that the two streams leaving the first stage are at equilibrium and when they are recontacted in stage 2, no additional net mass transfer of solute occurs. Accordingly, a cocurrent cascade only provides increased residence time that is not necessary when the stages are ideal because equilibrium is reached between the streams after the first stage.

5.2.3 Crosscurrent

For the crosscurrent cascade shown in Figure 5.4b, the feed progresses through each stage, starting with stage 1 and finishing with stage N. When the total amount of solvent S is distributed in equal portions of S/N over N stages, the extraction factor for each stage becomes E/N. Application of eq. (5.10) by replacing E by E/N gives the concentration ratios in each of the N stages:

$$x_1/x_F = 1/(1 + E/N)$$
$$x_2/x_1 = 1/(1 + E/N)$$
$$\vdots \tag{5.15}$$
$$\vdots$$
$$x_N/x_{N-1} = 1/(1 + E/N)$$

Combining equations in (5.15) to eliminate all intermediate interstage variables, x_N, the final raffinate mass ratio is given by

$$\frac{x_N}{x_F} = \frac{1}{(1 + E/N)^N} \tag{5.16}$$

Thus, unlike the cocurrent cascade, the value of x_N decreases in each successive stage, and the distribution of fresh solvent over multiple stages gives an improvement compared to the separation in a single stage for a given solvent to feed ratio. For an infinite number of equilibrium stages, eq. (5.16) becomes

$$\frac{x_\infty}{x_F} = \frac{1}{\exp(E)} \tag{5.17}$$

Thus, even for an infinite number of stages, x_N cannot be reduced to zero.

5.2.4 Countercurrent

The best compromise between the objectives of high extract concentration and a high degree of extraction of the solute is the countercurrent arrangement. In the countercurrent arrangement in Figure 5.4c, the feed entering stage 1 is brought into contact with a solute-rich stream, which has already passed through the other stages, while the raffinate leaving the last stage has been in contact with fresh solvent. For a three-stage system, the material balances and equilibrium equations for the solute in each stage are as follows:

Stage 1: $\quad x_F F + y_2 S = x_1 F + y_1 S \tag{5.18}$

Stage 2: $\quad x_1 F + y_3 S = x_2 F + y_2 S \tag{5.19}$

Stage 3: $\quad x_2 F + y_S S = x_3 F + y_3 S \tag{5.20}$

Equilibrium: $$K = \frac{y_1}{x_1} = \frac{y_2}{x_2} = \frac{y_3}{x_3} \qquad (5.21)$$

Solving this set of equations, the ratio between raffinate and feed composition, $x_3/x_F = x_R/x_F$, is obtained by elimination of y_1, y_2 and y_3 from eqs. (5.18) to (5.21). Taking the mole fraction in the entering solvent $y_S = 0$, introduction of the extraction factor $E = K \cdot S/F$ and subsequent elimination of x_1 and x_2 from the resulting set of three equations gives

$$\frac{x_R}{x_F} = \frac{1}{1 + E + E^2 + E^3} = \frac{E-1}{E^4 - 1} \qquad (5.22)$$

If the number of countercurrent stages is extended from $n = 3$ to $n = N$, this becomes

$$\frac{x_R}{x_F} = \frac{1}{\displaystyle\sum_{n=0}^{N} E^n} = \frac{E-1}{E^{N+1} - 1} \qquad (5.23a)$$

As with the crosscurrent arrangement, the value of $x_N = x_R$ decreases in each successive stage. For the countercurrent arrangement this decrease is greater than for the crosscurrent arrangement. Therefore, the countercurrent cascade is the most efficient of the three linear cascades and normally preferred for commercial-scale operations because of economic advantages. Considering eq. (5.1), the value of E is an operational parameter set by the engineer. When designing a countercurrent extraction with multiple stages, it is important to choose the solvent and feed flows such that $E > 1$ for the solute that you would like to extract. For $E > 1$, the fraction remaining in the raffinate approaches zero with many stages, whereas for $E < 1$, the fraction remaining in the raffinate approaches $1 - E$.

Expressing N in eq. (5.23a) as a function of x_R / x_F gives a *Kremser equation for extraction*, similar as that for absorption (eq. (3.28)) and that for stripping (eq. (3.33)). The complement of x_R/x_F is the *fraction extracted*:

$$1 - \frac{x_R}{x_F} = 1 - \frac{E-1}{E^{N+1} - 1} = \frac{E^{N+1} - E}{E^{N+1} - 1} \qquad (5.23b)$$

similar to the fraction absorbed (eq. (3.29)) and the fraction stripped (eq. (3.34)).

5.3 Design of countercurrent extractions

5.3.1 Graphical McCabe–Thiele method for immiscible systems

The McCabe–Thiele analysis for immiscible extraction is very similar to the analysis for absorption and stripping discussed in Chapter 3. In order to use a McCabe–Thiele type of analysis we must be able to plot a single equilibrium curve, have the energy balances automatically satisfied and determine the operating line. The energy balances are automatically satisfied when it is assumed that the heat of mixing is negligible and the system is isothermal as well as isobaric. The operating line will be straight and the solvent and diluent mass balances are automatically satisfied when the carrier and solvent are assumed to be totally immiscible. In that case, the *solute-free carrier flow rate*, F', and *solute-free solvent flow rate*, S', remain constant.

$$F' = \text{solute-free carrier flow rate, kg s}^{-1} = \text{constant} \qquad (5.24)$$

$$S' = \text{solute-free solvent flow rate, kg s}^{-1} = \text{constant} \qquad (5.25)$$

Note again that these are flow rates of carrier and solvent only and do not include the total raffinate and extract streams. Equations (5.24)–(5.25) are the carrier and solvent mass balances that are automatically satisfied when the phases are immiscible. When the carrier and solvent are immiscible, *weight ratio units* are related to *weight fractions* as follows:

$$X = \frac{x}{1-x} \quad \text{and} \quad Y = \frac{y}{1-y} \qquad (5.26)$$

where X is kg solute per kg carrier and Y is kg solute per kg solvent. The operating equation is derived from the mass balance envelope shown in Figure 5.5. In weight ratio units, the steady-state mass balance is

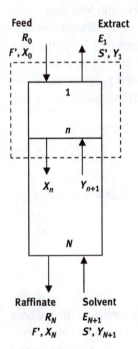

Feed
R_0
F', X_0

Extract
E_1
S', Y_1

1

n

X_n Y_{n+1}

N

Raffinate   Solvent
R_N E_{N+1}
F', X_N S', Y_{N+1}

Figure 5.5: Mass balance envelope of a countercurrent extractor.

$$F' X_n + S' Y_1 = S' Y_{n+1} + F' X_o \tag{5.27}$$

Solving for Y_{n+1} we obtain the operating equation:

$$Y_{n+1} = \left(\frac{F'}{S'}\right) X_n + \left[Y_1 - \left(\frac{F'}{S'}\right) X_0\right] \tag{5.28}$$

When plotted on a *McCabe–Thiele diagram* of Y versus X, this is a straight line with slope F'/S' and Y intercept $Y_1-(F'/S')\,X_0$, as illustrated in Figure 5.6. Note that since F' and S' are constants, the operating line is straight. For the usual design problem, F'/S' will be known as well as X_0, X_N and Y_{N+1}. Since X_N and Y_{N+1} are the concentrations of passing streams, they represent the coordinates of a point on the operating line. Equilibrium data for dilute extraction are usually represented as a distribution ratio

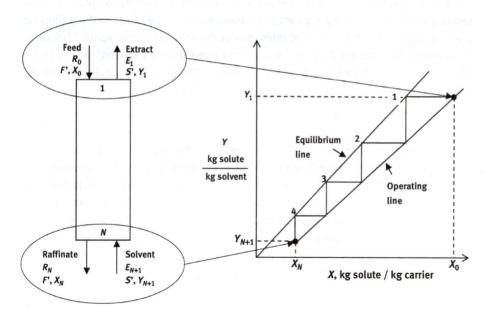

Figure 5.6: McCabe–Thiele diagram for dilute extraction.

$$K_B = \frac{y_B}{x_B} \tag{5.29}$$

in weight fractions or mole fractions. For very dilute systems, the distribution ratio will be constant, while at higher concentrations it often becomes a function of concentration. The distribution coefficient is nearly always temperature and pH dependent. Especially when the solute can be protonated (basic) or deprotonated (acidic), there is often a strong pH dependency in systems with an aqueous phase.

The McCabe–Thiele diagram shown in Figure 5.6 is very similar to McCabe–Thiele diagrams for stripping. This is because the processes are analogous in that both contact two fluid phases and solute is transferred from the X phase to the Y phase. In both analyses, we defined constant flows in each phase and used ratio units. The analogy breaks down when we consider stage efficiencies and sizing the column diameter, since mass transfer characteristics and flow hydrodynamics are very different for extraction and stripping. If the system is very dilute (<1 mol or weight percent solute), total flows will be constant and the McCabe–Thiele diagram can be directly plotted in fractions. Since adding large amounts of solute often makes the phases partially miscible, the assumption of complete immiscibility is often valid only for dilute systems.

Similar to absorption and stripping, a *minimum solvent flow rate* exists (see eq. (3.7) and Figure 3.6) for each feed flow rate, solute concentration, extent of extraction and operating temperature that corresponds to an infinite number of countercurrent equilibrium contacts between the extract and raffinate phases. As a result, a tradeoff exists in every design problem between the number of equilibrium stages and the solvent flow rate at rates greater than the minimum value. This minimum solute-free solvent flow rate S'_{MIN} is obtained from a mass balance over the whole extractor with an infinite number of stages. An overall mass balance over the column illustrated in Figure 5.5 gives the following result:

$$S' Y_{IN} + F' X_{IN} = S' Y_{OUT} + F' X_{OUT} \qquad (5.30)$$

The assumption of an infinite number of stages (in other words: a column of infinite length) means that the outgoing extract is in equilibrium with the incoming feed ($Y_{OUT} = K \cdot X_{IN}$ or $Y_1 = K \cdot X_0$, where α in eq. (3.13) is replaced by equilibrium constant K) allows the calculation of the solute weight ratio Y_{OUT} in the extract which is in equilibrium with the solute weight ratio X_{IN} in the feed. Reorganizing eq. (5.30) then gives the minimum solute-free solvent flow (compare with eq. (3.7)):

$$S'_{min} = F' \frac{X_{in} - X_{out}}{Y_{out, max} - Y_{in}} \text{ with } Y_{out, max} = K \cdot X_{in} \qquad (5.31)$$

For economical operation, a solvent flow rate greater than S'_{min} must be selected[2] for the extraction to be conducted in a finite number of stages. A reasonable starting value for S' is usually $1.5 \cdot S'_{min}$.

5.3.2 Analytical Kremser method for immiscible systems

If one additional assumption can be made, the Kremser equation can also be used for dilute extraction. Obviously, this assumption is that the distribution coefficient

---

2 For the determination of S_{min} in partially miscible systems, see Figure 5.10.

is constant. The dilute extraction model now satisfies all the assumptions used to derive the Kremser equations in Chapter 3, so these are applicable:

$$\text{Fraction of a solute, } i, \text{extracted} = \frac{x_{in} - x_{out}}{x_{in} - y_{in}/K} = \frac{E^{N+1} - E}{E^{N+1} - 1} \qquad (5.32)$$

In the case that the solvents are substantially immiscible and the distribution coefficient is constant, the required number of theoretical stages, N_{ts}, for countercurrent contact can be calculated directly from the Kremser equation:

$$N_{ts} = \frac{\ln\left[\left(\frac{x_{in} - y_{in}/K}{x_{out} - y_{in}/K}\right)\left(1 - \frac{1}{E}\right) + \frac{1}{E}\right]}{\ln E} \qquad (5.33)$$

5.3.3 Graphical method for partially miscible systems

For partially miscible ternary systems, especially when containing higher solute concentrations, stagewise extraction calculations are conveniently carried out with ternary equilibrium diagrams. In this section, triangular diagrams are used for the development and illustration of a graphical calculation procedure. Phase equilibrium may be represented as illustrated in Figure 5.7 on a triangular diagram,[3] where the dashed lines are the tie lines that connect the equilibrium phases of the equilibrium curve. On the equilibrium curve, all extract compositions lie to the right of the plait point, while all raffinate compositions lie to the left.

The problem posed is to determine the number of stages given the flow rates and compositions of the feed and solvent, and the desired raffinate composition. For this we consider a countercurrent, contactor with N stages for liquid–liquid extraction of a ternary system operating under isothermal, continuous, steady-state flow conditions, as shown in Figure 5.8. Stages are numbered from the feed end. Thus, the final extract is E_1 and the final raffinate is R_N. The feed consists of carrier and solute, but throughout the extraction process, this phase can also contain solvent up to the solubility limit. Besides solute, the solvent phase can also contain carrier due to the partial mutual miscibility. Equilibrium is assumed to be achieved at each stage, so that for each stage the extract and the raffinate are in equilibrium for all three components. Because most of the liquid–liquid equilibrium data are given in mass concentrations, we use the following symbol definitions:

The first step is to locate the *mixing point M*, which represents the *overall composition* of the combination of feed, F, and entering solvent, S in Figure 5.7.

---

3 See Appendix to this chapter how to read compositions from ternary diagrams.

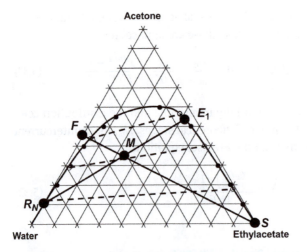

Figure 5.7: Location of product points in ternary diagram.
F = mass flow rate of feed to the extractor
S = mass flow rate of solvent to the extractor
E_n = mass flow rate of extract leaving stage n
R_n = mass flow rate of raffinate leaving stage n
$y_{i,n}$ = mass fraction of species i in extract leaving stage n
$x_{i,n}$ = mass fraction of species i in raffinate leaving stage n

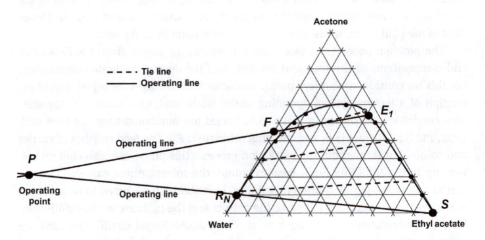

Figure 5.8: Location of operating point and construction of operating lines.

The overall composition is obtained from the overall mass balance and the three material balances over each component:

$$M = F + S \tag{5.34}$$

and

$$x_{i,M} M = x_{i,F} F + y_{i,S} S \qquad (5.35)$$

From any two of these $x_{i,M}$ values, point M is located as shown in Figure 5.7. Based on the properties of the triangular diagram, point M must be located somewhere on the straight line connecting F and S. With point M in place, the exiting extract composition E_1 is determined from the overall mass balances:

$$M = R_N + E_1 \qquad (5.36)$$

and

$$x_{i,M} M = x_{i,R_n} R_n + y_{i,E_1} E_1 \qquad (5.37)$$

Because the raffinate, R_N, is assumed to be at equilibrium, its composition must lie on the equilibrium curve of Figure 5.7. Therefore, if we specify the value $x_{solute,Rn}$, we can locate the point R_n. A straight line drawn from R_n through M will locate E_1 at the intersection of the equilibrium curve, from which the composition of E_1 can be read. Values of the flow rates R_n and E_1 can then be determined from the overall material balances above or from Figure 5.7 by the *inverse lever-arm rule*:

$$\frac{E_1}{M} = \frac{\overline{MR_N}}{\overline{E_1 R_N}} \qquad (5.38)$$

$$\frac{R_N}{M} = \frac{\overline{ME_1}}{\overline{E_1 R_N}} \qquad (5.39)$$

The second step involves the determination of the operating point and operating lines. Analogous to the McCabe–Thiele method, an operating line is the locus of passing streams. Referring to Figure 5.8, material balances around groups of stages from the feed end are:

$$F - E_1 = \cdots = R_{n-1} - E_n = \cdots = R_N - S = P \qquad (5.40)$$

Because the passing streams are differenced, point P defines a *difference point* rather than a mixing point. From the geometric considerations as applied to a mixing point, a difference point also lies on a straight line through the points involved. However, while the mixing point always lies inside the triangular diagram and between the two end points, the difference point usually lies outside the triangular diagram along an extrapolation of the line through two points such as F and E_1, R_N and S and so on. To locate the difference point, two straight lines are drawn through the point pairs (E_1, F) and (S, R_N), which were established during the first step and also shown in Figure 5.8. These lines are extrapolated until they intersect at point P. Figure 5.8 shows these lines and the difference point P. From point P straight lines can be drawn through points on the triangular diagram for any other pair of passing streams. Thus, we refer to the difference point as an *operating point*, and the lines drawn through pairs of points for passing streams and extrapolated to point P as *operating lines*. The operating point P lies on the feed or raffinate side of the triangular diagram in the illustration of Figure 5.8. Depending on the relative amounts of feed and solvent and the slope of the tie lines, point P may be located on the solvent or feed side of the diagram, and inside or outside the diagram.

The third step involves the construction of additional **tie lines** that connect opposite sides of the equilibrium curve, which is divided into the two sides by the plait point. Typically, a diagram will not contain all the tie lines needed. Tie lines may be added by centering them between the existing experimental or predicted tie lines. With these additional tie lines, the equilibrium stage construction can be started. Because the lines through point P represent the mass balances around each stage, Figure 5.8 can now be used to determine the number of required equilibrium stages. Equilibrium stages are stepped off on the triangular diagram by alternate use of equilibrium and operating lines, as shown in Figure 5.9. The approach is as follows. The extract stream E_1 comes from the first stage and is in equilibrium with raffinate stream R_1. This point must therefore be located at the opposite end of a tie line connecting to E_1 representing the equilibrium relation in the ternary diagram. The extract stream of the second stage E_2 passes R_1 and is therefore found by the intersection of the operating line, drawn through points R_1 and P. From E_2 we can continue by applying the equilibrium tie lines to locate R_2 and determine E_3 via the operating line. Continuing in this fashion by *alternating between equilibrium tie lines and operating lines*, we finally reach or pass the known point R_N. If the latter, a fraction of the last stage is taken.

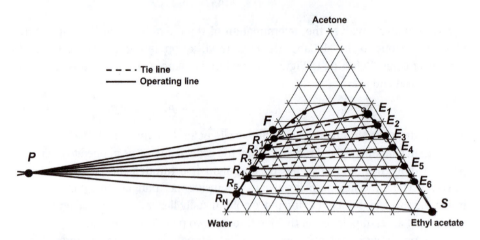

Figure 5.9: Determination of the number of equilibrium stages.

The graphical procedure just described presupposes that a solvent-to-feed ratio was chosen that allows separation with a finite number of stages. As with distillation, this solvent-to-feed ratio cannot be chosen completely arbitrarily. This can be illustrated graphically by considering the previous system in the construction that is shown in Figure 5.10. The points F, S and R_N are again specified, but E_1 is not because the solvent rate has not yet been specified. The first operating line is again drawn through the points S and R_N and extended to the left and right of the diagram. This line is the locus of all possible material balances determined by adding S to R_N. In order to minimize the amount of solvent, one needs to choose E_1 as high as possible. For a given feed F, the highest possible value is found when the second operating line through point F coincides with a tie line. In Figure 5.10, this is represented by the operating line drawn through point $P = P_{min}$ and point F extended to E_{max} at the intersection with the extract side of the equilibrium curve. Under this *minimum solvent-to-feed ratio condition* E_{max} should be in equilibrium with R_1 but the operating line dictates that the extract is in equilibrium with E_2. In this way, we end up with a situation where the entering flow F and the exit flow E_{max} are in equilibrium. Hence, an infinite amount of stages are required to achieve the desired separation.

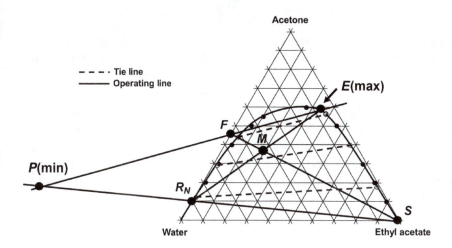

Figure 5.10: Determination of minimum solvent-to-feed ratio.

From the compositions of the four points, S, R_N, F and E_{max}, the mixing point M can be determined and the following material balances can then be used to calculate S_{min}/F:

$$F + S_{MIN} = R_N + E_{max} = M \quad \text{and} \quad x_{i,M} M = x_{i,F} F + y_{i,S} S_{min} \qquad (5.41)$$

The minimum solvent-to-feed ratio can also be determined graphically, as the ratio between the lengths of FM and MS:

$$(S/F)_{min} = \overline{FM}/\overline{MS} \qquad (5.42)$$

For the extraction to be conducted in a finite number of stages, a solvent flow rate greater than S_{min} must be selected. A reasonable value for S might be $1.5 \cdot S_{min}$. An important second condition is that the mixing point M must lie in the two-phase region. As this point is moved along the line SF toward S, the ratio S/F increases according to the inverse lever-arm rule. In the limit, a *maximum S/F ratio* is reached when $M = M_{max}$ arrives at the equilibrium curve on the extract side. At this point, all of the feed is dissolved in the solvent and no raffinate is obtained. To avoid this situation, a solvent to feed ratio S/F must be selected such that

$$(S/F)_{MIN} < (S/F) < (S/F)_{MAX} \tag{5.43}$$

In the example shown in Figures 5.8–5.10, this would result in an extreme S/F, because of the large miscibility gap between the solvent and the carrier. However, when the miscibility gap is smaller, operation close to $(S/F)_{max}$ can be realistic.

5.3.4 Efficiency of an ideal nonequilibrium mixer

The theory of liquid–liquid extraction presented in the previous sections is based on the assumption of equilibrium stages. However, liquid–liquid equilibrium typically is not instantaneously achieved, and when the contact time on a mixing stage is insufficient, it will operate off-equilibrium. The effectiveness of a mixing stage can be expressed in the overall stage efficiency. We will derive an expression to express the *overall efficiency* of an ideal nonequilibrium mixer in terms of mass transfer kinetics. Figure 5.11 shows stage n of a countercurrent extraction process with constant raffinate (R) and extract (E) flow. The composition in both phases is constant and equals the exit concentrations. In the stationary situation, no accumulation occurs and[4]

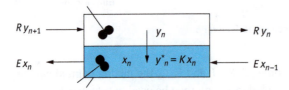

Figure 5.11: Mass transfer in a countercurrent mixer.

---

4 Compare this approach with that shown in Figure 4.7 and eq. (4.12) for a vapor–liquid tray and in Figure 4.16 and eq. (4.33) for a transfer unit in a packed column.

$$R\,y_{n+1} = R\,y_n + k_{oy}\,\rho_y\,A\,(y_n - y_n^*) \quad or \quad \frac{y_{n+1} - y_n}{y_n - y_n^*} = \frac{k_{oy}\,\rho_y\,A}{R} \tag{5.44}$$

In case of constants y_n and y_n^* the number of transfer units, defined in eq. (4.33), becomes

$$N_{oy} \equiv \int_{y_{in}}^{y_{out}} \frac{dy}{y_n^* - y} = \frac{y_{n+1} - y_n}{y_n - y_n^*} \quad \text{hence } N_{oy} = \frac{k_{oy}\,\rho_y\,A}{R} \tag{5.45}$$

Combining eq. (5.45) with the definition of the overall efficiency E_0 (the ratio of the actual difference in composition and the maximum difference obtainable, see eq. (4.10)) results in

$$E_0 = \frac{y_{n+1} - y_n}{y_{n+1} - y_n^*} = \frac{y_{n+1} - y_n}{y_{n+1} - y_n + y_n - y_n^*} = \frac{N_{oy}}{N_{oy} + 1} \tag{5.46}$$

This equation allows the estimation of E_0 from known values of the raffinate flow R, mass transfer coefficient k_{oy} and interfacial area A.

5.4 Industrial liquid–liquid extractors

Extractors are usually classified according to the methods applied for interdispersing the phases and producing the countercurrent flow pattern. To maximize the dispersion of one phase in the other, and to minimize backmixing, extractors are equipped with trays, packings or mechanical moving internals. The location of the principal interface depends upon which phase is dispersed. When the light phase is dispersed, the interface is located at the top of the extractor. When the heavy phase is dispersed, the interface is located at the bottom. The solvent can be the heavy or light phase, and dispersed or continuous. Usually the phase that is fed at the lowest rate (normally solvent) is the dispersed phase.

5.4.1 Mixer–settlers

As shown in Figure 5.12, mixer–settler systems involve a mixing vessel for phase dispersion that is followed by a settling vessel for phase separation. They are widely used in the chemical process industry because of reliability, flexibility and high capacity. Although any number of mixer–settler units may be connected together, these extractors are particularly economical for operations that require high throughput and few stages. Pump circulation, air agitation or mechanical stirring can achieve dispersion. Intense agitation in the dispersion vessel leads to high rates of mass transfer and close approach to equilibrium. However, because the resulting dispersion can be difficult to

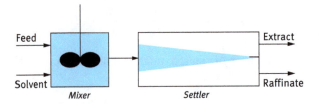

Figure 5.12: Mixer–settler.

separate, designs of mixer–settler systems must be carefully balanced between dispersion intensity and time of settling. Scale-up and design of mixer–settlers is relatively reliable because they are practically free of interstage backmixing and stage efficiencies are high. The main disadvantages of mixer–settlers are high capital cost per stage and large inventory of material in the vessels. In large industrial units, the settlers usually represent at least 75% of the total volume.

5.4.2 Mechanically agitated columns

Many modern differential contactors employ rotary agitation for phase dispersion. The best-known commercial rotary-agitated contactors are shown in Figure 5.13. The *Scheibel column* is designed to simulate a series of mixer–settler extraction units. An impeller agitates every alternate compartment with self-contained mesh-

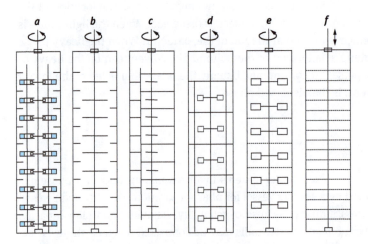

Figure 5.13: Mechanically agitated columns: Scheibel column (a), rotating disk contactor (b), asymmetric rotating disk contactor (c), Oldshue–Rushton multiple mixer column (d), Kuhni column (e) and Karr reciprocating plate column (f).

type coalescers between each contacting stage. The *rotating disk contactor* (RDC) uses the shearing action of a rapidly rotating disk to interdisperse the phases. RDCs have been used widely throughout the world for propane deasphalting, caprolactam purification and sulfolane extraction for aromatics/aliphatic separation. The *Oldshue–Rushton column* consists essentially of a number of compartments separated by horizontal stator-ring baffles. *Kuhni contactors* are similar to the Scheibel columns and have gained considerable commercial application. A baffled turbine impeller promotes radial discharge within a compartment. The principal features are the use of a shrouded impeller to promote radial discharge within the compartments.

A more energy-efficient way to obtain phase dispersion in a column is reciprocating or vibrating of plates in a column. Reciprocating of plates requires less energy and has the same effect in terms of mixing patterns and uniform dispersion. The difference between the different reciprocating–plate columns that have been built for industrial use lies in the plate design. The open-type *Karr column* is the best known. This type of column has gained increasing industrial application in the petrochemical, pharmaceutical and hydrometallurgical industries.

5.4.3 Unagitated and pulsed columns

Despite their low efficiency, unagitated columns are widely used in industry because of their simplicity and low cost. They are particularly suitable for processes requiring few theoretical stages and for corrosive systems where the absence of mechanical moving parts is advantageous. The three main types of unagitated column extractors are shown in Figure 5.14. *Spray columns* are the simplest in construction but suffer from very low efficiency because of poor phase contacting and excessive backmixing in the continuous phase. They generally provide no more than the equivalent of one or two equilibrium stages and are typically used for basic operations, such as washing and neutralization.

Packed columns have better efficiency because of improved contacting and reduced backmixing. The used packings in extraction are similar to the ones used in distillation and include random and structured packings. It is important that the packing material should be wetted by the continuous phase to avoid coalescence of the dispersed phase. The main functions of the packing elements are to reduce backmixing in the continuous phase and promote mass transfer due to jostling and breakup of dispersed phase drops. Because the packing elements reduce the cross-sectional area for flow and decrease the velocity of the dispersed phase, the column diameter for a given rate will always be greater than for a spray tower. However, normally a packed column is preferred over a spray column because the reduced flow capacity is less important than the improved mass transfer.

The *sieve tray extractor* resembles sieve tray distillation. If the light phase is dispersed, the light liquid flows through the perforations of each plate and is dispersed into drops, which rise through the continuous phase. The continuous phase flows horizontally across each plate and passes to the plate beneath through a

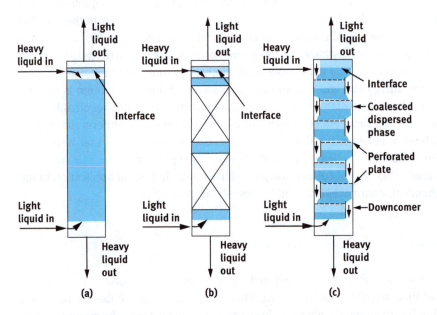

Figure 5.14: Unagitated column extractors: (a) spray column, (b) packed column and (c) perforated plate column.

downcomer. If the heavy phase is dispersed, the column is reversed and upcomers are used for the continuous phase. Perforated-plate columns are operated semistagewise and are reasonably flexible and efficient.

An increased efficiency of sieve-plate and packed extraction columns is obtained by *applying a sinusoidal pulsation* to the contents of the column. The well-distributed turbulence promotes dispersion and mass transfer while tending to reduce axial dispersion in comparison with the unpulsed column. A pulsed-plate column is fitted with horizontal perforated plates that occupy the entire cross section of the column. The total free area of the plate is about 20–25%. Pulsed-packed columns (Figure 5.15) contain random or structured packing. The light and dense liquids passing countercurrently through the columns are acted on by pulsations transmitted hydraulically to form a dispersion of drops. The pulsation device is connected to the side of the column, usually at the base, through a pulse leg. Typical operating conditions are frequencies of 1.5–4 Hz with amplitudes of 0.6–2.5 cm.

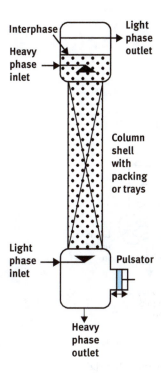

Interphase

Heavy phase inlet

Light phase outlet

Column shell with packing or trays

Light phase inlet

Pulsator

Heavy phase outlet

Figure 5.15: Pulsed packed column.

5.4.4 Centrifugal extractors

In centrifugal extractors, centrifugal forces are applied to reduce contact time between the phases and accelerate phase separation. The units are compact and a relatively high throughput per unit volume can be achieved. Centrifugal extractors are particularly useful for *chemically unstable systems* such as the extraction of antibiotics or for systems in which the phases are slow to settle. Advantages include short contact time for unstable materials, low space requirement and easy handling of emulsified materials or fluids with small density differences. The disadvantages are complexity and high capital and operating costs. Centrifugal extractors have been widely used in the pharmaceutical industry and are increasingly used in other fields. For example, when the low space requirement is an important criterion, such as in off-shore applications, centrifugal separators may be beneficial.

Figure 5.16 shows a schematic of the first differential centrifugal extractor used in industry. This *Podbielniak extractor* can be regarded as a perforated plate

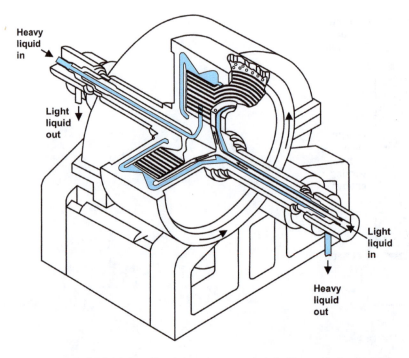

Heavy
liquid
in

Light
liquid
out

Light
liquid
in

Heavy
liquid
out

Figure 5.16: Podbielniak centrifugal extractor, reproduced with permission from [2].

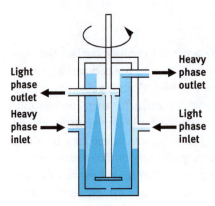

Light
phase
outlet

Heavy
phase
outlet

Heavy
phase
inlet

Light
phase
inlet

Figure 5.17: CINC extractor.

column wrapped around a rotor shaft. This extractor consists of a drum rotating around a shaft equipped with annular passages at each end for feed and raffinate. The light phase is introduced under pressure through the shaft and then routed to the periphery of the drum. The heavy phase is also fed through the shaft but is channeled to the center of the drum. Centrifugal force acting on the phase density difference causes dispersion as the phases are forced through the perforations. Other manufacturers of centrifugal extractors are Robatel, Westfalia, Alfa-Laval and CINC. In recent literature on centrifugal extraction, the CINC is mostly used, see Figure 5.17. It is an annular centrifugal contactor and incorporates mixing in the annular zone between the static wall and the rotating centrifuge, and separation in the centrifuge. Where the Podbielniak is conceptually multistage and typically two to five stages are achieved, depending on the mass transfer characteristics, the CINC and other annular centrifugal separators such as the Robatel are conceptually limited to a single stage and for multistage countercurrent operation, one device per stage is needed.

5.4.5 Selection of an extractor

The large number of extractors available and the number of design variables complicate selection of a contactor for a specific application. Some important criteria that should be taken into consideration during contactor selection are:
- Stability and residence time
- Settling characteristics of the solvent system
- Number of stages required
- Capital cost and maintenance
- Available space and building height
- Throughput
- Experience with the type of extractor

The preliminary choice of an extractor for a specific process is primarily based on the consideration of the system properties and number of stages required for the extraction. Choosing a contactor is still both an art and a science. Although cost ought to be a major balancing consideration, in many actual cases experience and practice are the deciding factors. Mechanically agitated devices show some efficiency advantages but whether these advantages come at the expense of higher cost, must be evaluated by designing for the specific system at hand. Some

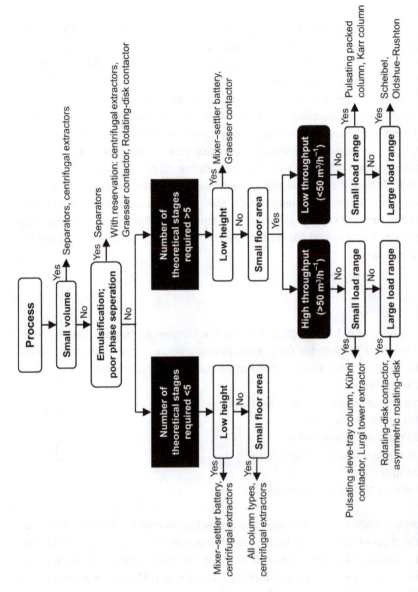

Figure 5.18: Selection scheme for commercial extractors (adapted from K.-H. Reissinger and J. Schroeter, I. Chem. E. Symp. Ser. 54(1978) 33–48).

guidance for the *selection of commercial extractors* is provided by the selection scheme in Figure 5.18. An overview of typically observed performances in various columns is given in Figure 5.19. Note that the performance in columns can be greatly affected by the chemical nature of the two-phase system, and that small impurities might have a large effect on the performance. For example, due to the emulsifying behavior of an impurity, the total capacity can be reduced significantly because the flooding point is reached much earlier. This is also why for the RDC and the Karr columns both an area and a line are presented. This reflects the variety of experiences with different chemical systems.

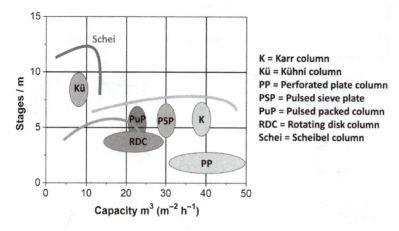

K = Karr column
Kü = Kühni column
PP = Perforated plate column
PSP = Pulsed sieve plate
PuP = Pulsed packed column
RDC = Rotating disk column
Schei = Scheibel column

Figure 5.19: Typical extraction column performances. The capacity is in total throughput, thus the carrier phase plus the solvent phase.

APPENDIX

How to read compositions in ternary diagrams

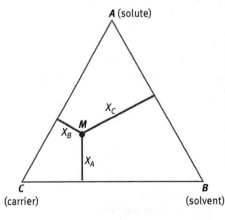

The three vertices A, B and C of the equilateral triangle represent the composition of a pure component: $X_A = 1$ in A, $X_S = 1$ in B and $X_C = 1$ in C, respectively.

Any point between two vertices indicates a binairy mixture, a ternary mixture is given by a point M inside the triangle.

Taking tha altitude as unity, the mole fraction of A is given by the shortest distance from M to the side opposite vertex A.
The same rule applies to the values of X_S and X_C.

Alternatively, the mole fractions can be found as three segments on one side (e.g., BC) by drawing lines through M parallel to the other two sides (in this case AB and AC).
Taking the length of a side as unity, the length of each segment represents a mole fraction, as shown in the diagram to the right.

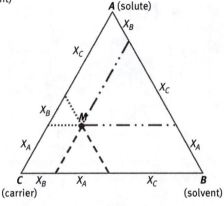

The third method is more accurate than the former two if one component (B) is the minor constituent.
The ratio of x_A and x_C is given by CM_{AC}/AM_{AC}.
Point M_{AC} is found by drawing a line from the vertex representing the minor component (B) through M to the opposite side AC.
All compositions are then calculated from:

$$\frac{X_B}{X_A + X_C} = \frac{M_{AC}M}{MB} \text{ or } X_B = \frac{M_{AC}M}{M_{AC}B}$$

$$\text{with } \frac{X_A}{X_C} = \frac{M_{AC}C}{M_{AC}A} \text{ and } X_A + X_B + X_C = 1$$

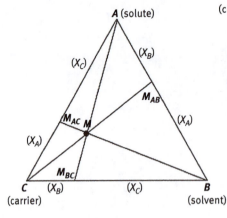

Nomenclature

A, B, C	Components	
A	Interfacial area, eq. (5.43)	m^2
E	Extraction factor, eq. (5.1)	–
E, F, R, S	Extract, feed, raffinate, solvent flow, Figure 5.4	amount s^{-1}
E_0	Overall efficiency, eq. (5.45)	–
F', S'	Solute-free feed flow, solute-free solvent flow, eqs. (5.24)–(5.25)	amount s^{-1}
k_{oy}	Mass transfer coefficient, eq. (5.43)	–
K_i	Distribution coefficient of component i, eq. (5.2)	–
M	Overall flow, eqs. (5.34)–(5.35)	amount s^{-1}
N_{ts}	Number of theoretical stages, eq. (5.33)	–
N	Stage number	–
N	Number of stages	–
N_{oy}	Number of transfer units, eq. (5.44)	–
P	Operation point (net flow) eq. (5.40)	amount s^{-1}
R/ρ_y	Raffinate volume flow rate, eq. (5.44)	$m^3 \, s^{-1}$
x, y	Mole or weight fraction	–
X, Y	Mole or weight ratio, eq. (5.26)	–
β_{12}	Selectivity of component 1 over component 2, eq. (5.3)	–
γ	Activity coefficient, eq. (5.4)	–

Indices

A,B, i	Components
E, F, R, S	Extract, feed, raffinate, solvent flow
$1, 2, .., n, N$	Stage number

Exercises

1 A water solution containing 0.005 mole fraction of benzoic acid is to be extracted by pure toluene as the solvent. If the feed rate is 100 mol h^{-1} and the solvent rate is 100 mol h^{-1}, find the number of equilibrium stages required to reduce the water concentration to 0.0001 mole fraction benzoic acid. Operation is isothermal and countercurrent. The equilibrium represented by:
mole fraction benzoic acid in water = 0.446 * *mole fraction benzoic acid in toluene.*
Compare the McCabe–Thiele method with the analytical Kremser method.

2 The system shown in the following figure is extracting acetic acid from water using toluene as the solvent. The temperature shift is used to regenerate the solvent and return the acid to the water phase. The distribution coefficient of acetic acid between the toluene and water phase ($K_D = y_A/x_A$) amounts to 0.033 at 25 °C and 0.102 at 40 °C.

The indicated number of stages refers to equilibrium stages.

a. Determine y_1 and y_{N+1} for the extraction column.

b. Determine R' and x_N' for the regeneration column.

c. Is this a practical way to concentrate the acid?

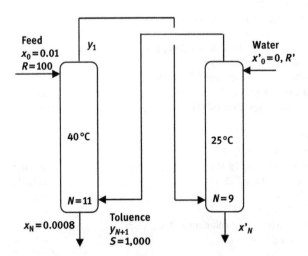

3 Phenol is a toxic chemical that is used in the chemical industry for the production of resins. It has a solubility in water of approximately 8 g/ 100 g water, and for environmental reasons it is important to treat process waters. Consider a 1,000 kg aqueous process water stream with 8 wt% phenol that is to be treated. Because of its immiscibility with water, heptane is considered as solvent. The heptane is phenol free. The distribution coefficient of phenol at low concentrations is constant and has a value of $K_D = 0.2$ (wtfr. phenol in heptane/wtfr. phenol in water).

Calculate the solvent requirements for removing 99% of the phenol in:

a. a single batch extraction;

b. two crosscurrent stages using equal amounts of heptane;

c. two countercurrent stages;

d. an infinite number of crosscurrent stages;

e. an infinite number of countercurrent stages.

4 An 11.5 wt% mixture of acetic acid in water is to be extracted with 1-butanol at atmospheric pressure and 25 °C and countercurrent operation. We desire outlet concentrations of 0.5 wt% in water and 9.5 wt% in butanol. Inlet butanol is pure. Find the number of equilibrium stages required and the ratio of water to 1-butanol by using the McCabe–Thiele diagram design method. The acetic acid equilibrium distribution is given by $K = 1.613$ (wtfr. in butanol/wtfr. in water). Required steps:

a. Convert the streams and equilibrium data to mass ratio units.
b. Plot the equilibrium curve.
c. Plot the operating curve to determine the water to 1-butanol ratio.
d. Step off the equilibrium stages.

5 About 1,000 kg per hour of a 45 wt% acetone in water solution is to be extracted at 25 °C in a continuous countercurrent system with pure ethyl acetate to obtain a raffinate containing 10 wt% acetone. Using the following equilibrium data, determine with the aid of a ternary diagram
a. the minimum flow rate of solvent;
b. number of stages required for a solvent rate equal to 1.5 times the minimum;
c. flow rate and composition of each stream leaving each stage.

	Acetone wt%	Water wt%	Ethylacetate wt%
Extract	60	13	27
	50	4	46
	40	3	57
	30	2	68
	20	1.5	78.5
	10	1	89
Raffinate	55	35	10
	50	43	7
	40	57	3
	30	68	2
	20	79	1
	10	89.5	0.5

	Raffinate wt% acetone	Extract wt% acetone
Tie line data	44	56
	29	40
	12	18

6 Under the umbrella of "green chemistry", chemicals produced by fermentation are also considered. This means that instead of fossil oil as feedstock, microorganisms are applied to transform a biobased feedstock into valuable chemicals. For example, organic waste can be used to make new chemistry out of waste. An example is the production of small carboxylic acids, volatile fatty acids. The smallest of them is formic acid with formula $HCOOH$. With extraction, the formic acid can be removed from the fermentation broth. In this exercise, octanol is applied. The ternary equilibrium data is given in the table:

Raffinate phase weight fractions			Extract phase weight fractions		
Octanol	Water	Formic acid	Octanol	Water	Formic acid
0.0001	0.9998	0	0.667	0.333	0
0.0006	0.975	0.025	0.596	0.361	0.043
0.0017	0.948	0.05	0.521	0.393	0.086
0.0041	0.918	0.078	0.442	0.431	0.127
0.009	0.884	0.107	0.358	0.477	0.465
0.02	0.838	0.142	0.265	0.537	0.198

a. Using this data, construct the ternary diagram, and determine for a 10 wt% formic acid in water solution, what would be the maximum concentration of formic acid that you can obtain in the extract phase when you use pure octanol to recover 90% of the formic acid.

b. Determine the minimum solvent to feed ratio.

c. How many stages are required in an extraction process with a realistic solvent to feed ratio?

7 Caprolactam is industrially recovered and purified by extraction with benzene. The flow diagram of this extraction section is shown in the following figure.

In the first column, caprolactam is extracted from the aqueous phase. The initial concentration amounts to 60 wt% caprolactam. The objective of this extraction is to obtain an aqueous raffinate stream with only minimal amounts of residual caprolactam.

a. Determine the minimal required solvent flow for the first column.

b. Determine the caprolactam concentration (wt%) and the amount (ton h^{-1}) of the extract stream when 1.5 times the minimal required benzene stream is used.

c. When 15 ton h^{-1} water is used for the extraction in the second column, determine the final concentration of caprolactam in the water after both extractions.

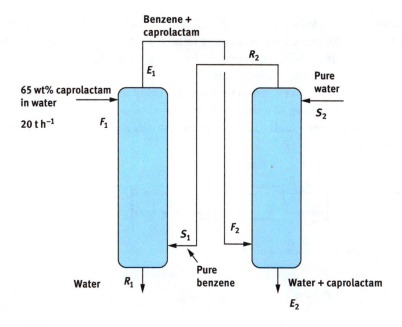

Equilibrium data at 20 °C are as follows:

Extract			Raffinate		
Caprolactam wt%	Water wt%	Benzene wt%	Caprolactam wt%	Water wt%	Benzene wt%
0	0	100	0	99.9	0.1
1.5	0.1	98.4	9.5	90.3	0.2
4.0	0.2	95.8	18.6	81.0	0.4
8.2	0.4	91.3	28.2	71.1	0.7
10.9	0.7	88.4	37.0	61.6	1.3
16.2	1.2	82.6	45.9	51.0	3.0
23.6	2.4	74.0	53.4	39.5	7.1

8 Using the selection scheme of Figure 5.18, suggest the extraction equipment that should be considered for:
   a. extraction of large volumes of copper ore leached with only two to three stages required;
   b. recovery of caprolactam from aqueous carrier stream in chemical plant;
   c. recovery of penicillin from a fermentation broth.

9 An aqueous stream F of 0.11 kg s^{-1} comprises pyridine and water in equal weights. This stream should be purified in a countercurrent extraction process

(see Figure 1). The concentration of pyridine in the raffinate R_N should be reduced to 5 wt% (or less). Pure benzene is to be used as solvent S. The ternary equilibrium diagram of water–pyridine–benzene is given in Figure 2.

a. Calculate the minimum solvent stream, S_{min}.
b. The solvent stream S is chosen to be 0.11 kg s^{-1}. Determine the value of flow E_1.
c. Determine the number of equilibrium separation stages, N.
d. Determine the size and composition of the extract leaving the second stage, E_2.

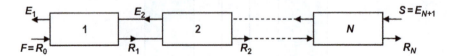

Figure 1: N-stages countercurrent extraction.

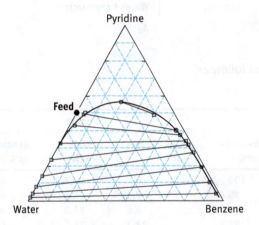

Figure 2: Equilibrium diagram water-benzene-pyridine, compositions in wt%.

10 In the past decade, due to governmental regulations, a large number of biodiesel production facilities have been built in Europe. In a typical process, triglycerides from plant oils are converted into fatty acid methyl esters (or biodiesel); see the reaction scheme. From the scheme you can see that also glycerol is formed as a byproduct of biodiesel.

Such transesterification reactions are typically equilibrium-limited and there-fore in batch processes, typically a sixfold (molar basis) of methanol is applied. The process is essentially biphasic and the excess of the methanol distributes over the glycerol phase and the biodiesel phase. Most of the methanol is in the polar phase, but a water wash is necessary to remove unconverted methanol from the biodiesel phase; see process scheme.

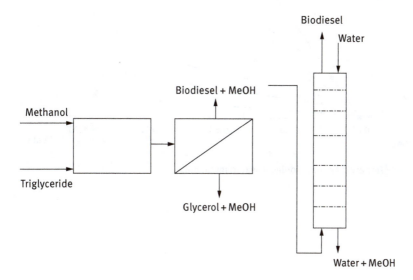

Figure 1: Process flow scheme.

a. In Figure 2, the ternary diagram for biodiesel/water/methanol is given. Explain, on the basis of the ternary diagram, whether or not you think water is a suitable solvent to wash methanol from biodiesel.
b. A new factory should produce 50 tons of biodiesel per hour, furthermore:
    1. The product stream contains 1/9 weight fraction of methanol.
    2. The water used for washing initially does not contain any methanol.
    3. The aqueous extract contains 50 wt% methanol.
    4. The maximum allowed raffinate concentration is 1 wt% methanol.
       How many m³ wash water does the plant need per day?
c. How many extraction stages are necessary to reach the desired purity of the raffinate?
d. Which separation technology do you advise to remove the methanol from the wash water?

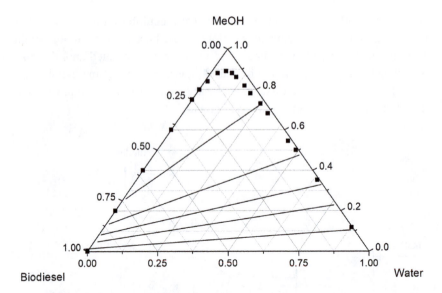

Figure 2: Ternary diagram: biodiesel/water/methanol.

Chapter 6
Adsorption and Ion Exchange

6.1 Introduction

Adsorption and ion exchange are *sorption* operations, in which targeted components in a fluid phase are selectively transferred onto insoluble particles. Adsorption processes leverage the natural tendency of molecules in liquid or gas phase to "stick" to the surface of a solid material. As a result of this natural tendency, *adsorption* of one or more components, *adsorbates*, occurs at the surface of a microporous solid, *adsorbent*, *as shown in* Figure 6.1. An adsorbent is an example of a mass-separating agent, used to facilitate the separation. As the molecules in fluid phase spontaneously bind to adsorbent, heat is released. Adsorption/desorption can be considered as a dynamic equilibrium process, where the spontaneous adsorption releases heat to the environment and heat needs to be supplied to initiate the desorption (Figure 6.1).

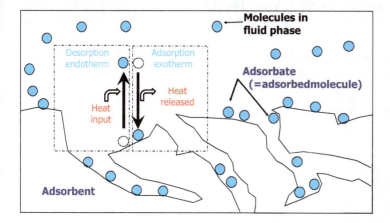

Figure 6.1: Schematic of the dynamic adsorption/desorption equilibrium at the adsorbent surface. The desorption endotherm extracting heat from the surroundings and adsorption exotherm releasing heat to the surroundings are highlighted.

Adsorption processes can be classified into two groups based on the nature of the bond formed, namely physisorption and chemisorption. Physisorption involves intramolecular forces facilitated by Van der Waals, electrostatics or other interactions. Intramolecular forces function between molecules without changing the chemical identity of interacting molecules. In contrast, chemisorption involves formation and destruction of intermolecular chemical bonds such as ionic or covalent bonds. In physisorption, the attractive forces binding the adsorbate are weaker than those of

https://doi.org/10.1515/9783110654806-006

intermolecular chemical bonds, allowing the adsorption to be reversed by either raising the operating temperature or by reducing the concentration or partial pressure of the adsorbate. Changing operating temperature or absorbate concentration alters the dynamic equilibrium, that is, the relative rate of adsorption and desorption processes. In chemisorption, the attractive forces binding the adsorbate are chemical in nature, therefore, stronger than the bonds formed in the physisorption process. Moreover, chemisorption can be more selective due to the chemical nature of bonds formed. However, the selectivity comes at a cost. The chemisorption involves high temperatures, which may lead to product degradation or unwanted changes in product chemistry. Consequently, physisorption is more commonly used in industrial practice.

The ability to manipulate the dynamic adsorption/desorption equilibrium enables combining selective adsorption with desorption or regeneration step that releases adsorbed molecules from adsorbate surface as shown in Figure 6.2. *In a sorption process, the desorption or regeneration step is a key step for designing a reversible process*. *Reversibility is only possible due to dynamic equilibrium characterized by weak binding energies characteristic to physisorption*. It allows recovery of adsorbates when they are valuable and permits reuse of the adsorbent for further cycles. The downside of the need to regenerate the adsorbent is that the overall process is necessarily cyclic in time. Only in a few cases, desorption is not practical, and the adsorbate must be removed by thermal destruction, another chemical reaction or the adsorbent is simply discarded.

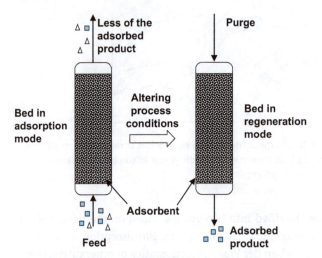

Figure 6.2: Schematic of an adsorption/desorption process. A feed stream carrying two molecules represented by squares and triangles enter a bed in the adsorption mode. The target product is adsorbed on a solid adsorbent surface while the bed is in the adsorption mode. Once the process conditions are changed, for instance, the bed temperature is increased, and the adsorbed product is released with a purge stream.

Adsorption processes may be classified as purification or bulk separations, depending on the concentration in the feed fluid of the components to be adsorbed. Nowadays, adsorption is the most widely used nonvapor–liquid technique for molecular separations throughout a wide range of industries. Application should be considered in cases when: (i) separation by distillation becomes difficult or expensive, for instance, for mixtures with very close boiling points or azeotropes and (ii) a suitable adsorbent exists. Suitable adsorbent should have proper selectivity and capacity (discussed in detail in Section 6.2.1). Moreover, it should be easily regenerated, and it should not cause thermal damage to the products during the adsorption/desorption cycle. A representative list of industrial applications is given in Table 6.1.

Table 6.1: Examples of industrial adsorptive and ion exchange separations.

Separation	Application	Adsorbent
Gas bulk separations	Normal paraffins, isoparaffins, aromatics	Zeolite
	N_2/O_2	Zeolite
	O_2/N_2	Carbon molecular sieve
	CO, CH_4, CO_2, N_2, NH_3/H_2	Zeolite, activated carbon
	Hydrocarbons from vent streams	Activated carbon
Gas purification	H_2O removal from cracking gas, natural gas, air, etc.	Silica, alumina, zeolite
	CO_2 from C_2H_4, natural gas, etc.	Zeolite
	Organics from vent streams	Activated carbon, silicalite
	Sulfur compounds from organics	Zeolite
	Solvents and odors from air	Activated carbon, silicalite
Liquid bulk separations	Normal paraffins, isoparaffins, aromatics	Zeolite
	p-Xylene/*o*-xylene, *m*-xylene	Zeolite
	Fructose/glucose	Zeolite, ion exchange resins
Liquid purifications	H_2O removal from organic solvents	Silica, alumina, zeolite
	Organics from water	Activated carbon, silicalite
	Odor and taste components from drinking water	Activated carbon
	Product decolorizing	Activated carbon
	Fermentation product recovery	Activated carbon, affinity agents
Ion exchange	Water softening	Polymeric ion exchange resins
	Water demineralization	
	Water dealkalization	
	Decolorization of sugar solutions	
	Recovery of uranium from leach solutions	
	Recovery of vitamins from fermentation broths	

Ion exchange is a chemisorption process involving a chemical ion complex reaction. In ion exchange processes, positive (*cations*) or negative (*anions*) charged ions in an aqueous solution are exchanged with "displaceable ions of same charge but different chemical identity" on a solid *ion exchanger*. Ion exchangers are often made up of polymeric materials with polymeric backbones functionalized to capture ions. In water softening, the calcium ions in solution are exchanged with sodium ions weakly attached to a cation polymer exchanger surface as shown in the following reaction:

$$Ca^{2+}_{aq} + 2\,NaR_S \Leftrightarrow CaR_{2,S} + 2\,Na^+_{aq} \tag{6.1}$$

where R_S is the polymeric backbone of the ion exchanger. The exchange of ions is reversible, which allows extended use of the ion exchange resin. Regeneration is usually accomplished with concentrated acid, base or salt solutions.

Industrial applications of ion exchange range from the purification of low-cost commodities such as water to the purification and treatment of high-cost pharmaceutical derivatives as well as precious metals. Table 6.1 illustrates various applications of ion exchange. Among these applications, the most widely known application is water treatment. Other major industrial applications are the processing and decolorization of sugar solutions and the recovery of uranium from relatively low-grade mineral acid leach solutions.

6.2 Adsorption fundamentals

6.2.1 Industrial adsorbents

The role of the adsorbent is to *provide the selectivity and capacity required* for the separation of a mixture. Selectivity is the preference of an adsorbent for a target molecule over other molecules present in the process fluid. For a binary mixture containing molecules A and B, the selectivity for target molecule A, α_A is the ratio of partition coefficient of A, K_A, over the partition coefficient of undesired molecule B, K_B. This concept is illustrated in Figure 6.3 and it is expressed as follows:

$$\alpha_A = \frac{K_A}{K_B} \tag{6.2}$$

The absorbent capacity refers to the maximum amount of adsorbed molecules possible for a given mass of adsorbent. Capacity is often thought as the number of available binding sites per adsorbent mass. It is proportional to the absorbent surface area. In other words, the larger the surface area of an absorbent, higher the number of actives sites and consequently the capacity of an absorbent.

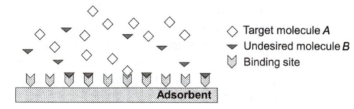

Figure 6.3: Schematic illustrating selectivity and capacity in adsorption/desorption process. The target product A and undesired molecule B dissolved in a solvent are brought in contact with the adsorbent surface. Both A and B adsorb however with different partitioning coefficients. Selectivity is the ratio of partition coefficient for A over the partition coefficient of unwanted molecule B. In this illustration for pedagogical purposes, consider that we have a snapshot of the dynamic equilibrium, out of ten A molecules, four of them are adsorbed resulting in a partitioning coefficient of $K_A = 4/6$. For undesired molecule B, one out of ten molecules are absorbed; hence, $K_B = 1/9$. Thus, the selectivity is $K_A/K_B = 6$. The number of binding sites per mass of adsorbent is defined as capacity.

In the majority of adsorptive separation processes, the selectivity is provided by the physical adsorption equilibrium. Because the adsorption forces depend on the nature of the adsorbing molecule as well as on the nature of the surface, different substances are adsorbed with different affinities. Table 6.2 gives a simple classification scheme, where *equilibrium-controlled adsorbents* are primarily divided into *hydrophilic* and *hydrophobic* surfaces. Adsorbents with hydrophilic surfaces will preferentially attract polar molecules, in particular water, and adsorbents with hydrophobic surfaces attract nonpolar molecules.

Table 6.2: Classification of commercial adsorbents.

Equilibrium selective		Kinetically selective	
Hydrophilic	**Hydrophobic**	**Amorphous**	**Crystalline**
Activated alumina	Activated carbon	Carbon molecular sieves	Small-pore zeolites
Silica gel	Microporous silica		
Al-rich zeolites	Silicalite, dealuminated mordenite and other silica-rich zeolites		
Polymeric resins containing–OH groups or cations	Other polymeric resins		

Besides high selectivity, high capacity is a desirable feature of absorbents since the adsorption capacity determines how much absorbent is required to absorb a given concentration of adsorbed molecules; consequently, the size and therefore the cost of adsorbent bed. To achieve a high adsorptive capacity, commercial adsorbents are made from microporous materials with a large specific area, typically in the range of 300–1,200 m^2 g^{-1}. To visualize how large these typical surface areas are, it is informative to think that a basketball court is approximately 436 m^2. Hence, 1 g of adsorbent powder can have the approximate surface area as high as three basketball courts. Such a large area is made possible by combining particle porosity, ε_{part}, between 30 and 85 vol% with average pore diameters from 1 to 20 nm. For a series of N cylindrical pores of diameter d_p and length L (Figure 6.4a), the surface area-to-volume ratio is defined by dividing the total area of pores ($S_{pore} = N\pi d_{pore}L$) with the total volume ($V_{pore} = N\pi d_{pore}^2 L$):

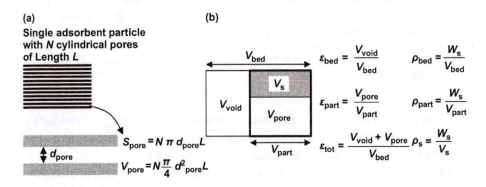

(a)
Single adsorbent particle with N cylindrical pores of Length L

(b)

$\varepsilon_{bed} = \dfrac{V_{void}}{V_{bed}}$ $\rho_{bed} = \dfrac{W_s}{V_{bed}}$

$\varepsilon_{part} = \dfrac{V_{pore}}{V_{part}}$ $\rho_{part} = \dfrac{W_s}{V_{part}}$

$\varepsilon_{tot} = \dfrac{V_{void} + V_{pore}}{V_{bed}}$ $\rho_s = \dfrac{W_s}{V_s}$

$S_{pore} = N\pi d_{pore}L$

$V_{pore} = N\dfrac{\pi}{4} d_{pore}^2 L$

Figure 6.4: (a) Schematic of pores in a single adsorbent particle. (b) Schematic representation of a bed with adsorbent particles and related definitions of porosities ε and densities ρ (W_s is weight of the solid adsorbent).

$$\frac{S_{pore}}{V_{pore}} = \frac{N\pi d_{pore}L}{N\frac{\pi}{4} d_{pore}^2 L} = \frac{4}{d_{pore}} \tag{6.3}$$

If the particle porosity is ε_{part} and the particle density is ρ_{part}, the *specific surface area S*, in area per unit mass of adsorbent, becomes

$$S = \frac{4\,\varepsilon_{part}}{\rho_{part}\, d_{pore}} \tag{6.4}$$

Thus, the specific surface area of a porous solid with $\varepsilon_{part} = 0.5$, $\rho_{part} = 1$ g cm^{-3} and $d_{pore} = 2$ nm becomes 1,000 m^2 g^{-1}. The most important properties of various microporous industrial adsorbents are listed in Table 6.3. Included are representative values of the mean pore diameter, d_{pore}, particle porosity, ε_{part}, particle density, ρ_{part}, and specific surface area, S. According to the definition of the IUPAC, the pores in adsorbents fall into three categories:
1. Micropores $d_{pore} < 2$ nm
2. Mesopores $d_{pore} = 2$–50 nm
3. Macropores $d_{pore} > 50$ nm

Table 6.3: Properties of industrial adsorbents.

Adsorbent	Pore diameter (nm) d_{pore}			Particle porosity	Particle density	Specific area
	Micropores $d_{pore} < 2$ nm	Mesopores	Macropores $d_{pore} > 50$ nm	ε_{pore}	ρ_{pore} g cm^{-3}	S m^2 g^{-1}
γ-Alumina		3–6	200–600	0.5	1.2–1.3	200–400
Silica gel	2	25		0.5–0.7	0.6–1.1	200–900
Activated carbon		2–4	800	0.4–0.6	0.5–0.9	400–1,200
Zeolite 5A	1		30–1,000	0.2–0.5	0.7	
Carbon molecular sieve	0.4–0.5		10–100		0.9–1.0	100–300
Polymeric adsorbents		25–40		0.4–0.6	0.4–0.6	100–700

In a *micropore*, the target molecule inside the pore never escapes from the confining influence of pore walls, while in *mesopores* and *macropores*, the molecules away from walls are free from this influence. This confining influence can alter the transport mechanism of target molecules within pores. Macropores provide very little surface area relative to the pore volume and hardly contribute to the adsorptive capacity. Their main role is to provide a network of super highways to facilitate rapid transfer of molecules from bulk to the interior of the adsorbent particle. Indicated in Figure 6.4b, but not included in Table 6.3 is the bed porosity, ε_{bed}, which can be calculated from the bed and particle volume (V_{bed} and V_{part}, respectively) or from the bed and particle density (ρ_{bed} and ρ_{part}, respectively):

$$\varepsilon_{bed} = \frac{V_{bed} - V_{part}}{V_{bed}} = 1 - \frac{V_{part}}{V_{bed}} = 1 - \frac{\rho_{bed}}{\rho_{part}} \tag{6.5}$$

In industrial processes only four types of adsorbents dominate in usage: activated carbon, silica gel, activated alumina and molecular sieve zeolites. Among these four types, activated carbon is most often used for removing hydrophobic organic species from both gas and aqueous liquid streams. *Activated carbon* is produced in many different forms that differ mainly in pore size distribution and surface polarity, consequently in selectivity and capacity. For liquid-phase adsorption, a relatively large pore size is required, while the activated carbons used in gas adsorption have much smaller pores. Commercial carbons are used in the form of hard pellets, granules, cylinders, spheres, flakes or powders. *Silica gels* represent an intermediate between highly hydrophilic and highly hydrophobic surfaces. Most often, these adsorbents are used for removing water from various gases, but hydrocarbon separations are also sometimes feasible. *Activated alumina* is essentially a microporous (amorphous) form of Al$_2$O$_3$ that has quite a high affinity for water and is often used in drying applications for various gases. Like with silica gels, the water bond with the alumina surface is not as strong as with zeolites, so that regeneration of aluminas can often be accomplished at somewhat milder temperatures. As shown in Figure 6.5, activated alumina can have a higher ultimate water capacity, but the zeolites have a higher capacity at low water partial pressures. Thus, zeolites are typically chosen when very high water removal is necessary, while activated alumina is preferred if adsorbent capacity is more important.

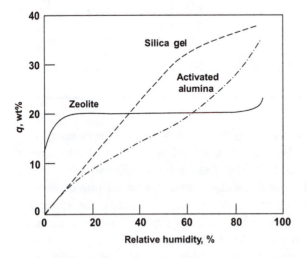

Figure 6.5: Comparison of water adsorption on various adsorbents.

Polymer-based adsorbents are presently used in operations for removing organic constituents from vent gas streams, such as the removal of acetone from air. As illustrated in Figure 6.6, these materials are usually *styrene–divinylbenzene copolymers*, which in some cases have been functionalized to give the desired adsorption properties.

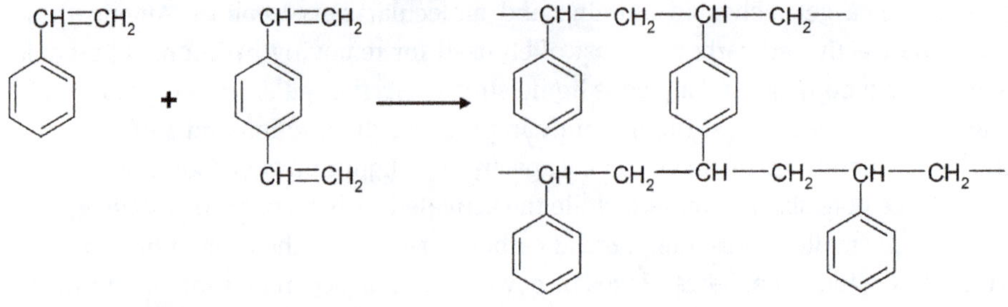

Figure 6.6: Structure of styrene–divinylbenzene.

In addition to adsorbent–adsorbate interactions represented in adsorption equilibrium, selectivity of adsorbents may originate from two important mechanisms:

- Exclusion of certain molecules in the feed because they are too large to fit into the pores of the adsorbent (*molecular sieving effect*)
- Differences in the diffusion of different adsorbing species in the pores of the adsorbent (*kinetic effect*)

Significant *selectivity* can be achieved with molecular sieving effect in zeolites. *Molecular sieve zeolites* are highly crystalline aluminosilicate structures with highly regular channels and cages illustrated in Figure 6.7. Zeolites are selective for polar, hydrophilic species, and very strong bonds are formed with water, carbon dioxide and hydrogen sulfide, while weaker bonds are formed with organic species. Molecular sieving is possible when a channel size is of molecular dimensions, restricting the diffusion sterically. As a result, small differences in molecular size or shape

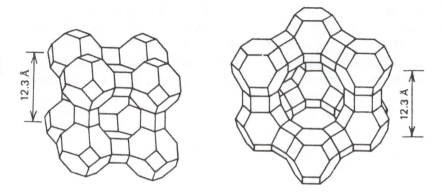

Figure 6.7: Schematic diagrams of structures of two common molecular sieve zeolites.

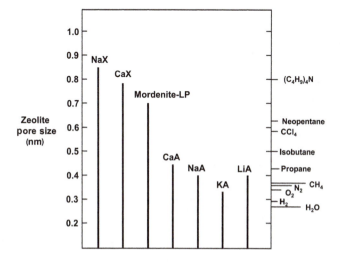

Figure 6.8: Comparison of molecular dimensions and zeolite pore sizes.

can lead to very large differences in diffusivity. In the extreme, certain molecules or a whole class of compounds may be completely excluded from the micropores. This is illustrated in Figure 6.8, where a range of molecular sizes is compared with the channel diameters of various zeolites. The most important examples of such processes are the separation of linear hydrocarbons from their branched and cyclic isomers using a 5A zeolite adsorbent and air separation over carbon molecular sieve or 4A zeolite, in which oxygen, the faster diffusing component, is preferentially adsorbed.

6.2.2 Equilibria

In adsorption, a dynamic equilibrium is established between the molecules in the fluid phase and molecules adsorbed on the solid surface as shown in Figure 6.9. Absorbate molecules constantly adsorb to the surface and desorb from the surface. When the rate of adsorption and desorption are in balance, a dynamic equilibrium is established. This equilibrium is commonly expressed in the form of an isotherm, a diagram showing the equilibrium adsorbed-phase concentration or loading as a function of the fluid-phase concentration, c_A, or partial pressure, p_A, at a fixed temperature. For pure gases, experimental physical *adsorption isotherms* are represented in amount of molecules adsorbed q_A versus partial pressure p_A or concentration c_A. According to their shapes, Brunauer classified adsorption isotherms into five types as shown in Figure 6.10. The simplest and the most commonly encountered isotherm is the *Langmuir type I isotherm*:

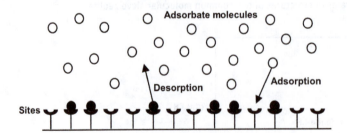

Figure 6.9: Schematic of the dynamic adsorption/desorption equilibrium fraction occupied $\theta_A = 6/14$; fraction not occupied $1 - \theta_A = 8/14$.

$$\theta_A = \frac{q_A}{q_{max,A}} = \frac{b_A p_A}{1 + b_A p_A} = \frac{b'_A c_A}{1 + b'_A c_A} \tag{6.6}$$

where θ_A is the fraction of sites occupied (see Figure 6.9); q_A is the amount adsorbed per unit volume or unit weight adsorbent; $q_{max,A}$ is the monolayer capacity, referring to maximum occupancy or $\theta_A = 1$; b_A is the Langmuir adsorption constant in reciprocal pressure unit; and b'_A is the Langmuir adsorption constant in reciprocal concentration unit.

Both types I and II are desirable isotherms as they exhibit strong adsorption at lower partial pressures or concentrations. This facilitates operation of an adsorption process at low pressures while enabling large loading. Figure 6.10 demonstrates the two limiting cases of a type I isotherm: linear behavior at low (relative) pressures ($b_A p_A \ll 1$) and maximum occupancy at higher concentrations ($b_A p_A \gg 1$). For $b_A p_A \ll 1$, the Langmuir isotherm reduces to a linear form and the following functional form of Henry's law,[1] called a *linear isotherm*, is obeyed:

$$q_A = q_{max,A}\, b_A\, p_A \tag{6.7}$$

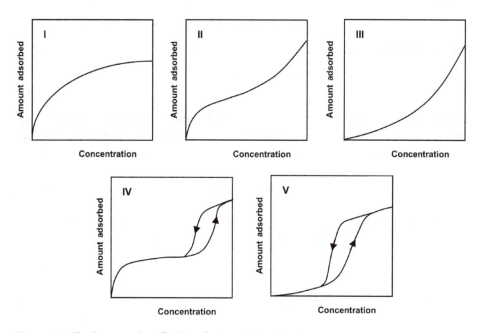

Figure 6.10: The Brunauer classification of adsorption isotherms.

---

1 Note that $q_{max}{\cdot}b_A$ equals the Henry coefficient H_A in eq. (1.9a).

The Langmuir isotherm can be derived based on four assumptions:
1. The adsorbed molecule or atom is held at definite, localized sites.
2. Each site can accommodate only one molecule or atom.
3. The energy of adsorption is constant over all sites.
4. There is no interaction between neighboring adsorbates.

Though the detailed derivation of eq. (6.6) follows from sound thermodynamic reasoning, it can also be obtained by realizing that *at equilibrium, the rates of adsorption and desorption are equal*. The rate of adsorption, r_{ads}, is proportional to the partial pressure, p_A as well as the fraction of empty sites, $1 - \theta$. The rate of adsorption can be written as $r_{ads} = k_{ads} \, p_A (1 - \theta)$, where k_{ads} is a first-order rate constant. Whereas the rate of desorption, r_{des}, is proportional to the fraction of occupied sites, θ. Consequently, r_{des} *can be written as* $r_{des} = k_{des} \, \theta$ where k_{des} is a zero-order rate constant.

At equilibrium, the difference between the adsorption and desorption rates vanishes:

$$r_{ads} - r_{des} = k_{ads} \, p_A (1 - \theta) - k_{des} \, \theta = 0 \tag{6.8}$$

Rearrangement of eq. (6.6) and introducing eq. (6.7)

$$b_A = \frac{k_{ads}}{k_{des}} \tag{6.9}$$

results in eq. (6.4), the *Langmuir equation* for adsorption from the gas phase. It is important to note that eq. (6.4) is in nondimensional form. However, in practice adsorption isotherms are often reported in dimensional form as given in eq. (6.9a):

$$q_A = q_{max,A} \frac{b_A \, p_A}{1 + b_A \, p_A} \tag{6.9a}$$

The adsorption constant b_A is temperature dependent in a way that is related to the activation energy of adsorption, ΔE_{ads}, through an Arrhenius-type equation:

$$b_A = b_A^0 \, e^{-\Delta E_{ads}/RT} \tag{6.10}$$

For an exothermic process, ΔE_{ads} is negative, and the adsorption constant decreases with increasing temperature.

The Langmuir model can faithfully represent a type I isotherm, since at low pressure it approaches Henry's law while at high pressure it tends asymptotically to the saturation limit. Although only few systems conform exactly to the Langmuir model, it provides a simple qualitative representation of the behavior of many systems. Therefore, it is widely used. For adsorption from a liquid phase, it is convenient to replace the (partial) pressure by concentration:

$$q_A = q_{max,A} \frac{b'_A c_A}{1 + b'_A c_A} \tag{6.11}$$

where b' is the Langmuir adsorption constant in reciprocal concentration units. The assumptions mentioned earlier for adsorption from the gas phase are hardly valid for adsorption from (aqueous) solutions, and the physical meaning of $q_{max,A}$ and b'_A is lost. In many cases, however, the equation is still applicable; both parameters can be treated as empirical constants, q_{max} and b_A, obtained by fitting the nonlinear equation directly to experimental data or by employing a linearized form of eq. (6.9a):

$$\frac{c_A}{q_A} = \frac{1}{q_{max}b_A} + \frac{1}{q_{max}} c_A \tag{6.12}$$

Using eq. (6.12), the best-fit straight line can be drawn through a plot of experimental data c_A / q_A versus c_A, giving a slope of $1/q_{max}$ and an intercept of $1/q_{max} b_A$.

The Langmuir model can be extended to *multicomponent* systems, reflecting the competition between species for the adsorption sites:

$$\frac{q_A}{q_{max,A}} = \frac{b_A p_A}{1 + b_A p_A + b_B p_B + \cdots} \tag{6.13}$$

and

$$\frac{q_B}{q_{max,B}} = \frac{b_B p_B}{1 + b_A p_A + b_B p_B + \cdots} \tag{6.14}$$

Since the total number of available adsorption sites should be equal for each component, thermodynamic consistency requires $q_{max,A} = q_{max,B}$. In practice, this requirement can cause difficulties when correlating data for sorbates of very different molecular sizes. For such systems, it is common practice to ignore this requirement, thereby introducing $q_{max,B}$ as an additional model parameter. The *equilibrium selectivity* of the adsorbent, α'_{AB} (see eq. (1.10)), now simply corresponds to the ratio of the equilibrium constants from the multicomponent Langmuir model:

$$\alpha'_{AB} = \frac{q_A/c_A}{q_B/c_B} = \frac{b_A}{b_B} \tag{6.15}$$

Because this selectivity is independent of composition, the ideal Langmuir model is often referred to as the *constant separation factor* model.

The equilibrium isotherms can be measured experimentally by the gravimetric approach. In the gravimetric approach, a known mass of absorbent is exposed to a given concentration of absorbate (or partial pressure if the adsorbate is in gas phase). The adsorbent mass is recorded continuously till it does not change anymore. Once the adsorbent mass stabilizes, the system is considered in equilibrium. The adsorbent weight change before and after contact is the amount of adsorbate absorbed. Repeating this procedure for several concentrations results in an experimental isotherm. More accurate methods such as N_2 adsorption can also be used to measure experimental isotherms. More often than not, the adsorption isotherms vary from type I Langmuir isotherms as shown in Figure 6.10. Despite its simplicity, Langmuir isotherms provide us a theoretical basis to understand other isotherms in Brunauer classification.

Type II–V isotherms shown in Figure 6.10 can be interpreted as deviations from the assumptions of Langmuir isotherm. For instance, type II and III isotherms do not reach a plateau at high concentrations as observed in type I. This can be seen as a deviation from assumptions 2–4 in the Langmuir isotherm deviation. More molecules adsorbed at a single site or favorably interactions between adsorbed molecules can increase the capacity. Type II and III isotherms can be considered as multimolecular adsorption isotherms, whereas type I is considered unimolecular adsorption. Type I and II are considered favorable isotherms due to high adsorption capacity compared to type III. Type VI and V isotherms show a distinct behavior. The isotherm appears different if the pressure/concentration gradually increased or decreased. As the pressure/concentration is increased, a steep increase in the amount of adsorbed molecules is observed. Interestingly, as the pressure/concentration is gradually decreased, the steep jump in adsorbed amount appears at lower pressure/concentration. This steep change is attributed to capillary condensation. In capillary condensation, adsorbate molecules in a nano/submicron size pore condense at lower pressures than the bulk saturation pressure. In bulk, adsorbate molecules interact exclusively with molecules of their own kind. However, inside a nano/submicron pore, adsorbate molecules on average interact more with the surrounding pore walls than molecules of their own kind. As the pore size decreases, this effect becomes more pronounced and it is represented in Kelvin equation as follows:

$$\frac{P_{sat}(d)}{P_{sat}} = \exp\left(-\frac{4\sigma\upsilon_L \cos\theta}{RTd}\right) \tag{6.16}$$

where $P_{sat}(d)$ and P_{sat} are the vapor pressure of liquid in pore of diameter d and the vapor pressure of liquid in bulk. Moreover, υ_L, σ, θ, R, T denote the interfacial energy of liquid in pore, molar volume of liquid, the contact angle liquid–pore material, gas constant and temperature in Kelvin, respectively. As the pressure is decreased, first the liquid adsorbed on external surfaces and larger pores is evaporated followed by the liquid in nano/submicron pores due to the Kelvin equation. This behavior gives rise to type VI and V isotherms.

6.2.3 Kinetics

In general, the rate of physical adsorption is controlled by mass transfer limitations rather than by the actual rate of equilibrium adsorption/desorption inside the adsorbent pores. As illustrated schematically in Figure 6.11, the adsorption of a solute onto the porous adsorbent surface requires the following steps:

1. *External mass transfer* of the solute from the bulk fluid through a thin film or boundary layer to the outer solid surface of the adsorbent.
2. *Internal mass transfer* of the solute by pore diffusion from the outer adsorbent surface to the inner surface of the internal pore structure.
3. *Surface diffusion* along the porous surface.
4. *Adsorption* of the solute onto the porous surface.

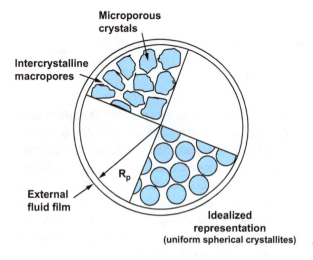

Figure 6.11: Resistances to mass transfer in a composite adsorbent pellet (adapted from [46]).

During the adsorbent regeneration, these four steps follow in reverse order, that is, step 4 to step 1. However, under most practical conditions the external film resistance, surface diffusion and adsorption are seldom rate limiting, so that internal mass transfer generally controls the adsorption/desorption kinetics. From this perspective, adsorbents may be divided into two broad classes: *homogeneous* and *composite*. Figure 6.12 illustrates that in homogeneous adsorbents (activated alumina and silica gel) the pore structure persists throughout the particle, while the composite adsorbent particles (zeolites and carbon molecular sieves) are formed by aggregation of small microporous microparticles. As a result, the pore size distribution in composite particles has a well-defined bimodal character with micropores within the microparticles connected through macropores within the pellet. Transport in a macropore can occur by bulk molecular diffusion, Knudsen diffusion, surface diffusion and laminar convection/Poiseuille flow. In liquid systems, bulk molecular diffusion is generally dominant, but in the vapor phase the contributions from Knudsen and surface diffusion may be large or even dominant. Knudsen diffusion becomes dominant when collisions with the pore wall occur more frequently than collisions with diffusing

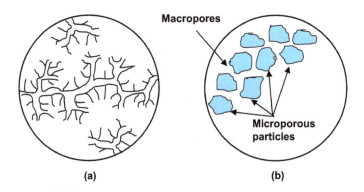

Figure 6.12: Homogeneous (a) and composite (b) microporous adsorbent particle (adapted from [46]).

molecules. In micropores, the diffusing molecule never escapes from the influence of the pore wall, and the Knudsen mechanism no longer applies. Diffusion occurs by jumps from site to site, just as in surface diffusion, and the diffusivity becomes strongly dependent on both temperature and concentration. The selectivity in a kineti- cally controlled adsorption process is dictated by both kinetic and equilibrium effects. This *kinetic selectivity* can be approximated when two species (A and B) diffuse inde- pendently and their isotherms are also independent. Under these conditions, selectiv- ity or the ratio of their uptakes at any time will be given by the Langmuir isotherm constants b_a, b_b and diffusion constants D_A, D_B of species A and B:

$$\alpha_{AB} = \frac{q_A/p_A}{q_B/p_B} = \frac{b_A}{b_B}\sqrt{\frac{D_A}{D_B}} \tag{6.17}$$

A typical example of kinetic selectivity is shown in Figure 6.13. The equilibrium ad- sorption and loading curves of oxygen and nitrogen on carbon molecular sieves commonly used in air separation are shown in Figure 6.13. The significant differ- ence between the uptake rates is caused by the fact that the pore diameter of the carbon is just very slightly larger than the diameters of the two gases. As a result, nitrogen has much more difficulty traversing a pore than oxygen. Thus, even though there is virtually no equilibrium selectivity, the operation in the kinetic re- gion results in very high selectivity for oxygen/nitrogen separation.

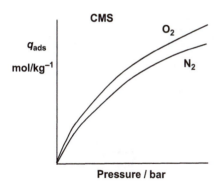 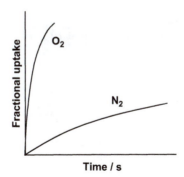

Figure 6.13: Equilibrium adsorption and uptake rates for carbon molecular sieves (CMS) used in air separation.

6.3 Fixed-bed adsorption

6.3.1 Bed profiles and breakthrough curves

In a fixed bed, a nearly solute-free liquid or gas effluent can be obtained until the adsorbent in the bed approaches saturation. A fixed bed is considered ideal under the following conditions: (i) external and internal mass transfer resistances are very small; (ii) plug flow is achieved instantaneously inside the bed, and the axial dispersion, that is, deviation from plug flow is negligible; (iii) the adsorbent is initially free of adsorbate. Under these *ideal fixed-bed adsorption* conditions, the instantaneous equilibrium results in a shock wave, the *stoichiometric front* (Figure 6.14), moving

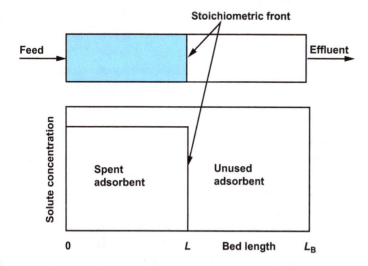

Figure 6.14: Stoichiometric equilibrium concentration front for ideal fixed-bed adsorption.

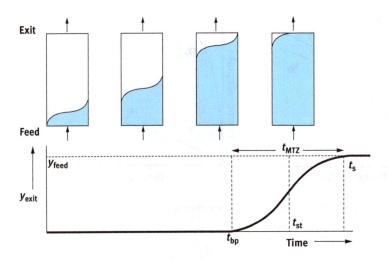

Figure 6.15: Breakthrough curve for real fixed-bed adsorption.

as a sharp concentration step through the bed. Upstream of the front, the adsorbent is saturated with adsorbate and the concentration of solute in the fluid is that of the feed. In the upstream region, the adsorbent is spent. Downstream of the stoichiometric front and in the exit fluid, the concentration of the solute in the fluid is zero, and the adsorbent is still adsorbate free. After a period of time called the *stoichiometric time*, the stoichiometric wave front reaches the end of the bed and the concentration of the solute in the fluid rises abruptly to the inlet value. Because no further adsorption is possible, this point is referred to as the *breakthrough point*. At this point, the bed can no longer separate the target compounds. The bed can regain its function through desorption/regeneration step.

In a real fixed-bed adsorber, internal and external transport resistances are finite and axial dispersion can be significant. These effects lead to a deviation from ideal behavior shown in Figure 6.14 and give rise to broader S-shaped concentration fronts like that shown in Figure 6.15. In such non-ideal systems, a well-defined mass transfer zone (MTZ) emerges in which the adsorbate concentration drops from the inlet to the exit value. This MTZ with gradually changing adsorbate concentration is significantly different from the sharp transition in ideal systems. This zone progresses through the bed as the run proceeds. At the breakpoint time t_{bp}, the zone begins to leave the column (Figure 6.15). If the run is extended, the effluent concentration rises sharply until the whole bed is saturated, t_{sat}, and the effluent concentration matches the inlet concentration. The S-shaped curve in Figure 6.15 is called the *breakthrough curve*, representing the outlet-to-inlet solute concentration in the fluid as a function of time from the start of the flow. The region between t_{bp} and t_{sat} is the *MTZ*, where adsorption takes place. At the breakthrough point t_{bp}, the leading point of the MTZ just reaches the end of the bed. For a flow rate Q, the useful capacity of the bed, Q_{uc}, and its total capacity Q_{tot} are, respectively:

$$Q_{uc} = \int_0^{t_{bp}} Q\,(c_f - c_{out})\,dt \qquad (6.18)$$

and

$$Q_{tot} = \int_0^{\infty} Q\,(c_f - c_{out})\,dt \qquad (6.19)$$

6.3.2 Equilibrium theory model

The steepness of the breakthrough curve determines the extent to which the capacity of an adsorbent bed can be utilized. Another complication is the fact that the steepness of the concentration profiles increases or decreases with time, depending on the shape of the adsorption isotherm. Therefore, determination of the breakthrough curve shape is very important in determining the length of an adsorption bed. For the fixed bed shown in Figure 6.16, the velocity of the concentration front can be derived by considering ideal plug flow of fluid through the column cross-sectional area, A, at a constant volume flow rate, Q, and constant temperature. As a first approximation, the so-called *equilibrium model* is used, which assumes that, at any point in the bed, equilibrium between the fluid phase and the adsorbent phase is achieved instantaneously. Under these conditions, the mass balance in a volume element of cross-sectional area A *and thickness* Δz of the bed with interparticle porosity ε_{bed} over a time step Δt can be written as follows:

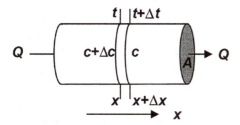

Figure 6.16: Concentration change in a thin slice of a fixed-bed adsorber.

accumulation insolution + accumulation in porous solid = flow in – flow out

or

$$A \Delta z \varepsilon_{bed} \frac{\Delta c}{\Delta t} + A \Delta z (1 - \varepsilon_{bed}) \rho_{part} \frac{\Delta q}{\Delta t} = Q(c) - Q(c + \Delta c) \qquad (6.20)$$

where ρ_{part} is the particle density (Figure 6.3) and q denotes the amount adsorbed per unit weight of adsorbent particles and c is the adsorbate concentration in solution. Dividing by $A\Delta z$ and taking the limit as $\Delta z \to 0$ gives

$$\varepsilon_{bed} \frac{dc}{dt} + (1 - \varepsilon_{bed}) \rho_{part} \frac{dq}{dt} = -\frac{Q}{A} \frac{dc}{dz} \qquad (6.21)$$

By introducing the superficial velocity $u_{carrier} \equiv Q/A$, the chain rule and the rules of implicit partial differentiation:

$$\frac{\partial q}{\partial t} = \frac{\partial q}{\partial c} \frac{\partial c}{\partial t} \quad \text{and} \quad \left. \frac{\partial z}{\partial t} \right|_c = - \frac{\left. \frac{\partial c}{\partial t} \right|_z}{\left. \frac{\partial c}{\partial z} \right|_t} \qquad (6.22)$$

the velocity v_c of propagation of a concentration c becomes

$$v_c \equiv \left. \frac{dz}{dt} \right|_c = \frac{u_{carrier}}{\varepsilon_{bed} + (1 - \varepsilon_{bed}) \rho_{part} \left. \frac{dq}{dc} \right|_c} \approx \frac{u_{carrier}}{(1 - \varepsilon_{bed}) \rho_{part} \left. \frac{dq}{dc} \right|_c} \qquad (6.23)$$

because usually $\varepsilon_{bed} \ll (1 - \varepsilon_{bed}) \cdot \rho_{part} \cdot dq/dc$. This equation gives the velocity v_c of any solute concentration c in the concentration wave front in terms of the superficial fluid velocity $u_{carrier}$, and the slope of the adsorption isotherm dq/dc at that concentration. In the analysis of adsorption and desorption processes, eq. (6.23) is important because it illustrates how the adsorption wave shape changes as it moves along the bed. For a linear isotherm, dq/dc is constant and all concentrations move at the same velocity v_c. If the adsorption isotherm is curved, regions of the front at a higher concentration move at a velocity different from regions at a lower concentration.

For these three conditions, Figure 6.17 illustrates the development of concentration gradients. For *unfavorable isotherms* (Figure 6.17a), the concentration gradients in the column at various times show that the wave front is *dispersive*, and leads to an adsorption zone which gradually increases in length as it moves through the bed. For the case of a linear isotherm, the zone goes through the bed unchanged. For *favorable isotherms* such as the Langmuir type (Figure 6.17c), concentration profiles become sharper and the wave front is *compressive* because points of high concentration in the adsorption wave move more rapidly than points of low concentrations. Since it is physically impossible for points of high concentration to overtake points of low concentration, the effect is for the adsorption zone to become narrower as it moves along the bed. The physical limit is a *shock wave* with a *shock wave velocity*:

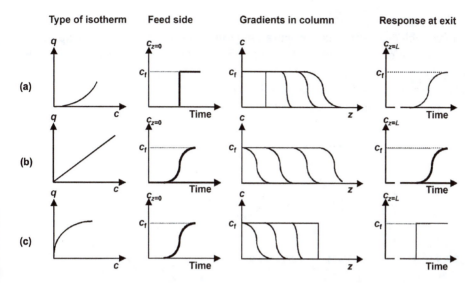

Figure 6.17: Development of an adsorption wave through a bed for (a) unfavorable, (b) linear and (c) favorable isotherms. The mass transfer zone at a certain time and increasing (a), constant (b) and decreasing (c) in time.

$$v_{sw} \approx \frac{u_{carrier}}{(1 - \varepsilon_{bed})\rho_{part}\frac{\Delta q}{\Delta c}} \tag{6.24}$$

where Δq and Δc are calculated between the feed and presaturation conditions. The time t_{sat} required saturating a freshly regenerated adsorbent of length L with a feed concentration c_f is the time of movement of this shock wave through the column:

$$t_{sat} = \frac{L}{v_{sw}} \approx \frac{L}{u_{carrier}}(1 - \varepsilon_{bed})\rho_{part}\frac{q^* - 0}{c_f - 0} \tag{6.25}$$

where q^* is the amount adsorbed in equilibrium with feed concentration c_f.

Note that eq. (6.25) can also be derived from a mass balance over the entire column (adsorbent weight W_s, plug flow rate Q) when the column is just saturated with $c = c_f$ at $t = t_{sat}$:

$$t_{sat}\, Q\, c_f = W_s\, q^* \tag{6.26}$$

6.3.3 Modeling of mass transfer effects

In nonideal systems commonly encountered in industrial practice, the finite mass transfer resistance and axial mixing in the column lead to departures from the idealized response predicted by the equilibrium theory in Section 6.3.2. In the case of a favorable isotherm, the concentration profile spreads in the initial region until a stable situation is reached in which the mass transfer rate is the same at all points along the wave front. Moreover, the wave speed exactly matches the shock velocity. This represents a stable situation and the profile propagates without further change in shape, as illustrated in Figure 6.18.

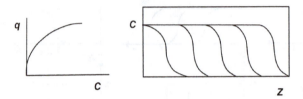

Figure 6.18: Effect of mass transfer on the adsorption wave through a bed for a favorable isotherm.

The distance required to approach the time-invariant stable pattern limit decreases as the mass transfer resistance decreases and the nonlinearity of the equilibrium isotherm increases. Calculating the concentration profile across the bed under these *constant pattern conditions* requires extension of the equilibrium model with mass transfer resistances outside and inside the adsorbent particle. Such a mass transfer model that includes all these mass transfer resistances requires simultaneous solutions of the mass balance equations within the particle as well as in the bed or for the bulk flow. This simultaneous solution is a formidable task, which can be avoided by using the *linear driving force (LDF) model*. In this model, which is widely used to simulate and design fixed-bed adsorption, the mass balance equation within the particle is eliminated by relating the overall uptake rate dq_A/dt in the particle to the bulk flow concentration c_A. This way the particle mass balance equation is eliminated from the model and the rate of mass transfer may be expressed in terms of a simple linear rate expression:

$$\frac{\partial \bar{q}_A}{\partial t} = f(q_A^*) = k \, (q_A^* - \bar{q}_A) \qquad (6.27a)$$

where $\bar{q}_A$ is the average or overall amount adsorbed in the particle; q_A^* is the amount adsorbed in equilibrium with the adsorbate concentration c_A; k is the first-order rate constant.

The rate constant k includes both external and internal transport resistances:

$$1/k = a/k_f + d_{part}^2/60D_{eff} \quad \text{(see, e.g., Seader and Henley for details)} \qquad (6.27b)$$

where k_f is the external mass transfer coefficient; a is the shape factor; $V_{part}/A_{part} = d_{part}/6$ (spherical particles); and D_{eff} is the effective coefficient of diffusion inside particle.

Incorporation of eq. (6.27 a&b) into the previously derived mass balance (eq. (6.21)) and taking into account axial dispersion (represented by the term with the coefficient of axial dispersion, D_{ax}) gives:

$$\varepsilon_{bed}\frac{\partial c_A}{\partial t} + k(1-\varepsilon_{bed})\rho_{part}\left(q_A^* - \bar{q}_A\right) = -u_{carrier}\frac{\partial c_A}{\partial z} + D_{ax}\frac{d^2 c_A}{dz^2} \qquad (6.28)$$

This equation provides the basic relation that can be integrated to calculate the effect of mass transfer resistances on the shape of breakthrough curves according to the LDF model. The *self-sharpening effect* as depicted in Figure 6.17c is balanced by axial dispersion; the resulting *constant pattern* is shown in Figure 6.18.

6.4 Basic adsorption/desorption cycles

Commercial adsorptions can be classified into two categories: bulk separations and purifications according to percentage of feed stream adsorbed. In bulk separation processes, 10 wt% or more of the feed stream are adsorbed. Processes in which considerably less than 10 wt% of the feed are classified as purification. Such a differentiation is desirable as different process cycles are used for the two categories. As schematically illustrated in Figure 6.19, there are three basic ways to influence the adsorbate loading on an adsorbent:
1. Changing temperature
2. Changing (partial) pressure in a gas or concentration in a liquid
3. Adding a component that competitively adsorbs with the adsorbate of interest

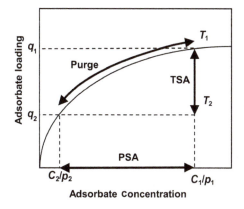

Figure 6.19: Schematic illustration of adsorption/desorption cycles.

A major difference in processes based on temperature change and those based on pressure or concentration change is the time required to change the bed from an adsorbing to a desorbing or regenerating condition. In short, pressure and concentration can be changed much more rapidly than temperature in practice. In other words, heating or cooling an adsorption bed takes more time that switching streams to purge a column. Consequently, *temperature swing* processes are limited almost exclusively to low adsorbate concentrations in the feed. For higher adsorbate feed concentrations, the adsorption time would become too short compared to the regeneration time. In contrast, *pressure swing* or *concentration swing* processes are much more suitable for bulk separations because they can accept feed, change to regeneration and be back to the feed condition within a reasonable fraction of the cycle time. The primary quantities of interest in regeneration are the required desorption time, t_{des}, to achieve a desired outlet concentration and the corresponding purge feed flow rate. Desorption of a species according to the Langmuir isotherm is equivalent to the adsorption of species with unfavorable isotherm. The input is now a negative step, which is shown in Figure 6.20 together with the concentration profiles in the bed and the desorption breakthrough curve. Because the high concentrations travel faster than the low concentrations, the breakthrough curve is now of a dispersive nature. The *desorption time*, t_{des}, is given by the bed height, L, divided by the propagation velocity of the wave front of the concentration at which the desorption takes place. According to the equilibrium theory, it follows from eq. (6.23):

$$t_{des} = \frac{L}{v_{c_{des}}} \approx \frac{L}{u_{carrier}} (1 - \varepsilon_{bed}) \rho_{part} \frac{dq}{dc}\bigg|_{c_{des}} \qquad (6.29)$$

For desorption with $c_{des} = 0$, $dq/dc|_{c=0}$ represents the initial slope of the isotherm.

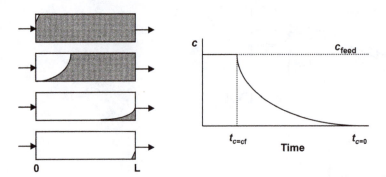

Figure 6.20: Dispersive wave front and breakthrough curve for desorption caused by a favorable isotherm.

6.4.1 Temperature swing

In a temperature swing or *thermal swing adsorption* (TSA) cycle, desorption takes place at a temperature much higher than the adsorption step. Desorption is promoted by increasing the temperature taking advantage of exothermic nature of adsorption. In this cycle, shown in Figure 6.21, a stream containing a minute amount of adsorbate at concentration c_1 is pumped through the adsorbent bed at temperature T_1. After this step, the bed temperature is raised to T_2, and desorption occurs by a continuous feed flow through the bed. In general, the regeneration is more efficient at higher temperatures. Heating, desorbing and cooling of the bed may range between few hours to over a day depending on the process size, making the desorption step often the cost-determining factor in the separation. Therefore, temperature swing processes are used almost exclusively to remove small adsorbate concentrations in order to maintain the process running for a significant fraction of the total process cycle time.

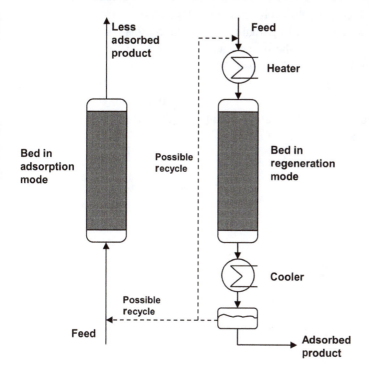

Figure 6.21: Temperature swing adsorption (TSA).

The most common application of TSA is drying. Zeolites, silica gel and activated aluminas are widely used in the natural gas, chemical and cryogenic industries to dry streams. Of these adsorbents, silica gel requires the lowest temperatures for regeneration. Other TSA applications range from CO_2 removal to hydrocarbon separations, and include the removal of pollutants, odors and contaminants with activated carbon.

6.4.2 Pressure swing

In a pressure swing adsorption (PSA) cycle, desorption step takes place at a pressure much lower than adsorption. In essence, this process takes advantage of the adsorption/desorption equilibrium isotherm shown in Figure 6.19. Decreasing the partial pressure of target compound drives the adsorption/desorption equilibrium toward desorption. The most common processing scheme has two, illustrated in Figure 6.22, or three fixed-bed adsorbers alternating between the adsorption step and the desorption steps. A complete cycle consists of at least three steps: adsorption, blow down and repressurization. During adsorption, the less selectively adsorbed components are recovered as product. The selectively adsorbed components are removed from the adsorbent by adequate reduction in partial pressure. During this countercurrent depressurization (*blow down*) step, the strongly adsorbed species are desorbed and recovered at the adsorption inlet of the bed. The repressurization step returns the adsorber to feed pressure and can be carried out with product or feed. PSA cycles operate at nearly constant temperature and require no heating or cooling steps. They utilize the exothermic heat of adsorption remaining in the adsorbent to supply the endothermic heat of desorption.

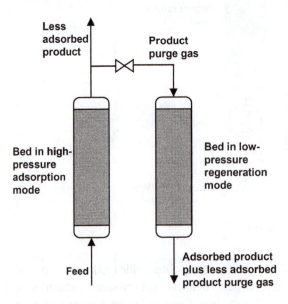

Figure 6.22: Pressure swing adsorption (PSA).

The principle application of PSA is for bulk separations where target compounds are present at high concentration. Fortunately, packed beds of adsorbent respond rapidly to changes in pressure. In most applications, equilibrium adsorption is used to obtain the desired separation. However, in air separation, PSA processes the adsorptive separation is based on kinetically limited systems where oxygen is preferentially adsorbed on 4A zeolite when the equilibrium selectivity favors N_2 adsorption. The major purification applications for PSA are for hydrogen, methane and drying. Air separation, methane enrichment and iso/normal separations are the principal bulk separations for PSA.

6.4.3 Inert and displacement purge cycles

In a purge swing adsorption cycle, the adsorbate is removed by passing a nonadsorbing gas or liquid, containing very little to no adsorbate through the bed. Desorption occurs because the partial pressure or concentration of the adsorbate around the particles is lowered. If enough purge gas or liquid is passed through the bed, the adsorbate will be completely removed. Just like PSA cycle, purge cycle takes advantage of the adsorption/desorption equilibrium isotherm shown in Figure 6.19. Once the purge removes the target compound, its partial pressure will decrease; consequently, desorption will be triggered. Most *purge swing* applications use two fixed-bed adsorbers to provide a continuous flow of feed and product (Figure 6.23). Its major application is for bulk separations when the target compounds are at high concentration. Applications

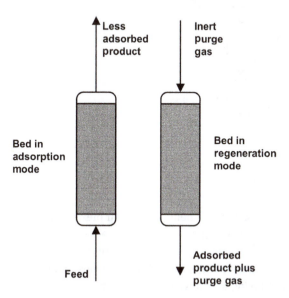

Figure 6.23: Inert purge cycle.

include the separation of normal from branched and cyclic hydrocarbons, gasoline vapor recovery and bulk drying of organics.

The displacement purge cycle differs from the inert purge cycle in that a gas or liquid that adsorbs about as strongly as the adsorbate is used to remove the adsorbate. Desorption is thus facilitated by adsorbate partial pressure or concentration reduction in the fluid around the particles in combination with *competitive adsorption* of the displacement medium. The use of an adsorbing displacement purge fluid causes a major difference in the process. Since it is actually absorbed, it will be present when the adsorption cycle begins and therefore contaminate the less adsorbed product. This means that the displacement purge fluid must be recovered from both product streams, as illustrated in Figure 6.24.

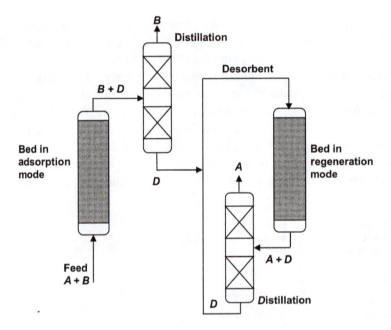

Figure 6.24: Displacement purge cycle.

6.5 Principles of ion exchange

Ion exchange is a chemisorption process with a chemical ion complex reaction. In ion exchange processes, positive or negative charged ions in an aqueous solution are exchanged with "displaceable ions of same charge but different chemical identity" on the surface of a solid *ion exchanger*. Water softening is the most commonly encountered ion exchange process. The ion exchange concept has been extended to completely remove inorganic salts from water in a two-step process called *demineralization* or *deionization*. In the first step, a cation resin is brought in contact with water to exchange hydrogen ions for cations such as calcium, magnesium and sodium. In the second step, the negatively charged ions sulfate, nitrate and chloride anions in water are exchanged with hydroxyl ions on an anion exchange resin. The hydrogen and hydroxyl ions exchanged in two steps react to produce water resulting in complete removal of inorganic ions from water.

6.5.1 Ion exchange resins

Naturally occurring inorganic aluminosilicates (zeolites) were the first ion exchangers used in water softening. Today, synthetic organic polymer resins based on styrene or acrylic acid-type monomers are the most widely used ion exchangers. These resin particles consist of a three-dimensional polymeric network with attached ionic functional groups to the polymer backbone. Ion exchange resins are categorized by the nature of functional groups attached to a polymeric matrix, by the chemistry of the particular polymer in the matrix. Strong acid and strong base resins are based on the copolymerization of styrene and a cross-linking agent, divinylbenzene, to produce the three-dimensional cross-linked structure as shown in Figure 6.25. In *strong acid cation exchange resins*, sulfonic acid groups in the hydrogen form exchange hydrogen ions for the other cations present in the liquid phase:

$$resin - SO_3^- \ H^+ + Na^+ \Leftrightarrow resin - SO_3^- \ Na^+ + H^+ \qquad (6.30)$$

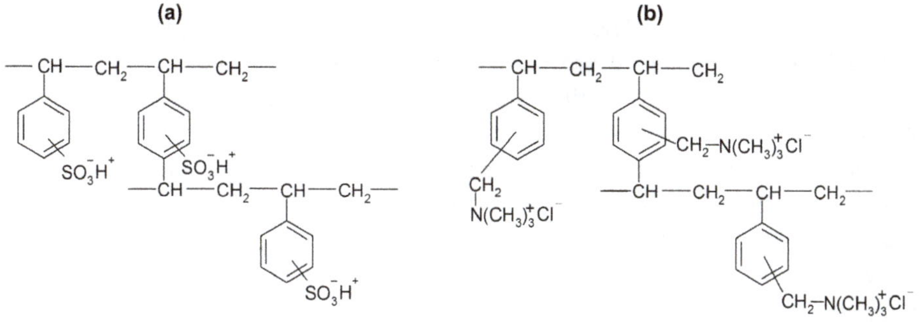

Figure 6.25: Styrene and divinyl benzene-based ion exchange resins: (a) sulfonated cation exchanger and (b) aminated anion exchanger.

It is not always necessary for the resin to be in the charged hydrogen form for adsorption of cations. Softening of water is accomplished by displacing sodium ions from the resin by calcium ions, for which the resin has a greater affinity:

$$2 \, resin - SO_3^- \ Na^+ + Ca^{2+} \Leftrightarrow (resin - SO_3^-)_2 Ca^{2+} + 2Na^+ \qquad (6.31)$$

In *strong base anion exchange resins*, quaternary ammonium groups are used as the functional exchange sites. They are used most often in the hydroxide form to reduce acidity:

$$\text{resin} - N^+ (CH_3)_3 \; OH^- + H^+ + Cl^- \Leftrightarrow \text{resin} - N^+ (CH_3)_3 \; Cl^- + H_2O \qquad (6.32)$$

During the exchange, the resin releases hydroxide ions as anions are adsorbed from the liquid phase. The effect is elimination of acidity in the liquid and conversion of the resin to a salt form. Ion exchange reactions are reversible. After complete loading, a *regeneration procedure* is used to restore the resin to its original ionic form. For strong cation and anion resins, this is typically done with dilute (up to 4%) solutions of hydrochloric acid, sulfuric acid, sodium hydroxide or sodium chloride.

Weak acid cation exchanger resins have carboxylic acid groups attached to the polymeric matrix derived from the copolymerization of acrylic acid and methacrylic acid. *Weak base anion exchanger resins* may have primary, secondary or tertiary amines as functional groups. The tertiary amine is most common. Weak base resins are frequently preferred over strong base resins for removal of strong acids in order to take advantage of the greater ease of regeneration.

6.5.2 Equilibria and selectivity

Ion exchange differs from adsorption in that one sorbate (a counterion) is exchanged for a solute ion, and the exchange governed by a reversible, stoichiometric chemical reaction equation. The exchange equilibria depend largely on the functional group type and the degree of cross-linking in the resin. The quantity of ions, acids or bases that can be adsorbed or exchanged by the resin is called the *operating capacity*. Operating capacities vary from one installation to another as a result of differences in composition of the stream to be treated. In ion exchange, significant exchange does not occur unless the functional group of the resin has a greater selectivity for the ions in solution than for ions occupying the functional groups, or unless there is excess in concentration as in regeneration. This is pictured in the following reaction where the cation exchange resin removes cations B^+ from solution in exchange for A^+ on the resin:

$$\text{resin} - SO_3^- A^+ + B^+ \Leftrightarrow \text{resin} - SO_3^- B^+ + A^+ \qquad (6.33)$$

For this case, we can define a conventional chemical equilibrium constant, determining the selectivity for B over A:

$$K_A^B = \frac{q_{B,\text{resin}}}{q_{A,\text{resin}}} \frac{m_{A,\text{liquid}}}{m_{B,\text{liquid}}} \qquad (6.34)$$

where m and q are the concentrations of the ions in the solution and the resin phase, respectively. A typical plot of K^B_A for univalent exchange is given in Figure 6.26. B is preferred by the exchanger if $K^B_A > 1$ while $K^B_A < 1$ B is less preferred and the ion exchanger preferentially absorbs species A. Selectivity for ions of the same charge usually increases with atomic weight (Li < Na < K < Rb < Cs), and selectivities for divalent ions are greater than for monovalents. For practical applications, it is rarely necessary to know the selectivity precisely. However, knowledge of relative differences is important when deciding if the reaction is favorable or not. Selectivity differences are marginally influenced by the degree of cross-linking of a resin. The main factor is the structure of the functional groups.

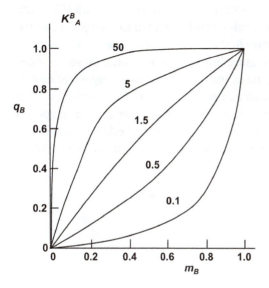

Figure 6.26: Equilibrium plot for univalent–univalent ion exchange.

6.6 Ion exchange processes

Ion exchange cycles consist of two principal steps, adsorption and regeneration. During adsorption, impurities are removed or valuable constituents are recovered. *Regeneration is in general of much shorter duration than the adsorption step.* Industrial systems may be batch, semicontinuous or continuous. They vary from simple one-column units to numerous arrays of cation and anion exchangers, depending on the required application and quality of the effluent. A single-column installation is satisfactory if the unit can be shut down for regeneration. However, for the processing of continuous streams, two or more columns of the same resin must be installed in parallel. Most ion exchanger columns are operated concurrently with the process stream and regeneration solution flowing down through the resin bed to avoid possible fluidization of the resin particles.

Deionization processes require two columns, one containing a cation exchanger and one an anion exchanger. The cation exchanger unit must be a strong acid-type resin and it must precede the anion exchange unit. Placing the anion exchanger first generally causes problems with precipitates of metal hydroxides. A column containing a mixture of cation and anion exchanger resins is called a mixed bed. The majority of the used resins consist of strong acid cation and strong base anion exchangers. *Mixed bed systems* yield higher quality deionized water or process streams than when the same resins are used in separate columns. However, regeneration of a mixed bed is more complicated than of a two-bed system.

To increase resin utilization and achieve high ion exchange reaction efficiency, numerous efforts have been made to develop continuous ion exchange contactors. Two such examples are shown in Figures 6.27 and 6.28. The *Higgins loop contactor* operates as a moving packed bed by using intermittent hydraulic pulses to move incremental portions of the bed from the contacting section to the regeneration section. In other continuous systems, such as the *Himsley contactor*, columns with perforated plates are used. The liquid is pumped into the bottom and flows upward through the column to fluidize the resin beads on each plate. Periodically, the flow is reversed to move incremental amounts of resin to the stage below. The resin at

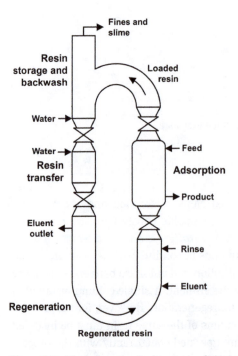

Figure 6.27: Higgins moving bed contactor (adapted from [15]).

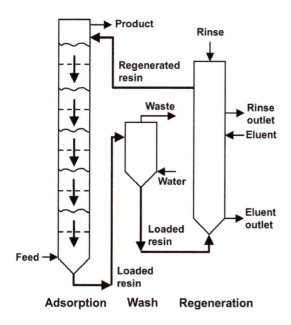

Figure 6.28: Himsley fluidized bed contactor (adapted from [10, 15]).

the bottom is lifted via the wash column to the regeneration column. A more recent approach involves the placement of a number of columns on a carousel that rotates constantly at an adjustable speed.

Nomenclature

a	Particle volume/surface area ratio ($=d_{part}/6$ for spheres), eq. (6.27b)	$m^3\,m^{-2}$
A, A_{part}	Area (column cross section, particle interface area)	m^2
b	Adsorption constant, eq. (6.4)	$m^3\,mol, Pa^{-1}$
c	Concentration	$mol\,m^{-3}$
d	Diameter	m
D_{ax}	Coefficient of axial dispersion, eq. (6.28)	$m^2\,s^{-1}$
D_{eff}	Effective diffusivity in the pores of a porous solid, eq. (6.27b)	$m^2\,s^{-1}$
k	First-order rate constant, eq. (6.27a)	s^{-1}
k_f	External mass transfer coefficient, eq. (6.27b)	$m\,s^{-1}$
K	Equilibrium constant	–
L	Column length	m
q	Amount adsorbed	$mol\,kg^{-1}\,(mol\,m^{-3})$
Q	Volume flow rate	$m^3\,s^{-1}$
S	Surface area in porous solid, eq. (6.2)	m^2
t	Time	s
u	Superficial velocity, eq. (6.21)	$m\,s^{-1}$

v	Interstitial velocity, eq. (6.23)	m s^{-1}
x, y	Mole fraction	–
V	Volume, Figure 6.3	m^3
W_s	Weight of dry solid (adsorbent)	kg
z	Length coordinate	m
α	Selectivity, eq. (6.17)	–
ε	Porosity, fractional void volume	–
ρ	Density	kg m^{-3}
θ	Occupancy, eq. (6.4)	–

Indices

A, B	Components
ads	Adsorption
bp	Break point
C	At concentration c
des	Desorption
f	Feed
max	Maximum (monolayer) capacity
part	Particle
sat	At saturation condition

Exercises

1   The following table gives the Langmuir isotherm constants for the adsorption of propane and propylene on various adsorbents.

Adsorbent	Adsorbate	q_m (mmol g^{-1})	b (bar^{-1})
ZMS 4A	C_3	0.226	9.770
	$C_3^=$	2.092	95.096
ZMS 5A	C_3	1.919	100.223
	$C_3^=$	2.436	147.260
ZMS 13X	C_3	2.130	55.412
	$C_3^=$	2.680	100.000
Activated carbon	C_3	4.239	58.458
	$C_3^=$	4.889	34.915

a.   Which component is most strongly adsorbed by each of the adsorbents?
b.   Which adsorbent has the greatest adsorption capacity?
c.   Which adsorbent has the greatest selectivity?
d.   Based on equilibrium conditions, which adsorbent is the best for the separation?

2 The following data were obtained for the adsorption of toluene from water on activated carbon and water from toluene on activated alumina. Fit both sets to the Langmuir-type isotherm and compare the resulting isotherms with the experimental data.

Toluene (in water) on activated carbon		Water (in toluene) on activated alumina	
C (mg L^{-1})	q (mg g^{-1})	c (ppm)	q (g 100 g^{-1})
0.01	12.5	25	1.9
0.02	17.1	50	3.1
0.05	23.5	75	4.2
0.1	30.3	100	5.1
0.2	39.2	150	6.5
0.5	54.5	200	8.2
1	70.2	250	9.5
2	90.1	300	10.9
5	125.5	350	12.1
10	165.0	400	13.3

3 An aqueous amount of V m^3 contains 0.01 kg m^{-3} of nitrobenzene. We want to reduce this to 10^{-6} kg m^{-3} by adsorption on activated carbon. The solution is fed to an ideally stirred tank containing m_C kg adsorbent; see outline. At these low concentrations, the equilibrium adsorption isotherm appears to be linear

$$q = H' \cdot c$$

V m³ wastewater

C_{in}

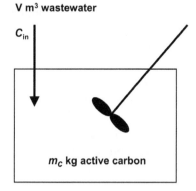

m_C **kg active carbon**

with q is the amount adsorbed, kg kg^{-1};
c is the concentration of nitrobenzene, kg m^{-3};
H' is the Henry constant = 675 m^3 kg^{-1}.
How many kg of activated carbon is required per m^3 of treated water?

4 Nitrobenzene in an aqueous effluent should be reduced from its initial concentration of 0.02 kg m^{-3} to a value of not more than $2 \cdot 10^{-5}$ kg m^{-3}. Nitrobenzene can be removed from water by adsorption on active carbon. The amount q of nitrobenzene adsorbed in equilibrium with its concentration c in water is given by the following Langmuir isotherm:

$$q(c) = \frac{510 \cdot c}{1 + 4,550 \cdot c} \quad \text{with } c \text{ in kg m}^{-3} \text{ and } q \text{ in kg kg}^{-1}$$

Two different process schemes (see outlines below) are considered to carry out the required reduction:
– A packed column (see process scheme a)
– A batch process in an ideally stirred tank (see process scheme b)

The packed column contains W_{pc} kg carbon and is fed with wastewater at a rate of $\Phi_f = 0.015$ m^3 s^{-1}.

The rate of convective transport through the bed is such that local adsorption equilibrium is established throughout the column. The concentration profile is flat (plug flow), as long as the column is not saturated, and the outlet concentration is zero.

The length of the column $L = 2.0$ m, its diameter $D = 0.6$ m, the bed porosity $\varepsilon_{bed} = 0.6$ and the particle density $\rho_p = 800$ kg m^{-3}.

a1. Calculate from a component balance the amount of carbon per unit volume of wastewater (W_{pc}/V in kg m^{-3}).
a2. How much wastewater (V in m^3) is purified when the column is about to breakthrough?

After saturation of the column, it is regenerated with a water flow of the same size in which the concentration of nitrobenzene amounts to $2 \cdot 10^{-5}$ kg m^{-3}.

a3. How long does it take to regenerate the column?

In the second process, an ideally stirred tank, V m^3 wastewater (as calculated in part $a1$) is treated. This batch is treated with W_{batch} kg active carbon. After equilibration, the resulting concentration of nitrobenzene should be $2 \cdot 10^{-5}$ kg m^{-3}.

b1. Calculate the excess amount of carbon to be used in the batch process compared to that required in the continuous process; in other words, calculate W_{batch} / W_{pc}.
b2. Comment on the answer of part b1.

a

b

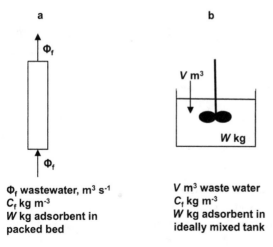

Φ_f wastewater, m³ s⁻¹
C_f kg m⁻³
W kg adsorbent in
packed bed

V m³ waste water
C_f kg m⁻³
W kg adsorbent in
ideally mixed tank

5   Derive the expression for the time required to saturate a fixed bed with $c = c_f$
(eq. (6.25)) starting from the overall mass balance given in eq. (6.26). List all the
necessary assumptions.

6   A couple of students got an assignment to determine experimentally the residence time of CO_2 during adsorption and desorption in a packed bed column.
They had to compare the results with a theoretical model based on local adsorption equilibrium.

   A column (L = 1 m, internal diameter D = 5 mm) was filled with activated
carbon and its exit provided with a CO_2 detector. The temperature was kept constant at 20 °C.

   Flushing with pure helium pretreated the column. At time τ_0 a CO_2–He mixture, containing 4 vol% CO_2, was fed to the column.

   Some time after saturation, at time τ_1, the feed was switched to pure He.
The concentration of CO_2 in the effluent was measured continuously and is
given schematically (not to scale) in the plot below:

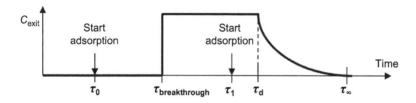

Additional data

The particle density of the carbon $\rho_{particle}$ = 790 kg m^{-3} and the bed porosity ε_{bed} = 0.40. The adsorption of CO_2 at 20 °C is given by the Langmuir-type adsorption isotherm:

where monolayer capacity q_m = 1.39 mol kg^{-1}, the adsorption constant b = 0.068 m^3 mol^{-1} and the concentration c is in mol m^{-3}.

In all experiments (flushing with He, adsorption with He/CO_2, desorption with He), the total gas flow Φ_V was maintained at 4·10^{-6} m^3 s^{-1} at 20 °C and 1 bar. In this equipment and at these conditions, adsorption and desorption are isothermal processes. R = 8.31 J mol^{-1} K^{-1}.

The following problems are to be addressed:

a. Explain the form of the breakthrough curve qualitatively.
b. Calculate the breakthrough time, for example, $\tau_{breakthrough}$ − τ_0, using eq. (6.25).
c. Some students calculated the breakthrough time from the ratio of the maximum amount to be adsorbed and the flow rate of carbon dioxide:

$$\tau_{breakthrough} - \tau_0 = \frac{W_s \times q(c_f)}{\Phi_v^* c_f}$$

When calculated properly, this gives the same result as in part *b*.
Explain.
d. Calculate the desorption time, $\tau_\infty - \tau_1$.

7 An aqueous waste stream contains small amounts of acetone (1) and propionitrile (2). These contaminants can be captured by adsorption on active carbon. The respective adsorption isotherms of both contaminants at 20 °C are of the Langmuir type:

$$q_1 = \frac{0.190 \cdot c_1}{1 + 0.146 \cdot c_1} \quad \text{and} \quad q_2 = \frac{0.173 \cdot c_2}{1 + 0.0961 \cdot c_2}$$

with q in mol kg^{-1} and c in mol m^{-3}. Simultaneous adsorption of acetone and propionitrile from the waste stream can be described by the "extended Langmuir" model.

a. Calculate q_1, m and q_2, m.
b. Calculate the equilibrium loading of both contaminants when the waste stream is led through an active carbon column and contains 40 mol acetone m^{-3} and 34 mol propionitrile m^{-3}.
c. Calculate the ratio of the velocities by which both contaminants flow through the column.

8 A commercial ion exchange resin is made of 88 wt% styrene (MW = 0.104 kg mol^{-1}) and 12 wt% divinyl benzene (MW = 0.1302 kg mol^{-1}). Estimate the maximum ion exchange capacity in equivalents kg^{-1} resin when a sulfonic acid group (MW = 0.0811 kg mol^{-1}) has been attached to each benzene ring.

9 A continuous stream of a soil/water mixture (0.3 m^3 s^{-1}) containing 15 mol m^{-3} of a heavy metal M^+ is treated with an ion exchange resin to remove the heavy metal by exchange against the sodium Na$^+$ ions present on the resin. The concentration M^+ has to be reduced from 15 to 1 mol m^{-3}. The concentration M^+ on the incoming regenerated ion exchange resin equals 200 mol m^{-3}. The regeneration liquid contains 3,000 mol m^{-3} Na$^+$ and no M^+.

a. Determine the equilibrium diagrams for extraction and regeneration when the total ion concentrations in the resin, soil/water mixture and regeneration solution are given by

Resin:	$[C^+] = [Na^+] + [M^+] = 2,400$ mol m^{-3}
Soil/water:	$[C^+] = [Na^+] + [M^+] = 30$ mol m^{-3}
Regeneration liquid:	$[C^+] = [Na^+] + [M^+] = 3,000$ mol m^{-3}

   and the equilibrium follows the ideal behavior with an equilibrium constant $K = 5$.

b. Determine the minimal required ion exchanger stream.

c. Determine the number of equilibrium stages and concentration M^+ in the outgoing ion exchange resin for 1.2 times the minimal ion exchanger stream.

d. Determine the minimal value of the regeneration liquid stream to obtain a regenerated resin with 200 mol m^{-3} residual M^+.

e. How many equilibrium stages are required with 1.2 times the minimal regeneration stream?

Chapter 7
Drying of Solids

7.1 Introduction

Drying is usually defined as the removal of a volatile liquid from solids, slurries or solutions by evaporation to yield solid products. The separation may be carried out in a mechanical manner without phase change or by evaporation through the supply of heat. This *thermal drying* is discussed in the following sections. Thermal drying consists of two steps. First, heat is supplied to the solid material to evaporate the moisture out of the product. Second, the vapor is separated from the product phase and, if necessary, condensed outside of the dryer. If possible, mechanical predrying is installed upstream of thermal drying because solids handling is made easier and liquid separation without evaporation is less costly. Figure 7.1 illustrates a possible position of predrying and thermal drying in an overall process.

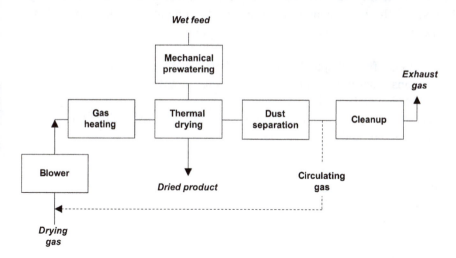

Figure 7.1: Predrying and drying steps in an overall process.

In industry, drying is applied to many types of materials, which differ tremendous in properties and in the required specifications. A few examples are
- building materials like oxides and clays;
- granular materials like sand and salts;
- foodstuffs like fruit juices, coffee, milk;
- sheet-like materials such as paper, leather;
- fabrics.

https://doi.org/10.1515/9783110654806-007

Also, there are many reasons why a drying step is required. Some examples are:
- to avoid unnecessary transport of water;
- to improve the handling properties of a powder;
- to preserve food;
- to recover remaining traces of a solvent;
- to meet the specifications (for storage or as a reaction agent).

Heat may be transferred to the solid in direct dryers by *convection*, in indirect dryers by *conduction* and in either case by *radiation*. The properties of the wet feed and the required specifications of the dried product are of major importance in the choice of the appropriate method of heat transfer and, subsequently the type of dryer. A comprehensive treatment of all kinds of combinations requires more space than available and is thus outside the scope of this textbook. The reader will find these in specialized books such as mentioned in the reference section.

This chapter will be restricted to convective thermal drying of wet solids, mainly by the use of heated air. A common liquid to be removed is water; hence, we will often refer to the liquid part as *moisture*. It should be noted that the theoretical derivations of the fundamental drying equations equally apply to other volatile liquids, except where the use of psychrometric charts is involved; see Appendix to this chapter.

The content of this chapter is organized along the following lines:
- definitions of specific terminology in drying technology;
- identification of drying mechanisms;
- derivation of simplified rate equations to estimate required drying times;
- drying methods and drying equipment.

7.2 Humidity of carrier gas

7.2.1 Definitions

Let us consider a stream of heated air at temperature T_f and vapor (water) concentration c_f being used to convey heat to a wet surface. After evaporation of liquid, the moisture is subsequently removed as vapor. For low vapor concentrations and low drying rates, the drying rate Φ_{vap} is proportional to the driving force $c_{sat}(T_s) - c_f$ and the rate expression becomes

$$\Phi_{vap} = k_g(c_{sat}(T_s) - c_f) \qquad\qquad (\text{mol s}^{-1}\ \text{m}^{-2}) \quad (7.1)$$

where T_s is the temperature of wet surface (K); c_{sat} is the saturation vapor (water) concentration at T_s (mol m^{-3}); c_f is the vapor (water) concentration in *feed* gas (air) (mol m^{-3}); k_g is the mass transfer coefficient (m s^{-1}).

Two special cases are to be considered. In case the amount of air is large compared to the amount of evaporated moisture, the dynamic equilibrium temperature of the wetted surface is called the *wet bulb temperature* T_{wb}. This is the situation existing in a wet bulb thermometer, where the temperature decrease of the liquid surface balances the heat required for evaporation; see the Appendix to this chapter for details. In the stationary situation, the drying rate is balanced by the rate of heat conduction from the heated air at T_f to the wetted surface (see also Figure 7A.1 in the appendix to this chapter):

$$\Phi_{vap} \cdot \Delta H_{vap} = h(T_f - T_{wb}) \qquad (Js^{-1}\, m^{-2}) \qquad (7.2)$$

where h is the heat transfer coefficient (W m^{-2} K^{-1}) and ΔH_{vap} is the molar heat of evaporation (J mol^{-1}).

Thus, *the wet bulb temperature indicates the maximum weight of vapor that can be carried by an amount of dry gas*, the driving force in the heat flux equation being $T_f - T_{wb}$.

On the other hand, when a stream of air Φ_{air} at T_f is mixed thoroughly and adiabatically with a lot of liquid, the air leaving the system in equilibrium with that liquid is at saturation temperature T_{sat}. By definition, the liquid is already at the eventual equilibrium temperature T_{sat}. By maintaining a constant amount of liquid at T_{sat}, it does not contribute to the enthalpy balance and the enthalpy decrease of the *carrier gas* equals just the heat of evaporation:

$$\Delta H_{vap}(c_{sat} - c_f)\, \Phi_{air} = C_p\, c_{air}\, M_{air}(T_f - T_{sat})\, \Phi_{air} \qquad (Js^{-1})$$

or

$$c_{sat} - c_f = \frac{\rho_{air}\, C_p}{\Delta H_{vap}}(T_f - T_{sat}) \qquad (7.3)$$

where C_p is the heat capacity of moist air (J K^{-1} (kg dry air)$^{-1}$) and $\rho_{air} = c_{air}.M_{air}$ is density of moist air (kg dry air (m^3 humid air)$^{-1}$).

The specific volume of moist air is calculated from the volume of a unit weight dry air plus the volume of vapor – at the average temperature $(T_f + T_{sat})/2$ – contained in that amount of air. Therefore, it is convenient to use *humidities* H_f, expressing moisture content per unit weight of *dry air*. The heat capacity C_p of moist air is then easily calculated from that of dry air and that of vapor contained in that air:

$$C_p = C_{p,\,dryair} + H_f.C_{p,\,vapor} \qquad (JK^{-1}kg^{-1}) \qquad (7.4)$$

The moisture content in a gas is directly related to the (water) vapor concentration c_f in that gas. The definition of *relative humidity* H_{rel} at a given temperature T is

$$H_{rel} = 100.\frac{c_f}{c_{sat}(T)} \approx 100.\frac{p_w}{p_{sat}(T)} \qquad (\%) \qquad (7.5)$$

It should be noted that the right-hand side of this equation is accurate only for ideal gases. The *absolute humidity* H_f is a function of just the vapor concentration at a given total pressure p_{tot}:

$$H_f \equiv \frac{c_f}{c_{air}}.\frac{M_w}{M_{air}} = \frac{c_f}{c_{tot} - c_f}.\frac{M_w}{M_{air}} \qquad (kg\, kg^{-1}_{\,dry\,air}) \qquad (7.6)$$

where M_w represents the molecular weight of (water) vapor. Usually $c_f \ll c_{tot}$ and c_{air} may then be replaced by c_{tot}. For not too high a temperature T, the maximum amount of water per unit weight of dry carrier gas is given by

$$H_{sat}(T) \approx \frac{c_{sat}(T)}{c_{tot}}\frac{M_w}{M_{air}} \qquad (7.7)$$

7.2.2 Air–water system: a special case

According to eq. (7.1), the drying rate of a completely wetted surface is proportional to the difference in vapor concentration at the LG interface and that in the vapor phase. Measurement of the wet bulb temperature provides a means to determine this difference because combination of eqs. (7.1) and (7.2) gives

$$c_{sat}(T_{wb}) - c_f = \frac{h}{\Delta H_{vap}\, k_g}(T_f - T_{wb}) \qquad (7.8)$$

Subtraction of eq. (7.6) from eq. (7.7) and elimination of $c_{sat} - c_f$ with eq. (7.8) leads to a similar expression in terms of humidities[1]:

$$H_{sat} - H_f = \frac{h}{\rho_{air} k_g}\frac{M_w}{\Delta H_{vap}}(T_f - T_{wb}) \qquad (7.9)$$

where H_{sat} should be evaluated at the wet bulb temperature T_{wb}. This equation applies to any kind of volatile liquid and requires empirical relations to evaluate appropriate values for the mass and heat transfer coefficients, k_g and h, respectively. The *saturation humidity* as a function of *adiabatic saturation temperature* T_{sat} follows from eq. (7.3):

$$H_{sat} - H_f = C_p\frac{M_w}{\Delta H_{vap}}(T_f - T_{sat}) \qquad (7.10)$$

---

1 It should be noted that in industrial applications, the change in enthalpy at evaporation often is expressed per unit *weight*, the *specific* heat of evaporation $\Delta H'_{vap}$ equals $\Delta H_{vap}/M_w$ (J kg^{-1}).

The following reasoning will show that for the *air–water* system, the ratio $h/\rho_{air}{\cdot}k_g$ happens to be equal to C_p. Both experimental data on the values of h/k_g and theoretical considerations, based on the analogy between mass and heat transfer, show that this ratio is constant. The *Chilton–Colburn transfer numbers* for heat, j_H, and mass, j_D, are defined as

$$j_H = Nu \; Re^{-1}Pr^{-1/3}$$
$$j_D = Sh \; Re^{-1}Sc^{-1/3}$$

(7.11)

with

$$Nu = \frac{hD}{\lambda}, \quad Sh = \frac{k_g d}{D_g}, \quad Re = \frac{\rho_{air}vd}{\eta}, \quad Pr = \frac{v}{a} = \frac{\eta\,\rho_{air}C_p}{\rho\;\lambda}, \quad Sc = \frac{v}{D_g}$$

The symbols used here have their usual meaning; for details see a textbook on transport phenomena. The analogy between heat and mass transport for geometrically similar cases and the same flow conditions at not to low a value of Re is

$$j_H = j_D$$

(7.12)

Using the definitions above, the ratio of Nusselt and Sherwood is given by

$$\frac{Nu}{Sh} = \frac{hD_g}{k_g\lambda} = \frac{h}{k_g\,\rho_{air}C_p}\frac{1}{Sc}\frac{Pr}{} = (Pr/Sc)^{-1/3} \Rightarrow \frac{h}{k_g} = \rho_{air}C_p(Sc/Pr)^{2/3}$$

(7.13)

The ratio h/k_g is proportional to the ratio of the thermal and mass diffusivity, which, in a stationary state, only depends on material properties. For air, *where Prandtl and Schmidt are about equal,* $\dfrac{h}{\rho_{air}\cdot k_g} \approx C_p$ and eq. (7.9) becomes

$$H_{sat} - H_f = C_p \frac{M_w}{\Delta H_{vap}}(T_f - T_{wb})$$

(7.14)

Comparing eqs. (7.10) and (7.14) reveals that in this special case the *wet bulb temperature equals the adiabatic saturation temperature*. The consequence is that wet bulb temperatures for air–water can be obtained from psychrometric charts.

7.2.3 Psychrometric charts

The calculation of the humidity H_f in eq. (7.10) from the experimental wet bulb temperature T_{wb} requires the saturation humidity H_{sat} to be known as a function of T_{wb}. The computational efforts involved in this evaluation can be avoided with the help of *psychrometric charts*. In a psychrometric chart at a given total pressure, the humidity H is plotted as a function of the temperature of a certain gas in a set of curved lines, each representing constant relative humidity. An additional feature of a psychrometric chart is the presence of a set of almost parallel straight lines, *adiabatic cooling lines*, giving the absolute humidity when gas is adiabatically cooled down. Figure 7.2 shows the essentials of a psychrometric chart:

- a curve representing 100% relative humidity, the boiling curve at constant total pressure;
- an adiabatic cooling line of a carrier gas at temperature T_f and humidity H_f, its slope given by eq. 7.9;
- the (dotted) wet bulb cooling line of the same gas, its slope given by eq. (7.8).

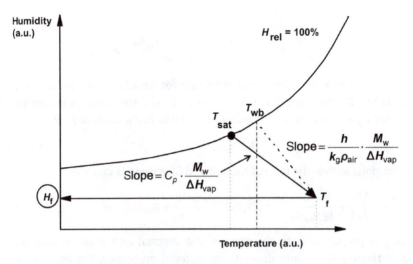

Figure 7.2: Illustration of the main features in a psychrometric chart. In case of the air–water system, the two slopes are equal and $T_{sat} = T_{wb}$.

In the air–water system, the adiabatic saturation temperature equals the wet bulb temperature. In other words, the wet bulb line and the adiabatic cooling line have the same slope. Thus, the psychrometric chart offers an elegant tool to facilitate the computations involved in solving eq. (7.10) for H_f. For air–water, the procedure to find the humidity H_f at 1 bar total pressure and temperature T_f is as follows:

- Find the wet bulb temperature on the line with 100% relative humidity (note that now $T_{wb} = T_{sat}$).
- Follow the adiabatic cooling line until the air temperature T_f is met.
- Read the humidity H_f on the vertical axis.

Reversibly, the wet bulb temperature T_{wb} – which governs the drying rate according to eq. (7.2) – can be found from known humidity H_f and feed temperature T_f.

Figure 7.3 gives a psychrometric chart for the air–water system at 1 bar total pressure. All adiabatic cooling lines have their endpoint on the 100% relative humidity line, where the water vapor in air is in equilibrium with liquid water. There,

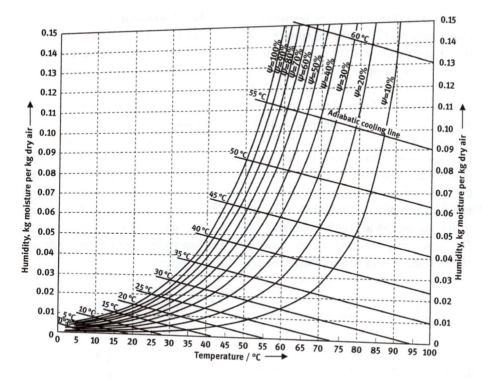

Figure 7.3: Psychrometric chart of air–water at 1 bar total pressure (adapted from [58]).

the saturation humidity H_{sat} can be read from the vertical axis and the wet bulb temperature $T_{wb} = T_{sat}$ from the temperature axis.

7.3 Moisture in solids

As the previous section dealt with properties of moist gases, this section is devoted to moisture in solid materials and focuses on
–   how is moisture bound in solid particles, and
–   how does the kind of moisture influence the drying rate.

7.3.1 Bound and unbound water

Many industrially important solids like activated carbons and molecular sieves are renowned for their highly porous character, average pore diameters ranging from microns down to molecular sizes. In pores that narrow the saturation vapor pressure of a liquid deviates from the equilibrium value for a bulk liquid at the same temperature. The drying rate is proportional to the difference in saturation vapor concentration at the external surface of the wet material and the vapor concentration in the carrier gas (cf. eq. (7.1)). Thus, to understand the relation between the properties of wet solids and their drying rate, it is necessary to look into the way a volatile liquid is held in the pores of a solid.

Two different forms of moisture in solids can be distinguished: *bound moisture* and *unbound moisture*. The liquid film at the external surface of solids and the liquid trapped in the spaces between solid particles exert a saturation vapor pressure in the same order of magnitude as that of bulk liquid. In case of water vapor, this kind of water is referred to as *unbound* or *excess* water; see Figure 7.4. All moisture that can be removed by drying is called *free moisture*, which comprises all unbound moisture and part of the bound moisture. The limiting moisture content by drying with a carrier gas at relative humidity H_1 is the equilibrium moisture content, w_{eq}. The critical moisture content, w_{crit}, separates the two regions with different drying mechanisms, as will be explained in Section 7.4.

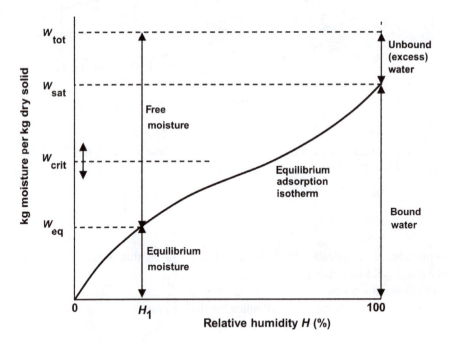

Figure 7.4: Graphical illustration of the various kinds of moisture.

Porous solids also adsorb water in (narrow) pores. In a filled pore, the liquid–gas interface is curved and the resulting saturated vapor pressure will be (much) lower than that of the bulk liquid. *The more curved the liquid–gas interface, the lower the saturation pressure.* This water will evaporate (much) slower than unbound water. The physical principles underlying the relation between saturated vapor pressure of a liquid in a narrow pore and pore radius will be explained in the following paragraphs.

The amount of moisture taken up by a porous solid depends on the actual vapor pressure. Generally, the higher the vapor pressure, the higher the uptake. The total pore volume limits the maximum uptake w_{sat} in the solid; see Figure 7.4. The removal of this moisture, drying, can only be accomplished at a vapor pressure below the saturated vapor pressure of partly or completely filled pores. At the same temperature, the saturated vapor pressure p_{sat} of a liquid in a pore is smaller than that of a bulk liquid p_{sat}^∞ because the liquid–gas interface in a pore is concave. To illustrate this effect, in Figure 7.5 the saturated vapor pressure of a liquid droplet in air and the saturated vapor pressure inside a cavity are compared with p_{sat}^∞ of a bulk liquid with a flat interface. The pressure in the liquid phase just below the flat interface, P_{liq}, equals the total pressure P in the gas phase. Thermodynamic considerations[2] show that if for some reason P_{liq} changes, the saturation pressure p_{sat} above the LG interface changes with P_{liq} according to

$$p_{sat} = p_{sat}^\infty \exp\frac{V_{liq}(P_{liq}-P)}{RT} \tag{7.15}$$

where V_{liq} is the molar volume of liquid. A reason for such a change is a change in the curvature of the LG interface: in a (very) small droplet (see left-hand side in Figure 7.5) the liquid pressure is higher to compensate for the force exerted by the surface tension y. The pressure difference across an LG interface is given by the *Laplace equation*

$$P_{liq} - P = y(1/r_1 + 1/r_2) \tag{7.16}$$

where r_1 and r_2 are the radii of the curved interface in two directions.

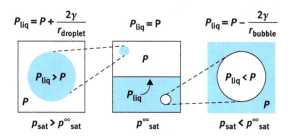

$$P_{liq} = P + \frac{2y}{r_{droplet}} \qquad P_{liq} = P \qquad P_{liq} = P - \frac{2y}{r_{bubble}}$$

$$P_{liq} > P \qquad P \qquad P_{liq} \qquad P_{liq} < P$$

$$p_{sat} > p_{sat}^\infty \qquad p_{sat}^\infty \qquad p_{sat} < p_{sat}^\infty$$

Figure 7.5: Saturation vapor pressure at curved LG interfaces.
Left: liquid droplet; center: flat interface for comparison; right: bubble in liquid phase.

---

2 See, for example, Atkins' Physical Chemistry, P.W. Atkins and J. de Paula, 7th Edn., Oxford University Press, 2002, Chapter 6.

The sign of the radius is positive for convex surfaces (see Figure 7.5, left-hand side) and negative for concave interface (Figure 7.5, right-hand side). In case of a liquid droplet in the gas phase, $r_1 = r_2 = + r_{droplet}$, whereas in a vapor bubble in liquid $r_1 = r_2 = - r_{bubble}$. Application of eqs. (7.15) and (7.16) to a liquid droplet in a gas phase and to a cavity (vapor bubble) in a liquid phase (see Figure 7.5) gives

$$\ln \frac{p_{sat}}{p_{sat}^{\infty}} = \frac{V_{liq}\gamma}{RT} \cdot \frac{2}{r_{droplet}} \quad \text{and} \quad \ln \frac{p_{sat}}{p_{sat}^{\infty}} = - \frac{V_{liq}\gamma}{RT} \cdot \frac{2}{r_{bubble}} \tag{7.17}$$

The increased liquid pressure inside the liquid droplet results in a higher saturated vapor pressure. Analogously, the pressure inside a small vapor bubble in a liquid phase is lower then the liquid pressure; consequently, the saturation pressure in a cavity is smaller than the equilibrium saturation pressure above a flat interface.

The next step is to see what vapor pressures exist in pores. To that purpose, we will discuss the filling and drying properties of a *hypothetical pore: a straight cylinder with zero contact angle*. Such a pore is characterized by two radii: $r_1 = \infty$ in the plane of drawing and $r_2 = -r_{pore}$ perpendicular to that plane. Increasing the vapor pressure p_{vapor} in such a pore (see right-hand side in Figure 7.6) will result in a higher amount adsorbed and the average thickness δ of the adsorbed film increases. Hence, the net radius of the pore, $r_{pore} - \delta$, decreases and so does the equilibrium saturated vapor pressure, as prescribed by eqs. (7.16) and (7.15). As soon as $p_{vapor} > p_{sat}$, or

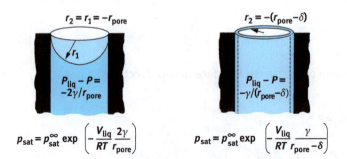

Figure 7.6: Saturation vapor pressure of a liquid in a cylindrical pore.
Left: filled pore, hemispherical LG interface, $r_1 = r_2 = -r_{pore}$; right: liquid film, cylindrical interface, $r_1 = \infty$, $r_2 = -(r_{pore} - \delta)$.

$$p_{vapor} \geq p_{sat}^{\infty} \exp\left(- \frac{V_{liq}}{RT} \frac{\gamma}{r_{pore} - \delta} \right) \tag{7.18}$$

the pore will be completely filled with liquid, instantaneously. The expression in eq. (7.18) is known as the *Kelvin equation* for a partly filled straight cylinder. This liquid–vapor equilibrium is not reversible. A slight decrease in p_{vapor} will not empty the pore because the LG interface is now hemispherical with radii equal to r_{pore}; see left-hand side in Figure 7.6. This interface is curved stronger than the interface in the partly filled pore. Consequently, the requirement for emptying the pore is more severe: $p_{vapor} < p_{sat}$, or

$$p_{vapor} < p_{sat}^{\infty} \exp\left(-\frac{V_{liq}}{RT}\frac{2\gamma}{r_{pore}}\right)$$ (7.19)

which is the form the Kelvin equation takes in case of a filled cylindrical pore.

The phenomenon that emptying a pore occurs at a lower vapor pressure than its filling is known as *hysteresis* and shows up as separated adsorption and desorption branches in adsorption isotherms as those presented in Figure 7.7. An important feature of the adsorption isotherm of a hypothetical solid, containing straight cylindrical pores of equal size, can be deducted. At low vapor pressure, a thin film will be present on the inside of the pores. At increasing vapor pressure, this adsorbed film will grow in thickness (following an isotherm equation relating vapor pressure and film thickness). As soon as the vapor pressure reaches the limiting value given by eq. (7.18), all pores will be completely filled instantaneously. This behavior is depicted in the right-hand branch of Figure 7.7a.[3]

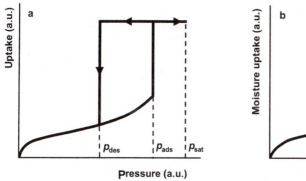
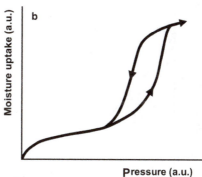

Figure 7.7: Water adsorption in porous solids. (a) Hypothetical porous solid with straight cylindrical pores of equal diameter and (b) adsorption–desorption branches of irregular shaped pores.

Reversely, starting from complete saturation, the moisture content with decreasing vapor pressure will follow the left-hand branch in Figure 7.7a. All pores are partially emptied as soon as the vapor pressure is below the limiting value indicated by eq. (7.19). In contrast with the hypothetical adsorption isotherm in Figure 7.7a, a real

---

3 Note that unlike the plot in Figure 7.4, where moisture content is plotted as a function of *relative humidity*, this isothermal graph gives moisture content as a function of *vapor pressure*.

adsorption isotherm of water on TiO_2 with a distribution of pore forms and sizes is shown in Figure 7.7b.

These examples illustrate that *the saturation pressure required for drying should be below the closing point p_{des} of the hysterisis loop to ensure a satisfactorily driving force for drying.*

7.4 Drying mechanisms

In the design of drying operations, an understanding of liquid and vapor mass transfer mechanisms is essential for quality control. Because no two materials behave alike, this understanding is usually obtained by measuring drying behavior under controlled conditions in a prototypic, pilot plant dryer. From these experiments, the drying rate is usually determined as the change of moisture content with time. For most moist solids, especially those having capillary porosity, the drying rate depends on the moisture content in a manner similar to that shown in Figure 7.8a, which gives moisture content as a function of time, or, related Figure 7.8b, showing the drying rate as a function of moisture content.

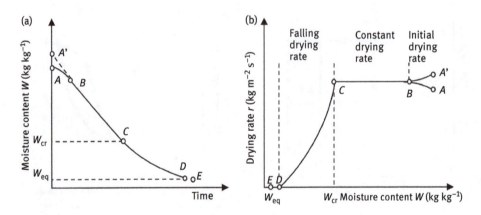

Figure 7.8: (a) Drying curve and (b) drying rate curve.

Three regions can be observed. In the *initial drying period AB*, the liquid in the solid has a lower temperature and it needs some time to get its stationary drying state. This initial heating period is usually quite short and can often be ignored. In the region BC, the drying rate is constant, whereas in the region CE the drying rate is decreasing. Point E represents the lowest possible moisture content of the solid, W_{eq}, in Figures 7.4 and 7.8 at the specified moisture content of the gas.

7.4.1 Constant drying rate

Initially, the surface of the solid is initially very wet and a continuous film of water exists on the drying surface. This water is entirely unbound and acts as if the solid were not present. Under these conditions, the rate of evaporation is independent of the solid and is essentially the same as the rate from a free liquid surface. The evaporation rate is controlled by the heat transfer rate to the wet surface. In the region *AB* (Figure 7.8), the mass transfer rate adjusts to the heat transfer rate and the wet surface reaches the wet bulb temperature T_{wb}. After reaching this stationary state temperature, the drying rate remains practically constant for a period of time, region *BC*. The resulting *constant drying rate* is obtained from eq. (7.1)

$$r_C = k_g(c_{sat}(T_{wb}) - c_f) \qquad (\text{kgs}^{-1}\text{m}^{-2}) \qquad (7.20a)$$

or, from eq. (7.2),

$$r_C = \frac{h(T_f - T_{wb})}{\Delta H_{vap}/M_w} \qquad (\text{kgs}^{-1}\text{m}^{-2}) \qquad (7.20b)$$

where ΔH_{vap} is the molar heat of evaporation (J mol^{-1}) and M_w the molar weight of (water) vapor (cf. note on page 7.4).

Given the driving force, the constant drying rate r_C in this first period depends on the mass transfer coefficient k_g and the heat transfer coefficient h (cf. eq. (7.20)) and thus on the velocity v of the carrier gas. Turbulent flow parallel to a *drying surface* will cause more turbulence than a well-developed flow parallel to a *flat plate*. Thus, the usual Sherwood and Nusselt relations predict too low a value for the transfer coefficients. An educated guess for this effect and applying average values for the physical constants of the air–water system at the temperature interval of interest (say 40–140 °C) lead to the following expression for the *heat transfer coefficient*:

$$h = 12(\rho v)^{0.8} \qquad (7.21)$$

with ρ being the density of moist air (kg moist air/unit volume of moist air) and v the velocity of moist air (m s^{-1}).

The drying rate remains constant as long as the external conditions are constant. Under these conditions all principles relating to simultaneous heat and mass transfer between gases and liquids apply. If the solid is porous, most of the water evaporated in the constant rate period – *free moisture* as defined in Figure 7.4 – is supplied from the interior of the solid. The constant rate period continues only as long as the water is supplied to the surface as fast as it is evaporated.

7.4.2 Falling drying rate

When the moisture content is reduced below a critical value, W_{cr}, in Figures 7.4 and 7.8, the surface of the solid dries out and a further decrease of the free moisture content starts taking place in the interior of the porous solid. This is called the *falling drying rate period*. In this period, the velocity of the carrier gas is less important because the transport of the moisture from within the solid to the outer surface. First, the wetted surface area decreases continuously until the surface is completely dry and the plane of evaporation slowly recedes from the surface. Heat for evaporation is transferred through the solid to the zone of evaporation. Then the drying rate is controlled by the internal material moisture transport and decreases with decreasing moisture content.

Liquid diffusion may become rate determining in the first part of this period and then the moisture content at the external surface is at the equilibrium value W_{eq}. Fick's second law for unsteady-state diffusion is applicable, but is difficult to solve because the liquid diffusivity varies with moisture content and humidity. Like in other diffusion problems, however, the general picture is still that the drying rate is directly proportional to the moisture content and inversely proportional to the square of the thickness of the drying layer.

In wet layers of small, granular particles (e.g., clays and minerals), water may get to the surface by *capillary flow* rather than by liquid diffusion. Now the drying rate is determined by both capillary flow and evaporation from the surface. Theoretical considerations (outside the scope of this chapter, but the reader is referred, e.g., Geankoplis [1] for elucidation of this point) lead to a rate model showing that the drying rate is inversely proportional to the thickness of the drying layer. The mechanism of evaporation is still the same as in the constant rate stage as long as a continuous liquid film across the pores is maintained, so gas velocity, humidity and temperature of the carrier gas also influence the overall drying rate. Therefore, finding experimental evidence for either mechanism is not too difficult.

The amount of moisture removed in the falling rate period may be relatively small but the required time is usually long. At the end of the falling rate period, the residual moisture in the solid is bound by sorption. Heat conduction through the solid and diffusion of vapor in the pores become rate determining. The drying rate decreases rapidly with decreasing moisture content and tends to zero as the (hygroscopic) equilibrium moisture content W_{eq} is approached.

7.4.3 Estimation of drying time

The form of the drying rate curve suggests that the determination of the time necessary to dry a solid until the required moister content is obtained, should be carried out separately for the first drying period (BC in Figure 7.8b) and for the second drying period (CD in Figure 7.8b). Because there is no way to predict the *critical moisture content*, which separates the two regimes, an experimental drying rate curve is required. Defining the drying rate r_C in terms of the change of moisture content W with time:

$$r_C = - \frac{w_0}{A} \frac{dW}{dt} \qquad (\mathrm{kgs}^{-1}\mathrm{m}^{-2}) \qquad (7.22)$$

where w_0 is the weight of dry solid (kg); A is the evaporating surface area (m^2); and the time t_C for the first drying period is obtained by integration of eq. (7.21) between the moisture content of the wet feed, W_0, and the critical moisture content W_{cr}:

$$t_C = \frac{w_0}{A} \frac{W_0 - W_{cr}}{r_C} \qquad (\mathrm{s}) \qquad (7.23)$$

The time necessary for drying from the critical moisture content W_{cr} until the desired moisture content W_{final}[4] is given by

$$t_F = - \frac{w_0}{A} \int_{W_{cr}}^{W_{final}} \frac{dW}{r_F} \qquad (\mathrm{s}) \qquad (7.24)$$

One way to develop the integral is to plot the reciprocal values of r_F as a function of the moisture content and to subsequently determine the area under this curve and between the two boundaries.

On the other hand, an analytical solution can be obtained when the dependence of r_F with moisture content is known. The shape of the falling drying period curve depends, partly, on the type of material to be dried. Some porous materials with relatively large external surface, like sheets of paper, show a linear change of drying rate with moisture content; see Figure 7.9:

$$t_F = a + bW \qquad (7.25)$$

Substitution in eq. (7.24) gives:

$$t_F = \frac{w_0}{bA} \ln \frac{a + bW_\sigma}{a + bW_{final}} = \frac{w_0}{bA} \ln \frac{r_C}{r_{F,\,final}} \qquad (7.26)$$

Also slope b in this equation can be eliminated; see Figure 7.9:

$$t_F = \frac{W_{cr} - W_{final}}{r_C - r_{F,\,final}} \frac{w_0}{A} = \ln \frac{r_C}{r_{F,\,final}} \qquad (7.27)$$

Introducing the *mean logarithmic drying rate* $r_{F,\,lm} = \dfrac{r_C - r_{F,\,final}}{\ln r_C / r_{F,\,final}}$, eq. (7.26) becomes

---

[4] The drying rate becomes zero approaching the equilibrium content W_{eq}; hence, $W_{final} > W_{eq}$.

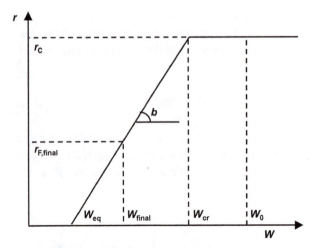

Figure 7.9: Elimination of slope b in eq. (7.26) from drying rate curve.

$$t_F = \frac{w_0}{A} \frac{W_\sigma - W_{final}}{r_{F,lm}} \qquad (7.28)$$

In either case the *total drying time* t_{tot} simply follows from

$$t_{tot} = t_C + t_F = \frac{w_0}{A} \left[\frac{W_0 - W_{cr}}{r_C} + \frac{W_{cr} - W_{final}}{r_{F,lm}} \right] \qquad (7.29)$$

7.5 Classification of drying operations

Drying methods and processes can be classified into several ways. One classification is as *batch*, where the material is inserted into the drying equipment and drying proceeds for a given period of time, or *continuous*, where the material is continuously added to the dryer and dried material continuously removed.

A batch dryer is best suited for small lots and for use in single-product plants. This dryer is one into which a charge is placed, the dryer runs through its cycle and the charge is removed. In contrast, continuous dryers operate best under steady-state conditions drying continuous feed and product streams. Optimum operation of most continuous dryers is at design rate and steady state. Continuous dryers are unsuitable for short operating runs in multiproduct plants.

Drying processes can also be classified to the physical conditions used to add heat and remove water vapor. The most frequently applied heat transfer mechanisms are *convection drying* and *contact drying*. In convection drying, the sensitive heat of a hot gas is supplied to the material surface by convection. The drying agent flowing past or through the body also removes the evaporated water and transports it from the dryer (Figure 7.1). For energy saving, partial recirculation of the drying medium is also used. Drying operations involving toxic, noxious or flammable vapors employ gas-tight equipment combined with recirculating inert gas systems having integral dust collectors, vapor condensers and gas preheaters (Figure 7.10). In contact drying, the heat is supplied to the wet material by conduction from the heated surface as bands, plates, cylinders or the dryer wall. The amount of heat transferred depends not only on the thermal conductivity of the heating surface but also on the heat transfer coefficient from the heating medium to the surface. Common heating mediums include steam, organic liquid and molten metals. Since all the heat for moisture evaporation passes through the material layer the thermal efficiency of contact drying is higher than convective drying where most of the heat is flowing over the material and wasted into the outlet air. During constant rate drying, material temperature is controlled more easily in a direct-heat dryer than in an indirect-heat dryer because in the former the material temperature does not exceed the gas wet bulb temperature as long as all surfaces are wet.

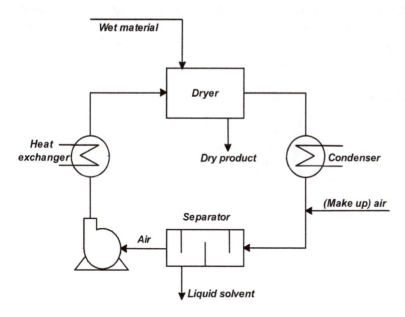

Figure 7.10: Convective drying in closed system.

7.5.1 Direct-heat dryers

In direct-heat dryers, steam heated, extended surface coils are used to heat the drying gas up to temperatures as high as 200 °C. Electric and hot oil heaters are used for higher temperatures. Diluted combustion products are suitable for all temperatures. In most direct-heat dryers, *more gas is needed to transport heat than to purge vapor*. Some of the most commonly used direct-heat dryers are listed below.

7.5.1.1 Batch compartment dryers

Direct heat batch compartment dryers are often called tray dryers because of frequent use for drying materials loaded in trays on trucks or shelves. Figure 7.11 illustrates a two-truck dryer. The compartment enclosure comprises insulated panels designed to limit exterior surface temperature to less than 50°C. Slurries, filter cakes, and particulate solids are placed in stacks of trays. Large objects are placed on shelves or stacked in piles. Unless the material is dusty, gas is recirculated through an internal heater as shown. Only enough purge is exchanged so as to maintain needed internal humidity. For inert gas operation, purge gas is sent through an external dehumidifier and returned. These dryers are economical only for single-product rates less than 500 t year^{-1}, multiple product operation and batch processing. Variable speed fans are employed to provide higher gas velocity over the material during early drying stages. To minimize dusting, the fans reduce velocity after constant drying when heat transfer at the material surface is no longer the limiting drying mechanism. Shallow tray loading yields faster drying, but care is needed to ensure depth uniformity and labor is increased.

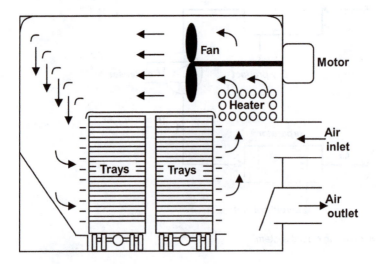

Figure 7.11: Two-truck tray dryer (adapted from [17]).

7.5.1.2 Belt dryers

In belt dryers shown in Figure 7.12, a loading device that is especially designed for the product is used to place the moist solid on the surface of a circulating belt,

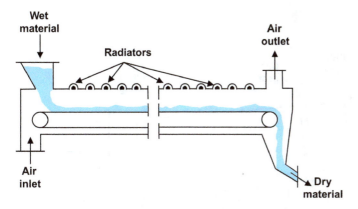

Figure 7.12: Radiation belt dryer (adapted from [54]).

which passes through a drying chamber that resembles a tunnel. The solid remains undisturbed while it dries. At the end of this chamber the material falls from the belt into a chute for further processing. In some installations the material falls onto another belt that moves in the opposite direction to the first one. Centrifugal or axial flow blowers are used to aerate the moist materials. The air stream can enter the solid from below or above. Belt dryers are ideal for friable, molded, granular, or crystalline products that require a long, undisturbed drying time. They are used in all branches of industry.

7.5.1.3 Rotary dryers

The direct-heat rotary dryer in Figure 7.13 is a horizontal rotating cylinder through which gas is blown to dry material that is showered inside. Shell diameters are 0.5–6 m. Batch rotary dryers are usually one or two diameters long. Continuous dryers are at least four and sometimes ten diameters long. At each end, a stationary hood is joined to the cylinder by a rotating seal. These hoods carry the inlet and exit gas connections and the feed and product conveyors. For continuous drying, the cylinder may be slightly inclined to the horizontal to control material flow. An array of material showering flights of various shapes is attached to the inside of the cylinder. Dry product may be recycled for feed conditioning if material is too fluid or sticky initially for adequate showering. Slurries may also be sprayed into the shell in a manner that the feed strikes and mixes with a moving bed of dry particles. Material fillage in a continuous dryer is 10–18% of cylinder volume. Greater fillage is not showered properly and tends to flush toward the discharge end.

Direct-heat rotary dryers are the workhorses of industry. Most particulate materials can somehow be processed through them. These dryers provide reasonably good gas contacting, positive material conveying without serious back-mixing, good thermal efficiency and good flexibility for control of gas velocity and material

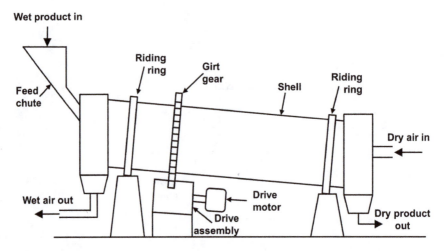

Figure 7.13: Direct heat rotary dryer (adapted from [59]).

residence time. Gas flow in these rotary dryers may be cocurrent or countercurrent. Cocurrent operation is preferred for heat-sensitive materials because gas and product leave at the same temperature. Countercurrent allows a product temperature higher than the exit gas temperature, which increases the dryer efficiency. To prevent dust and vapor escape at the cylinder seals, most rotary dryers operate at a slightly negative internal pressure.

7.5.1.4 Flash dryers
In flash dryers, materials are simultaneously transported and dried. The simplest form, illustrated in Figure 7.14, consists of a vertical tube in which granular or pulverized materials are dried while suspended in a gas or air stream. The available drying time is only a few seconds. Only fine materials with high rates of heat and mass transfer or coarser products with only surface moisture to be removed are used in such dryers. Solids that contain internal moisture can only be dried to a limited extend by this method. Flash dryers are well suited for drying thermally sensitive materials and are widely used for drying organic and inorganic salts, plastic powders, granules, foodstuffs and so on.

7.5.1.5 Fluidized bed dryers
These units are known for their high drying efficiency. A two-stage model is illustrated in Figure 7.15. The solid moves horizontally in a chute and the drying agent flows vertically through a perforated floor to fluidize the solid. These machines can operate continuously, because the solid is transported while suspended in the drying agent. Materials that can be suspended in the drying agent are usually

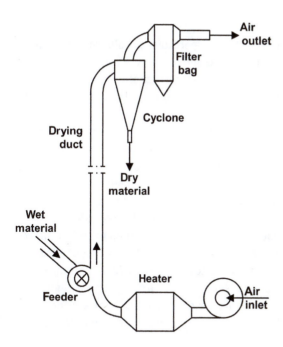

Figure 7.14: Flash dryer (adapted from [54]).

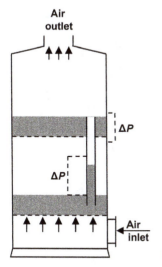

Figure 7.15: Two-stage fluid bed dryer.

powders, crystals and granular or short-fibered products that remain finely divided. Pastes and slurries are mixed with previously dried material and are then easily fluidized.

7.5.1.6 Spray dryers

Spray drying is used for the drying of pastes, suspensions or solutions. The moist material is sprayed into the drying agent and converted into a powder that is entrained by the gas stream. A spray dryer is a large, usually vertical chamber through which hot gas is blown and into which a solution, slurry, or pump able paste is sprayed by a suitable atomizer. The largest spray-dried particle is about 1 mm, the smallest about 5 μm. Because all drops must reach a nonsticky state before striking a chamber wall, the largest drop produced determines the size of drying chamber. Chamber shape is determined by nozzle or disk spray pattern. Nozzle chambers are tall towers, usually having height/diameter ratios of 4–5. Disk chambers are of large diameter and short. A spray dryer may be cocurrent, countercurrent or mixed flow. Cocurrent dryers are used for heat-sensitive materials. Countercurrent spray dryers yield higher bulk density products and minimize hollow particle production. Figure 7.16 shows an open-cycle, cocurrent, disk atomizer chamber with a pneumatic conveyor following for product cooling. Spray dryers are often followed by fluid beds for second-stage drying or fine agglomeration. Spray dryers are particularly suitable for drying solids that are temperature sensitive. Applications include coffee and milk powders, detergents, instant foods, pigments, dyes and so on.

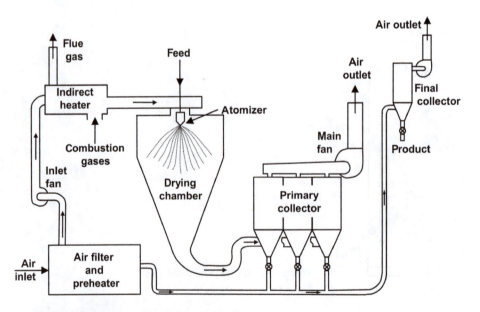

Figure 7.16: Spray dryer system (adapted from [17]).

7.5.2 Contact dryers

In contact dryers, most of *the heat is transferred by conduction*. Common heating media are steam and hot water, depending on the actual operating temperature. Based on dryer cost alone, contact dryers are more expensive to built and install than direct heat dryers. For environmental concerns, contact dryers are more attractive because they are more energy efficient and use only small amounts of purge gas. Dust and vapor recovery systems for contact dryers are smaller and less costly. In this paragraph some of the commonly encountered contact dryers are described.

7.5.2.1 Rotary and agitator dryers

The heat necessary for drying is transferred through the peripheral walls of these dryers in contact drying. Only a small amount of air is necessary to carry the moisture that is taken from the solid. Accordingly, the air velocity in these units is quite low. This is advantageous when drying materials that dust easily or form dust during drying. The rotary steam tube dryer is a horizontal rotating cylinder in which one or more circumferential rows of steam-heated tubes are installed. In agitator dryers (Figure 7.17), the vessel is stationary and the solid product is slowly agitated. Material holdup may be varied from a few minutes to several hours. Agitator speeds rarely exceed 10 min^{-1}, because at higher speeds the mechanical stress and power demand become intolerable.

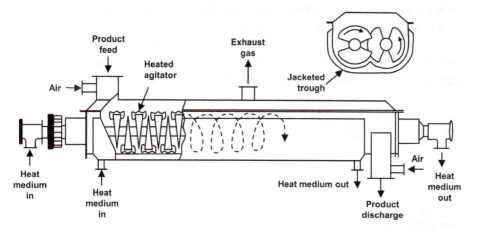

Figure 7.17: Indirect heat paddle-type agitator dryer.

7.5.2.2 Vacuum dryers

The principal differences in the vacuum dryers are their seals and the means to produce the vacuum. Drying under reduced pressure is advantageous for materials that are temperature sensitive or easily decomposed because the evaporation temperature is reduced. They are most often used to dry pharmaceutical products and foodstuffs. The simplest form of a vacuum dryer for batch drying is the vacuum shelf dryer. The moist solid lies on a heated plate. Improved heat transfer with higher efficiency is obtained in the vacuum tumble dryer of Figure 7.18, in which the moist solid is constantly agitated and mixed. These tumble dryers are also used for the final drying step during the production of various polyamides.

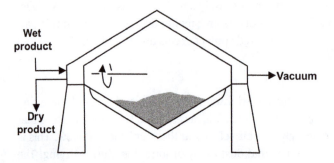

Figure 7.18: Tumble vacuum dryer.

7.5.2.3 Fluid bed dryers

Indirect-heat fluid bed dryers are usually rectangular vessels in which vertical pipe or plate coils are installed. Figure 7.19 is diagram of a two-stage, indirect-heat fluid bed incorporating pipe or plate coils heaters. The general design is used for several particulate polymers. Excellent heat transfer is obtained in an environment of intense particle agitation and mixing. Because of its favorable heat- and mass-transfer capabilities, flexibility for staging and lack of rotary seals, an indirect-heat fluid bed is an ideal vessel for vapor recovery and drying in special atmospheres.

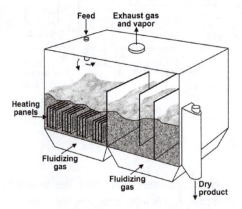

Figure 7.19: Indirect-heat fluid bed dryer.

7.5.3 Other drying methods

Some of the other less frequently used drying methods are radiation heating, dielectric drying and freeze-drying. In *radiation heating* the energy is supplied from an electromagnetic radiation source that is remote from the surface of the solid. Vibrations of molecules in the wet product, creating the thermal effect and evaporating moisture, absorb the electromagnetic radiation energy. Radiation drying is expensive because of the high cost of electrical power or combustion gas to heat the radiators. Since penetration depth of infrared waves is relatively small, it is mainly used for short drying times of thin product layers such as films, coatings and paints.

 · With *high frequency or dielectric drying*, the wet product forms a dielectric between the electrodes of a plate capacitor exposed to a high frequency electric field. The generated intermolecular friction causes the generation of heat inside the product. This is used to heat up the product and to evaporate the moisture. High frequency drying allows gentle thermal drying of the product. Substantial deformation and shrinkage cracks are avoided. Therefore, dielectric drying is employed for gentle drying of products such as fine wood, ceramic products, foods, pharmaceuticals and luxury goods.

 Freeze-drying is a vacuum sublimation drying process. At temperatures below 0 °C and under vacuum, moisture sublimes from a frozen wet product directly from the solid to the gaseous state. First, the wet product has to be frozen from −15 °C to approximately −50 °C. The freezing rate and final temperature essentially determine the drying time and final quality of the product such as structure, consistency, color and flavor. The frozen product is usually granulated, sieved and then charged to the dryer. Under vacuum it is then dried either discontinuously on heated plates or continuously while mixed and moved over a heated surface. Sublimed moisture vapor desublimes to ice on a cooling agent operated condenser. Drying rates are low because the low allowable rate of heat flow controls the process. Freeze or sublimation drying is carried out in a chamber freezer, vacuum disk dryer, vibrating film dryer, cascade dryer or spray freezing dryer. The high investment and operating costs of freeze-drying are only worthwhile for high-grade, thermally sensitive products. Certain important properties of the products are kept such as flavor, taste and color.

Nomenclature

A	Surface area	m^2
c	Concentration	$mol\ m^{-3}$
C_p	Heat capacity (of dry air)	$J\ K^{-1}(kg\ dry)^{-1}$
D_g	Coefficient of diffusion	$m^2\ s^{-1}$
D	(Characteristic) diameter	m
H	Humidity	$kg(kg\ dry)^{-1}$
H_{rel}	Relative humidity	$\%$
ΔH_{vap}	Molar heat of evaporation	$J\ mol^{-1}$
h	Heat transfer coefficient	$W\ m^{-2}\ K^{-1}$
k_g	Mass transfer coefficient in gas phase	$m\ s^{-1}$
M	Molar weight	$kg\ mol^{-1}$
p	Pressure	Pa
R	Gas constant	$J\ mol^{-1}\ K^{-1}$
r	Specific rate	$kg\ s^{-1}\ m^{-2}$
r	Radius	M
T	Temperature	K
V	Volume	m^3
v	(Superficial) velocity	$m\ s^{-1}$
w	Weight	kg
W	Moisture content	$kg(kg\ dry)^{-1}$
δ	(Film) thickness	M
γ	Surface tension	$N\ m^{-1}$
Φ	Molar flux	$mol\ s^{-1}\ m^{-2}$

Indices

m	Molar
f	Feed
s	Surface
sat	Saturation/saturated
h	Heat
vap	Evaporation/evaporated
wb	Wet bulb
w	Water
C	Constant (drying rate)
F	Falling (drying rate)
des	Desorption
ads	Adsorption

Exercises

1   Air of 20 °C and 1 bar with a relative humidity of 80% is heated to 50 °C. The heated air flows along a flat tray filled with completely wetted material. The heat transfer coefficient $h = 35$ W m^{-2} K^{-1} and the heat of evaporation $\Delta H_v = 2.45 \times 10^6$ J kg^{-1}.
Calculate r_c, the constant drying rate.

2   A wet solid with a moisture content $w = 0.5$ kg kg^{-1} dry occupies 0.1 m^2 on a flat tray. A laboratory experiment shows that the critical moisture content $w_c = 0.2$ kg kg^{-1} dry and the constant drying rate $r_c = 0.36$ g s^{-1} m^{-2}.
    How long will it take, under similar conditions, to reduce the moisture content of the wet solid on the tray, containing 2 kg solid (dry basis), from 0.4 to 0.2 kg kg^{-1} dry?

3   A wet solid is dried with air at 60 °C and a humidity $H_f = 0.010$ kg H$_2$O kg^{-1} dry air. The air flows parallel to the flat, wet surface of the solid at a velocity $v = 5$ m s^{-1}. Under these conditions the heat transfer coefficient h is given by $h = 14.3 \, (\rho v)^{0.8}$, where ρ is the density of the moist air in kg moist air per unit volume moist air. The surface area available for heat exchange with air is 0.25 m^2 and the heat of evaporation $\Delta H_v = 2.45 \times 10^6$ J kg^{-1}.
Calculate the total amount of water evaporating per unit time.

4   Calculate the relative change in saturation vapor pressure of water at 300 K when it is dispersed in droplets of 0.1 mm radius. $y_w = 0.073$ N m^{-1}.

5   Air at 20 °C and 1 bar with a relative humidity of 80% is available for drying a porous solid containing cylindrical pores with a diameter of 0.8 nm. The heat of evaporation $\Delta H_v = 44045$ J mol^{-1}.
    To what temperature should the air be heated at least to avoid that the pores remain filled with water? (Hint: the saturated vapor pressure obeys *Clapeyron.*)

6   Wet air (0.2 kg s^{-1}) of 30 °C and 2 bar with a relative humidity ($H_{rel} = p_{water}/p^0{}_{water}$) of 75% has to be dried by adsorption of water on porous alumina in a packed bed. The adsorption isotherm is given by Henry's law:

$$\frac{\text{kg water adsorbed}}{\text{kg dry alumina}} = K \times H_{rel} \quad \text{with} \quad K(30\,°C) = 0.350 \quad and \quad K(60\,°C) = 0.121$$

The process is conducted in an adsorption vessel with a length of 4 m and diameter of 0.5 m. Adsorption and regeneration take place at a constant temperature. The bed porosity $\varepsilon_{bed} = 0.32$ for a complete filling with the alumina particles of 2 mm diameter. The particle porosity $\varepsilon_{particle} = 0.58$. At the start of the drying process the water vapor pressure increases linearly until the maximum value of 75% relative humidity is reached. Furthermore it is given that the saturated vapor pressure of water, p^0_{water} equals 0.0418 bar at 30 °C and 0.199 bar at 60 °C. $M_w = 0.018$ and $M_{air} = 0.029$ kg mol^{-1}.

a. Calculate the maximum kilograms of dry alumina the adsorption vessel can contain.

b. When the first trace of water is detected at the column exit, only 70% of the column is used for adsorption. Draw the concentration profile in the adsorption vessel (relative humidity as function of the place).

c. How long did it take to achieve this situation of 70% column utilization?

d. Calculate the amount of water adsorbed to the alumina.

e. How long does it take to completely regenerate the column by heating the inlet air to 60 °C and reducing the pressure to 1 bar?

7 Silicagel is stored at room temperature (20 °C) in an environment with a relative humidity of 60%. The adsorption isotherm of this material is given in Figure 6.5. Before use the material is dried at 40 °C in air with a wet bulb temperature of 18 ° C. The air velocity is such that the heat transfer coefficient $h = 30$ W K^{-1} m^{-2}. The critical moisture content amounts 18 wt% and the heat of evaporation of water is 2.45·106 J kg^{-1}. Entrance effects can be neglected.

a. Determine the relative humidity of the drying air.

b. Calculate the constant drying velocity in wt% per second per unit of drying area.

c. How long does it take to reduce the moisture content to 18 wt%?

8 Filtration of a crystal suspension yields a cake that contains 10 wt% water. The residual water is removed from the filter cake in a belt dryer until a maximum moisture content of 0.1 wt%. This process can be modeled as a flat plate with a drying area of 2,500 m^2 with air of 80 ºC and a relative humidity of 10% flowing over it. Laboratory experiments have shown that the drying rate over the whole drying trajectory is constant. The density of this air is 1.3 kg m^{-3}. The heat of evaporation of water equals 2.45 × 10^6 J kg^{-1}.

a. Calculate the specific drying rate.

b. What should be the minimal value of the heat transfer coefficient?

c. Estimate the minimal velocity of the air stream.

9 A produced calcium sulfate filter cake (351 kg h^{-1}) needs to be dried from 30 wt% water per kg dry product (CaSO$_4$) till 5 wt%. This is done by distributing the wet cake in thin layers on several trays in a tray dryer. Air of 95 °C with a humidity of 0.04 kg moisture per kg dry air is led over these trays with a velocity of 2 m s^{-1}. Each tray has an area of 4 m^2 and the layers are sufficiently thin to assume a constant drying velocity of the whole drying trajectory. 2.45·× 10^6 J kg^{-1} water

and 1 kg m^{-3} can be taken for the heat of evaporation of water and the density of the air, respectively.

a. How high should the drying rate be to keep up with the rate of filter cake production?
b. How many trays of 4 m^2 are required?
c. When recycling part of the air from the dryer, till which temperature should the dryer with water vapor saturated drying air be cooled to reach again the initial moisture content of 0.04 kg water per kg dry air?

Appendix

Experimental determination of moisture content in air from wet bulb temperature

A liquid in equilibrium with its surrounding gas phase exerts the saturation pressure at the particular temperature of that liquid. A gas stream flowing along the surface causes evaporation taking place if and as long as the actual vapor pressure is below the saturation pressure at that temperature. The evaporation from the liquid surface results in a temperature decrease in a thin liquid surface layer. As a result, heat will be transported from the gas to the liquid phase. Figure 7A.1 shows the combined heat and mass transport. The change in molal enthalpy connected to the phase change is then balanced by heat transport from the gas to the liquid. The surface temperature T_s initially drops, decreasing Φ_{vap} and increasing Φ_h, until the stationary state is established. The so-called *wet-bulb temperature* T_{wb} is the steady-state temperature T_s, which is eventually reached when a gas is flowing at constant velocity along a wetted surface.

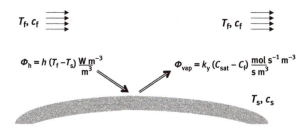

T_f, C_f ⇛

T_f, C_f ⇛

$$\Phi_h = h\,(T_f - T_s)\,\frac{W\,m^{-3}}{m^3}$$

$$\Phi_{vap} = k_y\,(C_{sat} - C_f)\,\frac{mol\,s^{-1}\,m^{-3}}{s\,m^3}$$

T_s, C_s

Figure 7A.1: Combined heat and mass transfer.

The higher the moisture content in this gas, the smaller the driving force for evaporation and the smaller the difference between the wet bulb temperature and the temperature of the fluid. By definition, the wet bulb temperature in a saturated gas is the same as the fluid temperature T_f.

In Figure 7A.2 the experimental setup is shown. The feed gas with unknown humidity H_f is entering at temperature T_f, also designated dry bulb temperature, which is measured with the first sensor. The second temperature sensor is positioned in a cylinder in such a way that no radial heat transfer resistance occurs between the sensor and the outer surface: the temperature in the cylinder is uniform. The outer surface is provided with a wick, which is kept wet, any water lost by evaporation from the surface is replaced with water at temperature T_f from a storage vessel connected to the wet wick. No conduction of heat is supposed to occur along the temperature sensors

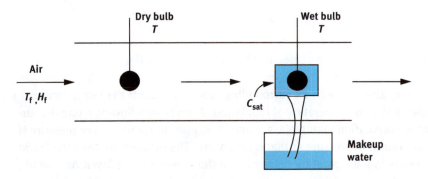

Figure 7A.2: Device for the measurement of wet bulb temperatures.

For the air–water system, in Section 7.2.2 the following relation between wet bulb temperature and humidity H_f has been derived:

$$H_{sat} - H_f = C_p \frac{M_w}{\Delta H_{vap}} (T_f - T_{wb}) \qquad (7.14)$$

This is the basic equation to calculate unknown moisture contents H_f from measured wet bulb temperatures T_{wb}. The graphical solution of this equation is shown in Figure 7A.3. Humidities can be converted to concentrations or partial pressures with eqs. (7.5)–(7.7).

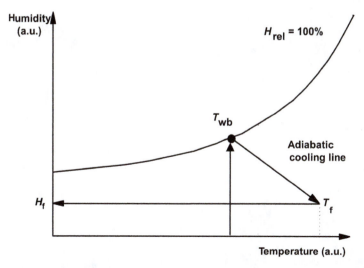

Figure 7A.3: Determination of air humidity from wet bulb temperature.

Chapter 8
Crystallization

8.1 Introduction

Crystallization is ubiquitous in nature from the formation of snowflakes to materialization of our teeth and bones. In industrial practice, it is estimated that 70% of all fine chemicals produced are solid. Consequently, crystallization, formation of a solid from a mother phase either a solution or melt, is among the most commonly encountered unit operations in the portfolio of industrial separations. Crystallization of a target molecule from a mother phase achieves two critical tasks simultaneously namely separation and purification. Moreover, formation of the crystalline phase purifies the target molecule as the crystalline form predominately made up target molecules consequently; it has high purity compared to mother phase. Once a target compound is crystallized in solid form, it can be separated from solvent by solid–liquid separation methods such as filtration or sedimentation owing to the density difference between the crystal and the mother phase as shown in Figure 8.1. The crystals separated from the solvent can be washed and dried depending on the product specifications. Large quantities of crystalline products are manufactured commercially such as table salt (sodium chloride), industrial salts (electrolysis brine), sucrose, battery materials (lithium salts), commercial polymers (poly(lactic acid), poly(vinyl acid)) and fertilizer chemicals (ammonium nitrate, ammonium phosphates, urea). In the production of organic chemicals such as pharmaceuticals, fine chemicals and polymers, crystallization is used to recover product, to refine intermediate chemicals and to remove undesired salts. Crystalline products coming from the pharmaceutical, fine chemical and dye industries are produced in relatively small quantities but represent a valuable and important industrial sector. Beyond the applications focused on manufacturing fine chemicals, crystallization is also used to separate and purify as in desalinization and water reuse.

Crystallization achieves separation through liquid-to-solid phase transition. Once crystallization occurs, solute molecules dissolved in solvent form a solid crystalline phase. Analogous to distillation or extraction, the new crystalline phase created from solution contains concentrated solutes. A unique feature of crystallization is that extremely high purities (>99% even 99.9%) can be achieved as in the crystalline phase. In melt crystallization, only one chemical species is considered, for instance, crystallization of a urea melt into urea crystals. In a solution crystallization process, the feed consists of a homogeneous solution of different chemical species, namely, dissolved solute and solvent. Impurities, that is, foreign substances different from solute or solvent, are always present in both melt and solution crystallization. In order to crystallize a solute, *supersaturation* in the mother phase (either solution or melt) is needed. A supersaturated solution carries more solute than the thermodynamic

https://doi.org/10.1515/9783110654806-008

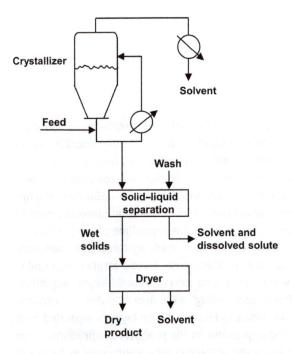

Crystallizer

Solvent

Feed

Wash

Solid–liquid
separation

Wet
solids

Solvent and
dissolved solute

Dryer

Dry
product

Solvent

Figure 8.1: Schematic of a crystallization process followed by solids processing sequence.

equilibrium value. Supersaturation is created either by cooling, evaporation of the solvent or addition of an anti-solvent to obtain the desired solute in a solid form. It is also possible to utilize these methods in tandem. The process of solid phase formation is termed as *crystallization*, and the operation occurs in a vessel called a *crystallizer* as shown in Figure 8.1. Crystallization consists of two fundamental phenomena, *nucleation* and *growth* (Figure 8.2). In nucleation, solute molecules dissolved in solvent come together and form the first-ordered solid structures called *nuclei*. The nucleation event involving only the solute molecules is called *homogeneous nucleation* (Figure 8.2). If the nucleation starts on the surface of a foreign object such as impurities or crystallizer walls, this is referred as *heterogeneous nucleation*. In addition to homogeneous and heterogeneous nucleation, a nucleation phenomenon with strong influence on crystal quality is *secondary nucleation*. In secondary nucleation, pieces of previously "chipped off" crystals are formed due to collisions of already nucleated crystals with the impeller, crystallizer walls or collisions between crystals. The chipped off crystals act as heteronucleants encouraging nucleation of dissolved solute on their surface (Figure 8.2). After nucleation, the nuclei grow by addition of more solute to existing nuclei and become larger until they are removed (harvested) from the crystallizer. Growing crystals can simultaneously go through aggregation and attrition as they grow to their final size. The performance of a crystallization process is also characterized by crystal quality parameters, *crystal morphology,*

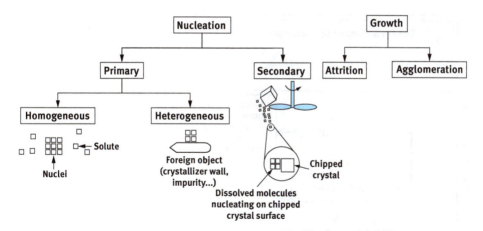

Figure 8.2: Schematic illustrating the fundamental phenomena occurring during crystallization process.

size distribution, polymorphism and purity. Consistently producing crystalline products with desired crystal quality parameters is essential in industrial practice. The aforementioned crystal quality parameters are also referred as four pillars of crystallization. These solid phase properties are especially important because crystallization is frequently the initial step in further solid processing sequence such as filtration, similar to that shown in Figure 8.1. After crystallization, the solids are normally separated from the crystallizer liquid, washed and discharged to downstream drying equipment. Since product size and suspended solids concentration are controlled to a large extend in the crystallizer, predictable and reliable crystallizer performance is essential for smooth operation of the downstream system.

The final crystallizer design is the culmination of the design strategy depicted in Figure 8.3. In the basic design stage, design and product specification along with thermodynamic and lab-scale kinetic data are used to make decisions on the solvent choice, process conditions, method of supersaturation generation, mode operation (continuous or batch), number of stages (single or multistage) and crystallizer type to be used. The decisions made are rationalized by calculations. After finalizing the crystallizer type, finally the final crystallizer design is concluded considering detailed cost analysis and large-scale kinetic data such as pilot-scale experimental data or computational operation scale simulations. The use of mathematical models for the design of industrial crystallizers has lagged behind compared to other unit operations because of the complexities associated with rationally describing and connecting fundamental crystallization phenomena (nucleation, growth) kinetics with the process configuration and mechanical features of the crystallizer. Yet significant progress has been done in recent years. During the operation, these complexities significantly influence the product quality and they are summarized in Figure 8.4. In the operation of a crystallizer, product quality parameters and process conditions are tightly

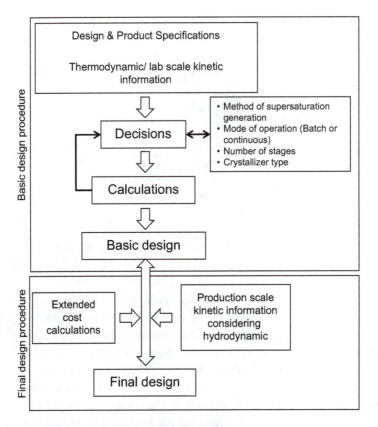

Figure 8.3: Crystallization system design strategy.

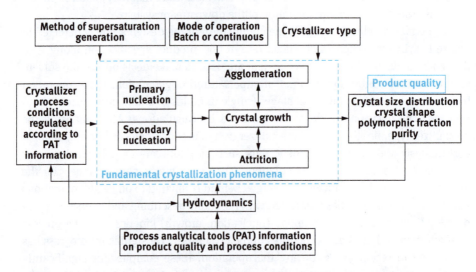

Figure 8.4: Schematic of a crystallization process illustrating interactions between fundamental crystallization phenomena, process conditions and process design/operation choices.

monitored through process analytical tools (PATs) such as in-line particle size analyzers. The information gathered from these sensors is collected and decisions on process conditions are made to ensure required product quality. The effectiveness of decisions made intimately related to our fundamental understanding of phenomena shown in Figure 8.4. The crystal quality is a result of the constituent phenomena, namely primary and secondary nucleation, growth, agglomeration and attrition as well as crystallizer design, method of supersaturation generation, hydrodynamics inside crystallizer and process conditions as illustrated in Figure 8.4 and it is an active area of research.

The design and analysis of crystallization processes of the continuous well-mixed suspension type have developed into formal design algorithms, which can now be applied in situations of industrial importance. Examples of specific process configurations that can be modeled rigorously include fines destruction, clear-liquor advance, classified-product removal, vessel staging and seeding. The basic requirement for these rigorous process configuration models is a population balance crystal size distribution (CSD) algorithm yet other crystal quality parameters also has to be optimized.

The main advantages of crystallization are the following. First, nearly pure solid product can be recovered in the desired crystal quality parameters such as size and morphology in one separation stage. With optimized design and operation while monitoring feed stream impurities, product purity in excess of 99.0% can be attained in a single crystallization, separation and washing sequence. This large separation factor is one of the reasons that make crystallization a desirable separation method. Moreover, crystallization can produce uniform crystals of the desired morphology, size distribution, polymorphic form and purity. During crystallization, these crystal quality parameters can be monitored and optimized so that the crystals have the desired crystal quality for direct packaging and sale. In industrial practice, it is a challenge to dictate the crystal qualities required due to fluctuations in process parameters such as feed streams' impurity levels and temperature. The major disadvantages of crystallization are that purification of more than one component and/or full solute recovery are not attainable in one stage. Thus, additional solid–liquid separation equipment (filtration, washing, drying …) is required to remove the solute completely from the remaining crystallizer solution as shown in Figure 8.1. Since crystallization involves processing and handling of a solid phase, the operation is normally applied when no alternative separation technique is discernable and as the final process prior to product meeting consumers.

Some important considerations that may favor the choice of crystallization over other separation techniques such as distillation are:

- *If the solute is heat-sensitive* and/or a high boiler that decomposes at the high temperatures required for distillation.
- *If the relative volatility* between the solute and contaminants is low or nonexistent and/or the existence of azeotropes between solute and contaminants.

- *If the desired state of the final product is solid*. After purification via distillation, solute must be solidified with additional processes such as flaking or prilling. This may introduce additional operating and capital cost, making crystallization the more convenient and economic choice.
- *If comparative economics favor crystallization over distillation*. If distillation requires high temperatures and energy usage, crystallization may offer economic incentives.

Precipitation, also referred as reactive crystallization, is an important separation method in production of many fine chemicals, in water and mineral processing. It is intimately related to crystallization since solutes dissolved in a solvent precipitate out. However, the precipitate may be amorphous and have a poorly defined shape and size. Precipitates are often aggregates of several species and may include salts or occluded solvent. Thus, precipitation serves as "rough cut" to either remove impurities or to concentrate and partially purify the product.

8.2 Crystal quality parameters

In crystallization processes, product requirements are a key issue in determining the ultimate success of the operation. Key parameters of crystal quality, also referred as pillars of industrial crystallization; are the size distribution (including mean and spread), the morphology (including habit or shape), crystal form (polymorphism, solvates, hydrates) and purity, as illustrated in Figure 8.5. The crystal quality parameters dictate product quality and are shaped by consumer demands and regulations.

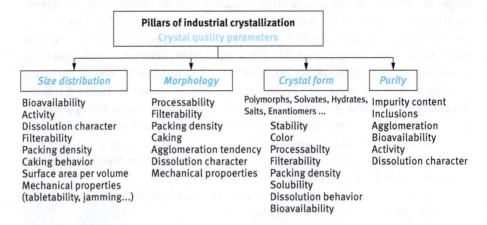

Figure 8.5: Schematic summary of crystal quality properties and the product properties each one influences.

Moreover, the desired crystal quality parameters may vary depending on the industry. For instance, in paint manufacturing, the purity requirements for the same crystalline material may be less relevant than food and pharmaceutical industry. This is due to the fact that food and pharmaceutical products are orally consumed; hence, the impurities are tightly regulated by public health agencies. The phenomena illustrated in Figure 8.4, namely, nucleation, growth, attrition and agglomeration influence more than one crystal quality parameters simultaneously and inter-connectively. Consequently, decisions made on crystallization design and operation conditions need to be optimized based on specific crystal quality parameters for each process.

8.2.1 Crystal size distribution

The CSD determines several important processing and product properties such as filterability, bioavailability, dissolution kinetics, packing density, caking behavior and surface area per volume. Most favored CSD is monodisperse distribution where all crystals are of the same size and dissolve at a known and reproducible rate. Critical phenomena that influence the CSD other than nucleation and growth are attrition (breakage) and agglomeration. Attrition of crystals is almost always undesirable because it is detrimental to crystal appearance and consumer perception. Broken or chipped off crystals are considered to be low quality or faulty by consumers. Moreover, it can lead to excessive small-sized crystals referred as fines. Agglomeration is the formation of a larger particle through two or more smaller particles sticking together. Just like attrition, agglomeration is also undesired as it created crystals with irregular appearance and it alters dissolution kinetics. Attrition often decreases the mean size while agglomeration increases the mean crystal size. Both processes lead to undesired broad size distributions.

Particles produced in a crystallization process have distribution of sizes that varies over a specific size range (Figure 8.6). A CSD is most commonly expressed as a population (number) distribution relating the number of crystals at each size to the size or as a mass (weight) distribution expressing how mass is distributed over the size range. The two distributions are related and affect many aspects of crystal processing and properties. For instance, a hydrophobic pharmaceutical crystal with large mean crystal size can take too long to dissolve and consequently taken out without performing its function. An average crystal size and spread of a distribution can be used to characterize a CSD. However, the average can be determined on any of several bases such as number, volume, weight and length. The *mean crystal size* L_D is most often used as a representation of the product size. The *coefficient of variation* (cv) of a distribution is a measure of the spread of the distribution about the characteristic size. It is often used in conjunction with dominant size to characterize crystal populations through the equation:

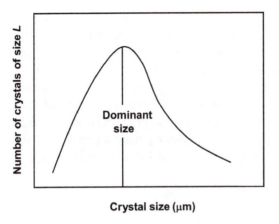

Figure 8.6: Various crystal morphologies of a molecule with hexagonal unit cell.

$$CV = \frac{\sigma_D}{L_D} \qquad (8.1)$$

where σ_D is the standard deviation of the distribution.

8.2.2 Morphology

Every chemical compound has a unique crystal shape that is a result of its unit cell and growth kinetic dictated by crystallizer conditions. The general shape of a crystal is referred to as *habit*. A characteristic crystal shape results from the unit cell (thermodynamically dictated regular internal structure of the solid) with crystal surfaces forming parallel to planes formed by the constituent units. The unique aspect of the crystal is that the angles between adjacent faces are constant. The surfaces (*faces*) of a crystal may exhibit varying degrees of growth but not the angles. With constant angles but different sizes of the faces, the shape of a crystal can vary enormously. This is illustrated in Figure 8.7 for three hexagonal crystals. The final crystal shape is determined by the relative growth rates of the crystal faces. Faster growing faces become smaller than slower growing faces and may disappear from the crystal altogether in the extreme. The relative growth process of faces and thereby final crystal shape are affected by many variables such as rate of crystallization, impurities, agitation, solvent used, degree of supersaturation and so forth. Especially the presence of impurities can alter the growth rates of crystalline materials significantly. Most common is a decrease in growth rate, which is considered to involve adsorption of the impurity onto the crystal surface. Once located on the surface, the impurity forms a barrier to solute transfer from the solution to the crystal. Another important effect associated with the presence of impurities is that they may change the crystal

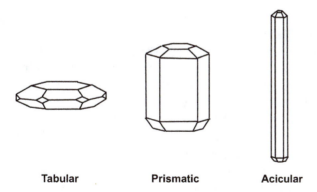

| Tabular | Prismatic | Acicular |

Figure 8.7: Illustration of a crystal size distribution showing the spread of distribution and its mean.

shape (habit). Habit alteration affects the appearance of the crystalline product, purity and its processing characteristics (such as washing and filtration) and is considered to result from unequal changes in the growth rates of different crystal faces.

8.2.3 Crystal form

A solute in a melt state or dissolved in a solvent can crystallize into different crystal forms with drastically different physical properties and appearance. These forms can be classified into two groups depending on the number of species involved in the final crystalline form namely, *polymorphism/allotropy* and *co-crystals*. In polymorphism, one species, that is, only the solute molecules are present in the final crystalline form. Polymorphism is defined as the same molecular building blocks arranged in different crystalline unit structures in three dimensions. The terminology allotropy is used exclusively to define atomic building block, whereas polymorphism is used for molecules according to IUPAC definition. Despite slight difference in definition, both polymorphs and allotropes show dramatically different physical properties. A commonly encountered example of allotropy is carbon, which can crystallize into graphite and diamond among other forms. Graphite is used in pencils and it is brittle and black in color, whereas diamond made of chemically identical atoms arranged differently in space exhibits extreme hardness as well as a shiny appearance. If more than one molecule is involved in the unit structure of crystallized solid, the crystalline form is called solvate, hydrate or co-crystal depending on the species taking part in the crystal structure formed. If the solvent molecules co-crystallize with the solute into one unit cell, the form is referred as solvate. If the solvent is water, the form is referred as hydrate. If a molecular species other than the solute co-crystallized into a crystalline form, then the form is called co-crystal.

Different polymorphs and co-crystals of the same compound exhibit very different physical properties such as melting point, solubility, tensile strength, ductility and processability. More recently, the polymorphism of pharmaceuticals has been subject of intense interest in the pharmaceutical industry. Organic molecules can exist in more than one polymorph. Among the all-possible metastable polymorphs, only one is the thermodynamically stable one referred as the most stable polymorph. Solution thermodynamics suggests that metastable polymorphs can transform spontaneously into more stable and, consequently, less soluble polymorphs. Reduced solubility brings down efficiency and disrupts the function of pharmaceuticals. In a notorious example, the unexpected transformation of anti-HIV drug Ritonavir into a five times less soluble polymorph forced a market recall, resulted in substantive economic losses and loss of the public's trust. In today's pharmaceutical industry where millions of potential compounds (43% of which are extremely water-insoluble) are synthesized by combinatorial methods every year, selectively crystalizing and identifying the most stable polymorph is a bottleneck limiting realization of potential treatments.

Once a new compound is synthesized, rigorous tests are carried out to identify the most stable polymorph. However, transition between polymorphs is a kinetic process. The transition time can be extremely very long and unexpected transitions after the product leaves the manufacturing line can occur as in the case of Ritonavir. Producing the most stable polymorph is still a major challenge for pharmaceutical industry.

8.2.4 Purity

The most basic phenomena influencing the purity are nucleation and growth. A crystal can nucleate heterogeneously on an impurity particle and the growing crystal can engulf this impurity in the crystal structure. The growth kinetics and mode also play a role whether impurities are incorporated inside the growing crystal. For instance, a crystal showing rough growth will grow with an irregular surface topology and can easily trap impurities or pocket of solvent known as *inclusions*. Attrition (breakage) and agglomeration can also influence purity. Attrition of crystals is almost always undesirable because the chipped off surface can exhibit rough growth trapping impurities and solvent pockets. Agglomeration of a larger particle through sticking of two or more smaller particles can also trap impurities or cause inclusion as shown in Figure 8.8. Just like attrition, agglomeration is also undesired particularly in case of inclusions if the solvent is not safe for consumption.

The effect on purity becomes even more important when crystallization is employed as a purification technique. Mechanisms by which impurities can be incorporated into crystalline products include adsorption on the crystal surface, solvent entrapment in cracks, crevices and agglomerates and inclusion of pockets of liquid. An impurity having a structure sufficiently similar to the material being crystallized can also be incorporated into the crystal lattice by substitution or entrapment.

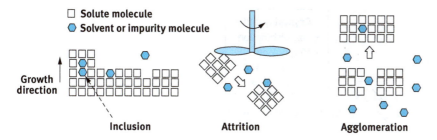

Figure 8.8: Illustration on how impurities are incorporated in crystals.

8.2.5 Interdependency among crystal quality parameters

The crystal quality parameters are dictated by the fundamental crystallization phenomena (Figure 8.2) involved in a process. The process design and operation conditions govern which of these fundamental phenomena are dominant under given conditions. Consequently, by controlling the process design and operation conditions, we can influence the crystallization phenomena active in a crystallizer, which in return dictates the crystal quality parameters. However, altering the process design and operation often affects more than one of the quality parameters in an interdependent manner. This interdependency is best illustrated with a closer look at polymorphic forms of a pharmaceutical compound commonly known as ROY, 5-Methyl-2-[(2-nitrophenyl)amino]-3-thiophenecarbonitrile (L.Yu et al JACS 122 (2000) 585). Altering the process conditions results in three different polymorphic form of ROY (Form A, B & C shown in Figure 8.9) with drastically different morphologies and size distributions. For instance, Form B is more aggregated than other forms, its inter-grown morphology facilitates undesired engulfment of solvent pockets. Form B is consequently of lower purity. The needle-shaped form C can cause problems in filtering forcing alterations in the process design as needle form often causes clogging in filtration process. The ROY case shown in Figure 8.9 is a good example of interdependency amongst crystal quality parameters. It demonstrates that the product quality parameters are intimately related to each other. Moreover, crystal quality parameters depend on the interplay among crystallization phenomena and process design. This network of interdependencies makes crystallization the most complex and difficult to control process among separation techniques (See references and further reading for more information).

8.3 Phase diagrams

Understanding the crystallization starts with exploring the equilibrium behavior or thermodynamics of solutions. Solution thermodynamics can be thought of as a

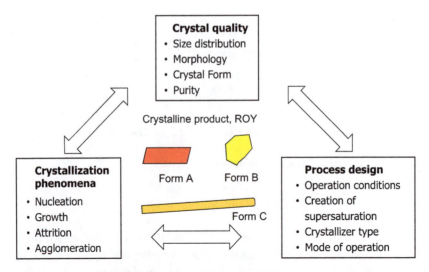

Figure 8.9: Interdependency of crystal quality, crystallization phenomena and process design.

balance between intermolecular interactions. By changing the thermodynamic state, we are changing how molecules interact with each other and their preferred phase. For instance, a mixture of compounds A and B will form a homogeneous one-phase solution if the intermolecular interactions between A and B at a thermodynamic state (at given Temperature and Pressure and composition) are much stronger than the interactions between A-A and B-B. In other words, A will dissolve in B and form a homogeneous solution as both A and B molecules prefer to be in the presence of each other rather than their own kind. This situation can change if the thermodynamic state parameters are altered. In crystallization, we change the thermodynamic state parameters (temperature and/or pressure); so that the system phase separates and forms two phases (solid and liquid) from one homogeneous solution or melt phase. In crystallization, at least one of the phases is crystalline with high purity.

Crystallization is the formation of a solid crystalline phase from a supersaturated mother phase. The success of crystallization process depends on the phase diagram and the thermodynamic state is chosen as operating conditions. A phase diagram can be thought of a map describing the thermodynamic conditions for stable phases. Figure 8.10 shows a *binary eutectic* at constant pressure. Points phase diagram, $T_{m,A}$ and $T_{m,B}$, correspond to melting temperatures of pure crystalline compounds of A and B. Curves $T_{m,A}$-E and $T_{m,B}$-E indicate equilibrium between the solution and the crystalline solid of A and B, respectively. Point E is the eutectic point where crystalline compounds A and B and solution phases exist in equilibrium. Eutectic phase diagrams are encountered in most inorganic compounds and

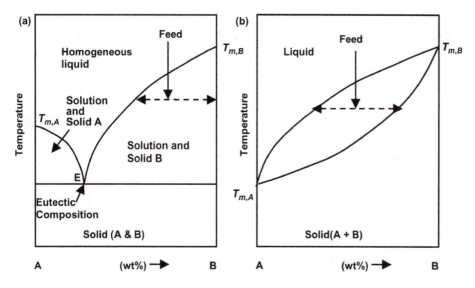

Figure 8.10: (a) Solid–liquid phase diagram of a eutectic system and (b) a binary solution system.

many organics produced in industrial crystallization. In a *binary eutectic*, shown in Figure 8.10a, a pure solid component is formed by cooling an unsaturated solution until solids appear. Continued cooling will increase the yield of pure component. At the eutectic point, both the components solidify and additional purification is not possible. Figure 8.10b shows another phase diagram for solid solutions commonly encountered in melt crystallization. These phase diagrams are interpreted as the two compounds A and B fit into each other's crystal lattice so that crystallization does not produce pure compound A or B.

For crystallization from solution, phase diagrams are often presented at constant pressure as shown in Figure 8.11. A solution with a given thermodynamic parameters (composition and temperature in this case) that sits below the solubility curve is called undersaturated. A solution with composition and temperature at the solubility line is referred as a saturated solution. If the thermodynamic parameters place a solution above the solubility line, this solution is denoted as a supersaturated solution. To crystallize a solution, thermodynamic parameters should be manipulated so that the solution resides in the supersaturated region. An undersaturated solution or a saturated will not crystallize unless the composition and temperature are changed. Thermodynamic systems spontaneously go to minimum energy state. Bringing the solution into a supersaturated state provokes crystallization as the suspension of crystals and supersaturated solution is thermodynamically favored compared to a supersaturated solution.

To illustrate utility of a phase diagram, we can focus onto a specific example of KNO_3 in Figure 8.11. An undersaturated homogeneous solution of KNO_3 can be prepared by adding a known amount of KNO_3 into a solvent. In this particular case, adding 110 gr of KNO_3 to 100 gr of water at 80 °C will place the solution at point A. At this point, all the solute molecules are dissolved in water and a homogeneous solution is observed. Upon decreasing the temperature to 40 °C creates a supersaturated solution. The supersaturated system at 40 °C will eventually crystallize forming a new ordered solid phase. At 40 °C, the supersaturated solution will crystallize as the suspension containing crystals and saturated solution is thermodynamically favored compared to a supersaturated solution at 40 °C. The crystals will grow and consume supersaturation till the solution becomes saturated. In other words, the supersaturated solution will phase separate into KNO_3 crystals and a saturated solution. On the phase diagram, this can be thought of as moving to point C from supersaturated point B. For a solution containing 100 gr of water and 110 gr of KNO_3, cooling the system from 80 °C to 40 °C will create a suspension containing the crystals and a saturated solution with solubility 65 gr/100 gr water. For 100 gr water solution, the amount of crystals then can be calculated from a simple mass balance. 110 gr of solute will be distributed between the crystals and solution. As the saturated solution can carry 65 gr/100 gr water, total crystal mass should be 110-65 gr, that is, 45 grams.

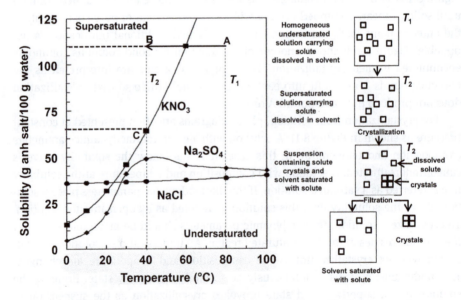

Figure 8.11: Solubility curve for various salts and an illustration of crystallization process when an undersaturated single phase solution is cooled from a high temperature, T_1 to a lower temperature T_2.

8.3.1 Predicting solubility curves

The design of a crystallizer is strongly influenced by the phase diagram. Phase diagram dictates the material and energy balance requirements of the design. Therefore accurate solid–liquid equilibrium data are essential to make the right choices in the crystallization process design. A saturated solution is a solution that is in thermodynamic equilibrium with the solid phase of its solute at a specified temperature. The saturation solubility of a solid is given by the fundamental thermodynamic relationships for equilibrium between a solid and a liquid phase. When the effect of differences between the heat capacities of the liquid and solid is neglected, the *saturation solubility* is given by:

$$\ln(\gamma_s x_s) = \frac{\Delta H_{melt}}{R} \left(\frac{1}{T} - \frac{1}{T_{melt}} \right) \tag{8.2}$$

where x_s is the solid solubility (mole fraction) in the solvent, γ_s is the liquid-phase activity coefficient, ΔH_{melt} is the enthalpy of melting, T_{melt} is the solid melting temperature and T is the system temperature in Kelvin. For ideal solutions, eq. (8.2) reduces to the *Van't Hoff relationship*:

$$\ln x_s = \frac{\Delta H_{melt}}{R} \left(\frac{1}{T} - \frac{1}{T_{melt}} \right) \tag{8.3}$$

With this equation, the solute solubility for an ideal solution can be calculated from the heat of fusion and the pure-solid melting temperature alone, as γ_s is unity for an ideal solution. In this case, the solubility depends only on the properties of the solute and is independent of the nature of the solvent.

8.3.2 Measuring solubility curves

In industrial practice, very few systems are ideal solutions and the solvent plays a critical role in determining the solubility. Moreover, impurities coming from both solvent and solute feeds can also influence solubility if they are present. Consequently, measuring solubility is essential for process design. The most basic approach is the gravimetric method. In this method, a known amount of solute crystals, m_c, and solvent, m_s, is brought in contact in a temperature-controlled jacketed vessel. The system is allowed to equilibrate over an extended amount of time at a constant temperature and pressure. The solute molecules in crystalline form dissolve into the solvent till the solvent is saturated. At this moment, system is considered to be in equilibrium. In other words, the rate of solute mass transfer from crystalline state to dissolved state and dissolved state to crystalline state equilibrates. After filtration, the change in mass of crystal, Δm_c, upon contact with solvent is recorded with analytical balance. The solubility measured by the gravimetric method is conventionally reported as $\Delta m_c/m_s$. However, gravimetric measurements work reasonably well for high solubility systems; it struggles with low solubility solutes due to difficulties in measuring small Δm_c with analytical balance. For low solubility systems, analytical chemistry methods such as high-performance liquid chromatography (HPLC), atomic absorption, mass spectroscopy and UV-vis spectroscopy are utilized to characterize the amount of solute dissolved in the solvent. To construct a phase diagram as shown in Figure 8.11, at constant pressure, the gravimetric method is repeated at different temperatures. The results are repeated and plotted with error bars indicating the repeatability of measurements.

8.4 Kinetics of crystallization

Phase diagrams provide information on which phases are thermodynamically favored under a given set of thermodynamic parameters, namely, pressure, concentration and temperature. However, they do not provide critical information essential for process design, that is, how long it would take to go to a thermodynamically favored situation from an unfavored one. This dynamic, that is, time information, is referred as the kinetics of crystallization. For pedagogical purposes, a phase map can be thought of as a geographical map without height information. It tells you where you are analogous to thermodynamics but no information is provided regarding how long it would take to get to point E from A. The point E can be on the top of a very high mountain, which makes the trip from A to E extremely long even unfeasible. For instance, at point A in Figure 8.12, the solute and solvent molecules form a homogeneous solution. This solution phase is thermodynamically favorable to crystalized two-phase system containing crystals and saturated solution. Once the system is cooled down to point E, the system is in a metastable supersaturated state. The first requirement of crystallization is to have supersaturation hence the system will eventually crystallize. Yet we do not know how long it would take to crystallize. To quantify and estimate this time required for crystallization, we need information on the crystallization kinetics and influence of driving force, that is, supersaturation to crystallize the system.

8.4.1 Metastable zone limit

The crystallization kinetics involves nucleation and growth. Along with other fundamental crystallization phenomena summarized in Figure 8.4, kinetics dictates crystal quality parameters. In simple terms, a collection of molecules come together and form nuclei then the nuclei grow into larger crystals. How fast this set of events will occur is called nucleation kinetics. It is intimately related to the driving force behind crystallization, supersaturation, defined as the deviation from thermodynamic equilibrium.

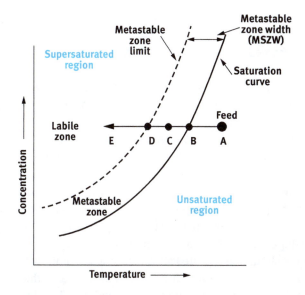

Figure 8.12: Schematic depiction of supersaturation generation, metastable and labile zones.

If a solution or melt are slowly cooled from temperature A to E in Figure 8.12, then the system goes from undersaturated region in the phase diagram into supersaturated region consisting of *metastable* and *labile* zones. As the system is cooled from A to E, it will reach point B first in the phase diagram. At this point, no spontaneous crystallization will take place, as there is no driving force for crystallization. Phase change may be in the supersaturated region when a certain degree of supersaturation is reached. Interestingly, a system can be supersaturated yet the driving force can be too low to initiate nucleation. This zone is referred as metastable zone. For instance, if the system is held in point C, it will not crystallize for an extended amount of time, (even years) despite the system being supersaturated. At point C, the crystallization kinetics is very slow with respect to observation time and the driving force is thought to be too small to trigger nucleation. If the system is further cooled to point D from C, at this point, supersaturation reaches a value that is large enough to initiate nucleation. At point D, the first nucleus is observed. The collection of points where the first nuclei are observed at different concentrations is referred as *metastable zone limit.* Cooling the system beyond point D to point E results in instant nucleation. This "sudden death" zone is known as the *labile zone*. The knowledge of metastable zone is essential in operation of a crystallizer. The preferred operation ensures that the crystallizer always stays in the metastable zone where there is sufficient supersaturation to nucleate and grow. Entry into labile zone is avoided as this leads to instant uncontrolled nucleation. Uncontrolled nucleation creates large variations in size and other crystal quality parameters.

8.4.2 Measuring metastable zone width

Engineers and scientists have observed the existence of metastable zone width (MSZW) since the early days of crystallization science. MSZW is illustrated in Figure 8.12. Nyvlt, using a very simple apparatus, conducted the first meaningful measurements of MSZW. The apparatus consists of 50-ml flask fitted with a thermometer and a magnetic stirrer, located in an external cooling bath. A solution containing known amounts of the solute and solvent are heated above the saturation temperature and cooled down at a slow steady rate. The experimentalist records the temperature at which the first crystal is observed. In early experiments of Nyvlt, the observation was done by visual inspection. More modern systems nowadays utilize automated procedures with microscopy or scattering techniques to detect the emergence of the first crystal. The difference between the measured temperature and saturation temperature is the maximum allowable undercooling at a corresponding cooling rate. MSZW measurements should be repeated statistically significant number of times due to stochastic nature of nucleation. It is important to realize that MSZW is a kinetic parameter not a thermodynamic one such as solubility. In other words, MSZW is depending on the system size and operation conditions. For instance, MSZW often decreases with increasing mixing rate, as MSZW is a kinetic parameter. Altering the mixing speed will not change thermodynamic parameters such as solubility under identical temperature and pressure.

8.4.3 Supersaturation

The *degree of supersaturation* is a measure of deviation from equilibrium and it can be expressed in many different ways. In crystallization, it is common to use *supersaturation ratio* formulated as the solute concentration relative to the concentration at equilibrium:

$$S = \frac{c}{c^*} \tag{8.4}$$

where c is the actual solution concentration and c^* the equilibrium saturation value. Another common expression is the *relative supersaturation* σ, given by the ratio of the difference between the solute concentration and the equilibrium concentration to the equilibrium concentration:

$$\sigma = \frac{c - c^*}{c^*} \tag{8.5}$$

Also the difference between the solute concentration and the concentration at equilibrium is termed *absolute supersaturation* Δc:

$$\Delta c = c - c^* \tag{8.6}$$

For melt crystallization, deviation from the melting temperature, T_m, of pure target compound is referred as *undercooling*, ΔT where the temperature of the melt is T:

$$\Delta T = T - T_m \tag{8.7}$$

Solution concentration may be expressed in a variety of units. For general mass balance calculations, kilograms anhydrate per kilogram of solvent or kilograms hydrate per kilograms of free solvent are most convenient. The former avoids complications if different phases can crystallize over the temperature range considered.

The importance of supersaturation in setting nucleation and growth rates is clarified in Figure 8.13, which schematically shows the influence of supersaturation on nucleation and growth. The key aspects in this figure are the qualitative relationships of the two forms of nucleation to growth and to each other. Growth rate and secondary nucleation kinetics are low-order (shown as linear) functions of supersaturation, while primary nucleation follows a high-order dependence on supersaturation. Design of a crystallizer to produce desired crystal quality parameters requires the quantification of nucleation and growth rates to externally controlled variables through supersaturation.

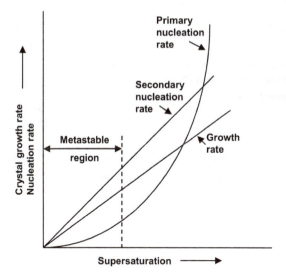

Figure 8.13: Influence of supersaturation on nucleation and crystal growth rates.

8.4.4 Theory of crystallization

For crystallization to occur, the state of solution or melt has to be shifted from a thermodynamically equilibrium undersaturated state to out-of-equilibrium supersaturated state. The degree of this deviation is termed as supersaturation. Once the mother liquor is supersaturated with respect to solute, the degree of supersaturation determines the fundamental crystallization phenomena and the rate of these processes. Consequently, supersaturation is a valuable tool to characterize and control crystallization process. To design crystallizers, we need to predict or measure crystallization kinetics as a function of supersaturation.

In thermodynamic equilibrium, the chemical potential of phases involved is equal. Thermodynamically this condition indicates that the chemical potential of the solute i in solution, $\mu_{i,\,\text{liquid}}^{*}$, is the same as the chemical potential of solute i in the solid phase, $\mu_{i,\,\text{solid}}^{*}$. Here the $*$ indicates the equilibrium and denotes to a point on the solid–liquid boundary in phase diagram such as point B in Figure 8.12.

$$\mu_{i,\,\text{solid}}^{*} = \mu_{i,\,\text{liquid}}^{*} \tag{8.8}$$

If an undersaturated solution or melt is brought out of equilibrium by external action (cooling, evaporation, chemical reaction or anti-solvent), the chemical potential of both the liquid and solid phase denoted by $\mu_{i,\,\text{liquid}}$ and $\mu_{i,\,\text{solid}}$, respectively, will change. For instance, a solution cooled from point B to point C in Figure 8.11 creates a deviation from equilibrium. The chemical potential derivation from the equilibrium, $\Delta\mu$, can be written as:

$$\Delta\mu = \mu_{i,\,\text{solid}} - \mu_{i,\,\text{solid}}^{*} = \mu_{i,\,\text{liquid}} - \mu_{i,\,\text{liquid}}^{*} \tag{8.9}$$

Chemical potential of a solute in a given state can be expressed as its chemical potential at a reference state plus the deviation from the reference state, a function of its molar activity a at constant temperature and pressure as:

$$\mu_{i,\,\text{state}} = \mu_{i,\,\text{state}}^{0} + RT \ln a \tag{8.10}$$

where $\mu_{i,\,\text{state}}^{0}$ is the chemical potential of a reference state, a is the activity of the solution. Using eq. (8.9) and eq. (8.10), the chemical potential difference $\Delta\mu$ can be written as:

$$\Delta\mu = RT \ln\left(\frac{a}{a^{*}}\right) = RT \ln\left(\frac{\gamma_s c}{\gamma_s^{*} c^{*}}\right) \tag{8.11}$$

where γ_s is the activity coefficient and c is the concentration. The $*$ denotes equilibrium. For relatively low supersaturation, $\gamma_s = \gamma_s^{*}, \Delta c/c^{*} \ll 1$ the equilibrium potential difference simplifies to:

$$\Delta\mu = RT \ln\left(\frac{c}{c^{*}}\right) = RT \ln S \approx RT\sigma \tag{8.12}$$

Therefore, at constant pressure and temperature for processes conducted at low supersaturation, the chemical potential change due to deviation from thermodynamic equilibrium can be approximated by the experimentally measurable and practical supersaturation relative supersaturation.

8.4.5 Nucleation

In crystallization, nucleation is the first step for formation of a solid phase from a mother phase (either liquid or melt). It sets the character of the crystallization process, and it is therefore the most critical component in relating crystallizer design and operation to crystal quality parameters. Nucleation can be classified into two mechanistically different categories:

- Primary nucleation
    - Homogeneous primary nucleation
    - Heterogeneous primary nucleation
- Secondary nucleation

In *primary nucleation,* a collection of randomly vibrating solute molecules come together and form precrystalline bodies or cluster that eventually give birth to crystalline nuclei. Existing crystals are not involved in the primary nucleation process. Primary nucleation is classified into two subcategories: *homogeneous and heterogeneous primary nucleation*. Homogeneous primary nucleation occurs in the bulk liquid phase in the absence of any foreign bodies. The formation of nuclei involves only the solute molecules. Heterogeneous primary nucleation occurs on impurity surfaces such as a dust particle or the vessel wall. Just like other, it is hypothesized that heterogeneous nucleation begins with adsorption of solute to the impurity surface where the solute diffuses and encounters other solute molecules to form a nucleus. In addition to primary nucleation, *secondary nucleation* plays a critical role dictating final crystal quality parameters. In secondary nucleation, solute molecules form nuclei on the surface of preexisting crystal. In essence, secondary nucleation is heterogeneous nucleation induced by existing crystals.

Homogeneous primary nucleation involves only solute molecules. In industrial practice, however, the working fluids often contain impurities. Moreover, the process equipment is often in contact with working fluids. In classical homogeneous nucleation theory, a large number of solute units (atoms, molecules or ions) can come together in a supersaturated solution to form clusters due to concentration fluctuations. To control the kinetics of crystallization consequently the crystal quality parameters, a theory connecting the supersaturation and rate of nucleation is valuable. Classic nucleation theory (CNT) provides such a framework.

CNT considers nucleation as a competition between surface and volume formation. If a solute molecule is added to a prenucleation cluster, both the volume and area increase. Volume and surface formation work against each other in terms of free energy. Formation of a volume decreases the free energy as solid phase is thermodynamically favored. In other words, crystalline phase is of lower free energy and thermodynamic systems tend to go to low energy state. This is also evident from the heat released when crystallization occurs. However, formation of a surface increases the free energy. In essence by increasing the surface area, the number of solute molecules increases interacting with solvent and solute molecules in solution. As the solute–solvent interactions are less favorable compared to solute–solute interactions, evidenced by heat released in crystallization, formation of a surface increases free energy.

Considering this balance between the energy involved in solid-phase formation, ΔG_v and in creation of the surface, ΔG_s of an arbitrary spherical crystal of radius r in a supersaturated fluid leads to the relationship:

$$\Delta G = \Delta G_v + \Delta G_s = \frac{4}{3}\pi r^3 \Delta\mu + 4\pi r^2 \gamma \tag{8.13}$$

where ΔG is the overall excess free energy associated with the formation of the crystalline body and $\Delta\mu$ is the free energy change associated with the phase change. The ΔG_v can be written as the volume of a spherical nucleus, $4\pi r^3/3$ multiplied by $\Delta\mu$. ΔG_s is represented as surface area of a spherical nucleus, $4\pi r^2$ multiplied by the interfacial energy between the crystalline nucleus and the surrounding solution, γ. γ is also referred as the solid–liquid interfacial tension. It is important to realize that the volume term is negative and proportional to r^3 whereas the surface area term is positive and proportional to r^2. The sign of terms is determined by convention. The proportionality of volume and surface terms with respect to radius of nuclei gives interesting deductions. For small nucleus sizes, the surface term dominates, whereas at larger sizes the volume term dominates. Once ΔG is plotted as a function of cluster radius or size of cluster, it is observed that ΔG first increases with increasing cluster size, reaches a maximum and then decreases. Within the framework of CNT, this behavior is interpreted as an energy barrier denoted by W^* in Figure 8.14. Through thermal fluctuations ever present above absolute 0 K, solute molecules form high-density collection of molecules so-called clusters or amorphous bodies. If these clusters grow above a critical size, they can form a nucleus and grow unhindered. This critical size is denoted as r^* or n^* and corresponds to the W^* in Figure 8.14. Any cluster smaller than the critical size dissolves back into its constituents. Considering the timescale of molecular diffusion, this situation should be seen as a dynamic equilibrium where clusters of all sizes constantly form and dissolve. Once in a while, the clusters above the critical size emerge and the probability to dissolve becomes negligible. This can be seen as an activated statistical process.

From a practical point of view, we would like to know how often clusters grow and how this frequency depends on the supersaturation. CNT allows for predicting nucleation rate as a function of supersaturation by making the assumption that the rate nucleation will happen is to connect W^* with an Arrhenius type relationship given below.

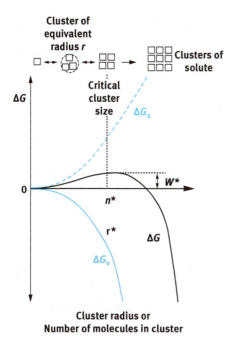

Figure 8.14: Illustration of free energy diagram for classic nucleation theory and formation of clusters consisting of solute molecules.

$$J = A \exp\left(-\frac{W^*}{k_b T} \right) \tag{8.14}$$

where J is the rate of nucleation giving the number of nuclei forming per volume and time, A is a prefactor with same units of J, k_b is the Boltzmann constant in Joule/K and T is the Kelvin scale temperature in units of K. $k_b T$ is the average kinetic energy of a molecules in Joules. In essence, $W^*/k_b T$ can be interpreted as relative height of the activation energy barrier with respect to kinetic energy available to a solute molecule due to thermal fluctuation. Equation (8.14) indicates that as W^* increases the probability that a cluster grows above the critical size consequently, the rate of nucleation decreases.

To connect the W^* to driving force behind crystallization, supersaturation; we first recall eq. (8.12) and replace it in eq. (8.13).

$$\Delta G = \Delta G_v + \Delta G_s = \frac{4}{3}\pi r^3 RT \ln S + 4\pi r^2 \gamma \tag{8.15}$$

Rewriting eq. (8.15) in terms of number of solute units in a cluster gives

$$\Delta G(n) = \Delta G_v + \Delta G_s = -nRT \ \ln S + c(vn)^{2/3} \gamma \tag{8.16}$$

where c is a constant accounting for cluster geometry and v is the molecular volume.

To find W^*, we take the derivative of eq. (8.16) and setting it to zero would lead to W^* in terms of n. This is expressed mathematically as

$$\frac{\partial \Delta G(n)}{\partial n} = 0 \text{ at } n = n^* \text{ and } \frac{\partial \Delta G(n^*)}{\partial n} = W^* \tag{8.17}$$

carrying out the operation suggested in eq. (8.17) for spherical molecules leads to

$$W^* = \frac{16\pi v^2}{3k^2 T^2} \frac{\gamma^3}{\ln^2 S} \tag{8.18}$$

and

$$n^* = \frac{32\pi v^2}{3k^2 T^2} \frac{\gamma^3}{\ln^3 S} \tag{8.19}$$

Replacing W^* in eq. (8.18) to eq. (8.14), connects the rate of nucleation to supersaturation

$$J = A \exp\left(-\frac{W^*}{k_b T}\right) = A \exp\left(-\frac{16\pi v^2}{3k_b^3 T^3} \frac{\gamma^3}{\ln^2 S}\right) \tag{8.20}$$

This is the simplest rate of expression for homogeneous primary nucleation. The high-order dependence on supersaturation is especially important as a small variation in supersaturation may produce an enormous change in nucleation rate and can lead to production of excessive fines when primary nucleation mechanisms are important. Experimentally, the solid–liquid interfacial tension is not easily measured consequently; a simplified version given below is used where all the molecular parameters are dumped to B, so-called the thermodynamic factor:

$$J = A \exp\left(-\frac{B}{\ln^2 S}\right) \tag{8.21}$$

A is called the kinetic parameter or frequency factor with approximately $A \approx 10^{30}$ nuclei/s·cm^3 for most organic compounds.

In *heterogeneous primary nucleation,* the crystal forms around a foreign object such as a dust particle or a crack in the vessel wall. The foreign object lowers the surface energy by altering γ in eq. (8.20) as well as A. For heterogeneous nucleation, rate of heterogeneous nucleation J_{HET} is defined as

$$J_{\text{HET}} = A_{\text{HET}} \exp\left(-\frac{16\pi v^2}{3k_b^3 T^3} \frac{\gamma_{\text{eff}}^3}{\ln^2 S}\right) \tag{8.22}$$

where A_{HET} and γ_{eff} are frequency factor and the effective interfacial tension. The frequency factor for heterogeneous nucleation is typically 4 to 5 orders of magnitude larger than the frequency factor for homogeneous nucleation. This greatly increases the rate of nucleation at much lower supersaturation values, which is of utmost importance in industrial systems that are never completely free of suspended solids, impurities and confining surfaces.

Measuring the primary rate of nucleation is challenging task. This is due to two fundamental challenges. Firstly, no experimental method offers atomistic resolution at the speed of molecules diffusing in solution. Secondly, nucleation is a rare event and consequently it is difficult to predict and capture the emergence of the first nuclei. Instead experimentalist measure a delay time called induction time. Once a solution is brought to a supersaturation, nucleation does not occur instantly. The time delay between the temperature reaching a desired value and the observation of the first crystal is called induction time (τ). Experimentalist measures the induction time in a given volume of solution and then connects this induction time to rate of nucleation, J, by the following equation $\tau = \frac{1}{JV}$ where V is the volume of the system. The rate of nucleation is defined as number of nuclei per time per volume. The rate of nucleation is measured by the following procedure: (a) a statistically significant number of vials (200 or more) are prepared by adding a known amount of solute to a known volume of solvent; (b) the vials are heated above saturation temperature and kept at fixed temperature till the solute is dissolved; (c) the vials are cooled down to a desired temperate very rapidly and the emergence of first crystals in vials are recorded as a function of time. The rate of nucleation is calculated either from the mean induction time or by constructing a cumulative induction probability function and fitting this function with $P(t) \approx exp\left(-\frac{t}{\tau}\right)$ to extract primary rate of nucleation.

Secondary nucleation is the formation of new crystals as a result of the presence of other solute crystals through mechanism such as initial breeding, contact nucleation and sheer breeding. In most industrial crystallizers, secondary nucleation is far more important than primary nucleation. This is due to the existence of crystals that can participate in secondary nucleation mechanisms in widely used continuous and seeded batch crystallizers. Moreover, these crystallizers typically operate at low supersaturation to obtain regular and pure product. The presence of crystals and low supersaturation can only support secondary nucleation and not primary nucleation. *Initial breeding* results from adding *seed crystals* to a supersaturated solution to induce nucleation. The presence of growing crystals can also contribute to secondary nucleation by several different mechanisms. *Contact nucleation* results from nuclei formed by collisions of crystals against each other, against crystallizer internals or with an impellor, agitator or circulation pump. *Sheer breeding* results when a supersaturated solution flows shear a crystal surface and carries along crystal precursors formed in the region of the growing crystal surface.

Many commercial crystallizers operate in the secondary nucleation range and contact nucleation is the most commonly encountered secondary nucleation mechanism in practice. Based on experimental observations, the contact nucleation rate was proposed to follow an exponential function of impact energy E minus a required minimum collision energy E_t to break small nuclei from the growing crystals:

$$B^o = k_N \exp(E - E_t) \quad E > E_t \tag{8.23}$$

Although this equation can be derived from fundamental principles, it is only accurate within an order of magnitude. For design purposes, the metastable limit can provide a useful *semiempirical approach* to correlate the effective or apparent heterogeneous and secondary nucleation rate:

$$B^o = k_n(c - c_m)^n \approx k_n(c - c^*)^n \quad c^* < c_m \tag{8.24}$$

where c is the solute concentration, c_m is the solute concentration at which spontaneous nucleation occurs and c^* is the solute concentration at saturation. The metastable limit must be determined through experimentation. Fortunately, c_m is very close to c^* for many inorganic systems. The parameters k_n and n must be evaluated from experimental data. The order for heterogeneous nucleation can range from 2 to 9. For secondary nucleation, the order is significantly lower and in the range of 0 to 3. Much of the experimental data are correlated by an empirical power-law function:

$$B^o = k'_n M_T^n G^m \tag{8.25}$$

where M_T = magma concentration (crystal weight per unit suspension volume) and G = crystal growth rate (m per unit time). The orders m and n can be determined from small-scale experiments; a reliable value of k'_n, however, should be determined from data for commercial crystallizers.

8.4.6 Crystal growth

Extending our current understanding of crystal growth is an on-going scientific challenge with immense industrial consequences just like nucleation. Growth can be expressed in a three of the following ways:
- linear advance rate of an individual crystal face;
- change in a characteristic dimension of a crystal;
- rate of change in mass of a crystal.

Adsorption layer or kinetic theories have proven to be quite fruitful in explaining crystal growth. It is postulated that there is an "adsorbed" layer of solute molecules, which is loosely held to the crystal face. These adsorbed units are free to move on the two-dimensional surface, but they have essentially lost one degree of freedom. Thus, the molecules must crystallize through *two-dimensional nucleation*. At low supersaturations, the energy required for two-dimensional nucleation is considerably less for normal nucleation. Thus, existing crystals can grow under conditions where three-dimensional nucleation will not occur. Crystal growth will usually occur at supersaturation levels lower than can be explained by the two-dimensional nucleation theory. This happens because growth is much faster at any imperfection where the surface is not a geometrically smooth plane. Since crystals are usually not geometrically smooth, growth usually proceeds by a "filling-in" process. Small crystals are more likely to be geometrically smooth than large crystals and therefore more likely to grow via two-dimensional nucleation. Large crystals are less likely to be geometrically smooth and are more likely to be damaged by the impeller or baffles. Thus, the large crystals can grow by healing kinks, pits and dislocations. These imperfections are also the reason that even two crystals of the same size may have different growth rates although all conditions appear to be the same. This is called *growth rate dispersion*. If one of the crystals happens to be more geometrically smooth than the other, it will have a lower growth rate.

The adsorption theory that explains the crystallization once a solute molecule is at the surface is often incorporated into mass transfer theories that describe the movement of the unit to the surface. In crystallization, the picture of the mass transfer process is somewhat different from mass transfer theories for other processes and different empirical expressions are used. Figure 8.15 shows a schematic of the mass transfer process that can be broken down in the following seven steps:

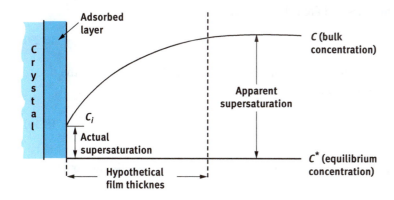

Figure 8.15: Schematic of crystal growth mass transfer process.

1. Bulk diffusion of dissolved solute through the film
2. Diffusion of dissolved solute in the adsorbed layer
3. Partial or total desolvation
4. Surface diffusion of solute units to growth site

5. Incorporation into lattice
6. Counter diffusion of solvent through the adsorbed layer
7. Counter diffusion of solvent through the film

Since it is difficult to include all these steps in a theory, usually a simplified model is used which includes step 1, combines steps 2 to 5 and ignores the last two steps. The mass transfer across the film is for step 1 which is given by:

$$\frac{dm}{dt} = k_f A (c - c_i) \tag{8.26}$$

where k_f is the film mass transfer coefficient and A the crystal surface area. Because m represents the mass of solute transferred across the film, m also equals the mass of solid deposited. Next, steps 2 to 5 are combined as a single surface "reaction" of "order" n with rate constant k_r through the empirical relation:

$$\frac{dm}{dt} = k_r A (c_i - c^*)^n \tag{8.27}$$

In general, it is extremely challenging to determine the interfacial concentration c_i. For a first-order surface reaction ($n = 1$), both equations can be combined to remove c_i in a similar way as shown in chapter 4 for gas/liquid contactors:

$$\frac{dm}{dt} = k_{OG} A (c - c^*) \tag{8.28}$$

in which k_{OG} is the overall rate constant. For crystals where the faces are growing at the same rate, the mass and area of the crystal can be written as

$$m = k_v L^3 \rho_c \quad A = k_A L^2 \tag{8.29}$$

where k_v and k_A are shape factors. Substituting into eq. (8.12), we obtain the *linear growth rate* G of the crystal, which is equal in all three dimensions:

$$G = \frac{dL}{dt} = \left(\frac{k_A}{3\rho_c k_V} \right) k_{OG} (c - c^*) \tag{8.30}$$

that can be simplified by the introduction of the combined growth rate constant k_G to:

$$G = \frac{dL}{dt} = k_G (c - c^*) \tag{8.31}$$

When the linear growth rate is independent of crystal size and the supersaturation is constant, this is reduced to very simple ΔL law, which will be used in the next section:

$$G = \frac{\Delta L}{\Delta t} \tag{8.32}$$

Because of the assumptions involved, it should not be surprising that many systems do not satisfy this law. In addition to supersaturation, temperature will also affect the crystal growth rate through diffusion and the crystal rearrangement process. If either mass transfer or surface reaction controls the temperature, dependence of growth kinetics can often be expressed in terms of an Arrhenius expression:

$$k_G = k_G^o \exp\left(-\frac{\Delta E_G}{RT}\right) \tag{8.33}$$

When both mechanisms are important, this equation is not valid and Arrhenius plots for crystal growth data tend to give curved instead of straight lines. When n is not a simple integer, it is not possible to solve explicitly for the growth rate. In these cases, a common approach is to introduce an empirical "overall growth-rate order" n to relate growth kinetics to supersaturation:

$$\frac{dm}{dt} = k_{OG}A(c - c^*)^n \tag{8.34}$$

yielding for the growth rate, a power-law function in which the constants k_G and n are determined by fitting experimental crystal growth data:

$$G = \frac{dL}{dt} = k_G(c - c^*)^n \tag{8.35}$$

Such an approach will only be valid over small ranges of supersaturation.

8.4.7 Population balance equations for the MSMPR crystallizer

An analysis of a crystallization system requires the quantification of nucleation and growth kinetics, and how they influence the crystal quality parameters such as CSD. There are a number of ways for describing the CSD for a crystallizer. Population balance is the major theoretical tool for predicting and analyzing CSDs. They allow quantification of the nucleation and growth processes by balancing the number of crystals within a given size range. A balance on the number of crystals in any size range, say L_1 to L_2, accounts for crystals that enter and leave the size range by convective flow into and out of the magma suspension with volume V_M and for crystals that enter and leave the size range by growth. Crystal breakage and agglomeration are ignored and it is assumed that crystals formed by nucleation are near to size zero. The number of crystals in the size range L_1 to L_2 is given by:

$$\Delta N = \int_{L_1}^{L_2} n dL \tag{8.36}$$

The population balance differs from a mass balance in that only the item balanced within a given size range is included, which are coupled to the more familiar balances on mass and energy. It is assumed that the population distribution is a continuous function and that a *characteristic dimension* L can describe crystal size, surface area and volume. Population balances for crystallizers are usually described by the *population density* n that gives the number of crystals dN in the size range L to $L + dL$ by:

$$n = \frac{dN}{dL} = \lim_{\Delta L \to 0} \frac{\Delta N}{\Delta L} \tag{8.37}$$

where ΔN is the number of crystals in size range ΔL per unit volume (clear liquor or slurry). The value of N depends on the value of L at which the interval ΔL is measured as shown in Figure 8.16. In practice, the number density is calculated from the relationship:

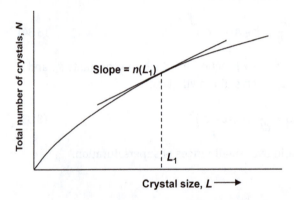

Figure 8.16: Schematic representation of the population density.

$$n = \frac{m_i}{k_v \rho_c \bar{L}_i^3 \, \Delta L_i} \tag{8.38}$$

where m_i is the mass of the residue on sieve i, $\bar{L}$ is the mean crystal size of the particles in that fraction of the sieve and ΔL_i is the size difference between the two sieves that produced mass fraction m_i.

Application of the population balance is most conveniently described for the idealized case of a continuous, steady-state *mixed-suspension mixed-product removal* (MSMPR) crystallizer as shown in Figure 8.17. Furthermore many industrial crystallizers are operated in a well-mixed manner and the equations derived are quite useful in describing their performance. Perfectly mixed crystallizers are highly constrained and the form of CSD produced by such systems is fixed. The assumptions made are that no crystals are present in the feed stream, all crystals are of the

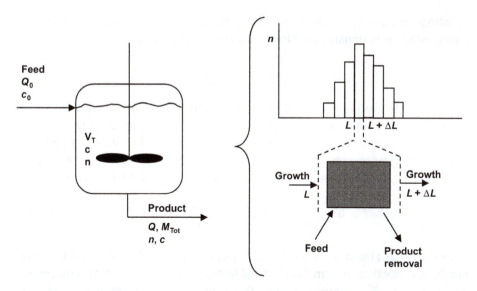

Figure 8.17: Schematic diagram of a simple, perfectly mixed crystallizer.

same size and crystal growth rate is independent of crystal size. For a well-mixed and constant slurry volume V_M in a crystallizer, the population balance for a time interval Δt and size range $\Delta L = L_2 - L_1$ becomes:

$$\begin{matrix}\text{number of} & & \text{number of} & & \text{number of} \\ \text{crystals growing} & = & \text{crystals growing} & + & \text{crystals in range} \\ \text{into range} & & \text{out of range} & & \text{leaving crystallizer} \end{matrix} \qquad (8.39)$$

which becomes in symbols:

$$n_L G V_M \Delta t = n_{L+\Delta L} G V_M \Delta t + Q \bar{n} \Delta L \, \Delta t \qquad (8.40)$$

where Q is the volumetric feed and discharge rate, G is the growth rate, defined in eq. (8.32), and $\bar{n}$ is the average population density.

As $\Delta L \to 0$, $\bar{n} \to n$, and, noting that $n_{L+\Delta L} = n_L + \frac{dn}{dL} \Delta L$, the population balance takes the form

$$G \frac{dn}{dL} = -\frac{Qn}{V_M} \qquad (8.41)$$

which can be written as:

$$G \frac{dn}{dL} + \frac{n}{\tau} = 0 \qquad (8.42)$$

by introducing a mean residence time τ defined as V_M/Q. This equation can be integrated using the boundary condition $n = n_0$ at $L = 0$ to obtain:

$$n = n_0 \exp\left(-\frac{L}{G\tau}\right) \tag{8.43}$$

If the logarithm of the number density n is plotted against the crystal size L, a straight line is produced with the negative slope $-(1/G\tau)$ and intercept $\ln n_0$. Under steady-state conditions, the total number production rate of crystals in a perfectly mixed crystallizer must be identical to the nucleation rate B^0. With the intercept of the ordinate n_0 for $L = 0$, we obtain the following for the nucleation rate B^0:

$$B^0 = \left(\frac{dN}{dt}\right)_{L=0} = \left(\frac{dN}{dL}\right)_{L=0} \left(\frac{dL}{dt}\right) = n_0 G \tag{8.44}$$

The slope $-(1/G\tau)$ and the intercept of the ordinate n_0 of the straight line in the number-density diagram can thus be used to determine the two kinetic parameters: nucleation rate B^0 and growth rate G. These two parameters determine the dominant crystal size L_D of the CSD. The dominant crystal size L_D is defined as the crystal size for which the maximum weight in the CSD is obtained. The crystal mass m_L within the range of L to $L + dL$ is obtained from the mass m_c of a single crystal multiplied by the number of crystals N (eq. (8.36)) within that population:

$$m_L = (\rho_c k_v L^3) N = \rho_c k_v n_0 L^3 e^{-\frac{L}{G\tau}} \tag{8.45}$$

The maximum weight is then obtained by setting the derivative of m_L to L, $\frac{dm_L}{dL}$, equal to zero and solving this equation for L. The resulting *dominant crystal size L_D* is then:

$$L_D = 3G\tau \tag{8.46}$$

In a similar way, the *suspension density* of total crystal (magma) mass M_T (kg crystals $/m^3$ suspension) is calculated from the mass of one crystal m_c with a size between L and $L + dL$, (eq. (8.29) and integrating this over the total number of crystals N_T (eqs. (8.36) and (8.43)):

$$M_T = \rho_c k_v n_0 \int_0^\infty L^3 e^{-\frac{L}{G\tau}} dL = 6\rho_c k_v n_0 (G\tau)^4 \tag{8.47}^2$$

---

1 Note that the total number of nuclei per unit volume $N = \int_0^\infty n \cdot dL = n_0 \int_0^\infty e^{-L/G\tau} dL = n_0 G\tau = B^0 \tau$

2 Note that $\int_0^\infty L^3 \exp\left(-\frac{L}{G\tau}\right) dL = (G\tau)^4 \int_0^\infty y^3 e^{-y} dy = 6(G\tau)^4$

which is easily converted into the *volumetric holdup of crystals* ϕ_T (m^3 crystals /m^3 suspension) by dividing with the crystal density:

$$\phi_T = 6k_v n_o (G\tau)^4 \tag{8.48}$$

8.5 Methods of supersaturation generation

Supersaturation is the driving force behind crystallization. The methods to produce supersaturation play a critical role in dictating crystal quality parameters and process design. There are five methods to produce supersaturation:
- Temperature change
- Evaporation of solvent
- Vacuum crystallization (cooling/evaporation)
- Chemical reaction
- Changing the solvent composition also known as anti-solvent crystallization

In industrial practice, these aforementioned methods can be used in tandem and/or sequential to achieve desired crystallization yields. The methods of supersaturation generation can be classified into two categories based on the phase of the mother liquor: crystallization from solution and crystallization from melt. In crystallization from solution, the mother liquor contains a solute dissolved in a solvent. The solute is crystalized using one of the approaches discussed in detail below namely, cooling, evaporation, vacuum cooling, precipitation and anti-solvent crystallization. Crystallization from solution is mostly applied as a single-stage separation technique to isolate high-quality solute crystals excluding the solvent and impurities. This is only possible if the solvent or impurity molecules do not fit into the unit cell of the solute. In crystallization from melt, there is no solvent involved. The melt of a solute is purified in a series of crystallization steps to get rid of impurities till the solute crystals are formed. In melt crystallization, the melt is cooled below its equilibrium melt temperature, T_m. The degree of supersaturation is referred as undercooling and denoted as $\Delta T = T_m - T$ (eq. 8.7).

8.5.1 Cooling

Cooling crystallization is applied to solutes with medium-to-high solubility solutes when the solubility curves strongly depend on the temperature. Usually the starting and final temperature of the process is chosen at the temperature point where the slope of the solubility curve is steepest. This ensures that maximum amount of solid is crystallized, that is, highest yield is achieved. The final temperature should be chosen as low as possible yet going below room temperature will require (i) additional cooling units consequently higher operational costs; and (ii) higher maintenance costs as cooled parts often cause blockage and aggregation. Choosing a final temperature close to room temperature such as 30 °C or higher is often convenient as such cooling operation can utilize ambient room temperature to cool the system. Operating a crystallization process at below ambient temperatures often results in low crystal growth rates and low heat/mass transfer rates consequently low liquid–solid separation.

In a cooling crystallizer, heat can be removed either directly or indirectly. In direct cooling, a refrigerant solution is introduced into the crystallizer directly. The refrigerant may be a gas, a liquid or a solid. It may be miscible or immiscible with the mother liquor. In indirect cooling, heat is removed via a heat exchanger through confining surfaces of the crystallizer often designed as a temperature controlled jacket surrounding the crystallizer. The heat removing surfaces are often prone to scaling or incrustation consequently; these scrappers can be added to remove the scaling from heat exchanging surfaces. Due to its noninvasive nature, indirect cooling is the more popular cooling method.

The main limitation of the cooling method is the yield that is determined by the solubility curve at the lowest rational temperature reachable. For high value products such as pharmaceuticals, the remaining solute in solution after cooling crystallization is often recovered by anti-solvent crystallization.

8.5.2 Evaporation

Evaporative crystallization creates supersaturation by evaporating the solvent consequently increasing the concentration of solute in solution. It is the preferred method of supersaturation generation for compounds with nearly flat solubility curves. Such solute–solvent systems cannot give appreciable yields with cooling crystallization. Once the solvent evaporates, the increasing supersaturation triggers nucleation and growth.

Evaporative crystallization is often carried out at the boiling temperature of the solvent. This method should not be utilized for heat sensitive compounds. Particularly, if heating the solution to the boiling temperature of the solvent chemically decomposes the solute, this method should be avoided. Scaling with retrograde solubility, such as calcium sulfate, can be challenging in evaporative crystallizers. Most evaporation units use low-pressure steam or by-product steam generated in other chemical processes.

8.5.3 Vacuum crystallization

Vacuum crystallization also referred as vacuum cooling and flash cooling crystallization is applied to systems exhibiting a nonflat solubility curve. It takes advantage of cooling and evaporative crystallization simultaneously. In this method, a hot saturated solution is fed to a vessel kept at a pressure below the vapor pressure of the solution. As the solution enters the chamber, part of the solvent evaporates to the gas phase and the liquid is adiabatically cooled down to the boiling temperature of the vessel pressure. In essence, the saturated solution is both evaporated and cooled down. The solution becomes supersatured due to solvent evaporation and cooling. The created supersaturation drives the nucleation and growth of the solute crystals.

Vacuum crystallization is another heat removal method for systems with scaling tendency. Vacuum crystallization methods often result in minimal scaling as the heat exchange occurs only in the confining crystallizer walls as oppose to a high area heat exchanger surface commonly used in cooling crystallization. Vacuum crystallization is also used for products that are not stable when heated to high temperatures to promote boiling in evaporative crystallization. It can be utilized for solutions of salts with high boiling point elevation or for solutions with large quantities of noncondensable gases.

8.5.4 Precipitation

Precipitation is an important industrial process applied to sparingly soluble compounds (solubility range 0.001 to 1 kg/m^3). In precipitation, a solute is made supersaturated so that dissolved solutes precipitate out. Precipitation differs from crystallization since the precipitate may be amorphous or crystalline with a poorly defined size and shape. Moreover, precipitate is often an aggregate and is usually not pure. Precipitation can be carried out in one step to remove a large number of compounds in a single precipitate. This is the easiest application and is most common. *Fractional precipitation* requires a series of steps with each step optimized to precipitate the desired component. It is not possible to do extremely sharp purifications by fractional precipitation. Precipitation and crystallization are often complementary processes since precipitation is used for a "rough cut" while crystallization is used for final purification. Precipitation is commonly used for the manufacture of pigments, pharmaceuticals and photographic chemicals. In the production of ultrafine crystalline powders, precipitation is often considered an attractive alternative to comminution, particularly for heat sensitive substances.

Like all crystallization processes, precipitation consists of three basic steps: supersaturation generation, followed by the generation of nuclei and the subsequent growth of these nuclei to visible size. The used equipment is often very similar to that used for crystallization. Precipitation operations usually have a mixing operation where various reagents are added to make the solution supersaturated. This is followed by a holding period, which allows for an induction period before nuclei form and a latent period before the supersaturation starts to decrease. Once the precipitate starts to grow, the supersaturation decreases often at a constant rate. Towards the end of the desupersaturation period, the rate decreases as the concentration approaches equilibrium. During this period, aging often occurs. During *aging*, small crystals redissolve and the solute is redeposited onto the larger precipitates. The result is a uniform crop of fairly large precipitate, which is relatively easy to separate from the solution by centrifugation.

Precipitation processes are conveniently classified by the method used to trigger reaction creating the supersaturation: mixing initiated reaction, salting out, anti-solvents and altering pH. Mixing reagents that react to form an insoluble precipitate produces many common commercial products. A representative example is the formation of gypsum by the addition of sulfuric acid to a calcium chloride solution:

$$CaCl_2 + H_2SO_4 \rightarrow CaSO_{4\,(solid)} + 2HCl$$

In aqueous solution, high inorganic salt concentrations also often cause the precipitation of solute. Anti-solvents or nonsolvents can be added to achieve that same effect. The solubility of many compounds is affected by pH.

8.5.5 Anti-solvent crystallization

In anti-solvent crystallization, an anti-solvent is mixed with a solution containing solute dissolved in solvent. The anti-solvent mix with the solvent and the resulting solvent mixture has low solubility for solute. Consequently, the solute becomes supersaturated in solvent mixture and eventually crystallizes. Since the anti-solvent addition also dilutes the solute concentration, the decrease in solute solubility due to anti-solvent should be large enough to overcome this dilution.

The supersaturation ranges created in anti-solvent crystallization are sensitive to nature of mixing. The highest supersaturations are created at the inlet of the anti-solvent and the created supersaturation gradually decreases away from the inlets. This situation gives raise to large supersaturation variations across crystallizer leading to large variations in crystal quality parameters as well as nucleation and growth kinetics across the crystallizer.

A commonly used variation of anti-solvent crystallization is salting out. In salting out, an easily soluble salt is added to an aqueous solution containing the compound to be crystallized. Crystallization is initiated due to deceased water activity in the presence of salt. Consequently the target solute becomes less soluble in salty aqueous solution and crystallizes. It is also possible to trigger crystallization by adding a salt carrying a common ion. For instance, an aqueous solution carrying an ionic compound AB can be forced to crystallize by addition of salt AC where A^+ is the common ion.

8.5.6 Melt crystallization

In some crystallization systems, the use of a solvent can be avoided. Melt crystallization is the process where crystalline material is separated from a melt of the crystallizing species by direct or indirect cooling until crystals are formed in the liquid phase. Melt crystallization can be considered as cooling crystallization conducted at very high concentrations.

Two basic techniques of melt crystallization are gradual deposition of a crystalline layer on a chilled surface or fast crystallization of a discrete crystal suspension in an agitated vessel. *Zone melting* relies on the distribution of solute between the liquid and solid phases to affect a separation. Normal freezing is the slow solidification of a melt. The impurity is rejected into the liquid phase by the advancing solid interface. The basic requirement of melt crystallization is that the composition of the crystallized solid differs from that of the liquid mixture from which it is deposited. The process is usually operated near the pure-component freezing temperature. High or ultrahigh product purity is desired in many of the melt purification processes. Single-stage crystallization is often not sufficient to achieve the required purity of the final product. Further separation can be achieved by repeating the crystallization step or by countercurrent contacting of the crystals with a relatively pure liquid stream.

Since there is no solvent, melt crystallization has the advantage that no solvent removal and recovery is required and contamination by the solvent is impossible. However, there is also no way to influence the melt properties (viscosity, diffusivity) and the chemicals being purified must be stable at the melting point. Melt crystallization becomes advantageous when the presence of a solvent would be detrimental or when highly pure products are desired. Currently, the interest in wider application of melt crystallization is stimulated by the energy-saving potential in large-scale processing because solvent evaporation is absent and the heat of fusion is several times less than the heat of vaporization.

8.5.7 Choosing a method of supersaturation generation

The choice of an appropriate method for a given process depends on the chemical identity of the target solute and solvent, the feed composition, thermodynamics, that is, solubility curve, physical properties of solute/solvent system as well as the crystal quality specifications. As a rule of thumb, cooling or evaporative crystallization is implemented if the aim is to obtain large crystals (mm to cm range) from solution. For fine particulate product, precipitation or anti-solvent crystallization is required. For ultrapure products, melt crystallization is the method of choice.

Figure 8.18 provides an overview of the choices in a systematic approach. The ultimate information required is the C^* the equilibrium concentration in g/g-solution

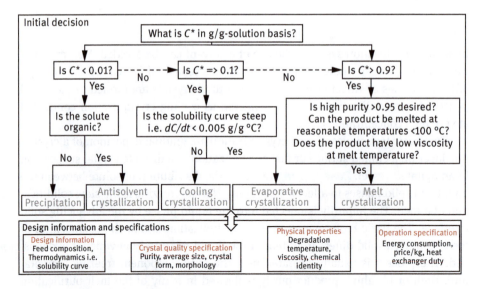

Figure 8.18: Diagram for method of supersaturation generation selection.

gbasis. Based on C^* value, design information particularly the solubility curve and product specifications, we can make an initial design decision on the method of supersaturation generation. One should of course consider that the final crystallizer design requires various iterations due to complexity of crystallization phenomena involved.

8.6 Basic design for industrial crystallizers

Moreover, they can be employed in single- or multistage crystallization or in batch operations. Generally, production rates over 50 ton/day justify continuous operation. The continuous operations tend to have higher yield and require less energy than batch, but batch is more versatile. Multistage operation is employed where evaporative requirements exceed the capabilities of a single vessel and/or energy costs dictate staging of the operation. Another reason for staging may sometimes be the production of more uniform and/or larger crystals. Operation of crystallizers in series generates CSDs having narrower size spread than the same volume of crystallizers in parallel. Batch crystallizers produce a narrower CSD than continuous well-mixed units. Although numerous design methods and kinetic theories exist to analyze specific crystallizer configurations, no clear-cut guidance is available for the choice between crystallizer types and/or modes of operation.

The design of an operational crystallizer can be summarized into two sections: basic design and final design as shown in Figure 8.19. The goal of basic design is to

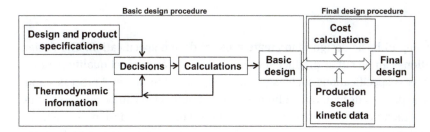

Figure 8.19: Diagram showing the crystallizer design workflow.

create a first guess for a feasibility study identifying the optimal crystallizer configurations. In the basic design, decisions on the crystallizer type, operation mode and auxiliary crystallizer equipment such as circulation devices are decided. These decisions are based on the thermodynamic information and specifications. A flow sheet of the design and initial design dimension are calculated from available design and product specification, that is, crystal quality parameters, production quantity, and feed composition information. Mass and energy balances are supported with heuristic scaling relationships for evaporation or cooling areas and impeller speeds to prevent.

The basic design procedure does not contain kinetic or hydrodynamic information. In the final design, detailed kinetic information such as MSZWs and hydrodynamic information is added to reach the final design. The final design procedure is the most challenging part of the design. This is due to the difficulties in representing the industrial crystallizers with complete hydrodynamic and kinetic equations. For instance, an industrial crystallizer can be as large as couple of meters in diameter and it can contain millions of crystals with sizes ranging between one micrometer and multiple millimeters. These crystals can collide with the confining walls and stirrers altering their size and other crystal properties. A detailed design procedure capable of predicting the product quality parameters as well the crystallizer scale and geometry requires a detailed description of the hydrodynamics and kinetics. This is simply a tall task to represent and/or simulate hydrodynamics of a system ranging six orders of magnitude in length scales accurately. Devising a detailed final design procedure is challenging, dependent to details of the crystallizer geometry. Consequently these procedures are slowly emerging in recent years. This relative lack of design tools compared to other separation techniques is a major drawback for wide spread use of crystallization. This is also the reason designing crystallizers are considered to be the most difficult design task in separation technologies. Currently, many iterations are required between basic and detailed design to pinpoint optimal design. The basic design procedure however provides a good first step to designing crystallizers.

8.6.1 Process and design specifications

The first step in basic design is to create a list of design specification. The design specifications can be either product specifications such as product quality parameters or process specifications such as yield, energy consumption and process stability. In basic process design, only a limited amount of this list can be addressed due to lack of kinetic and hydrodynamic information. For a basic design at least, the following design specifications need to be known (i) production capacity, (ii) average crystal size and (iii) purity. The production capacity dictates the design dimensions, mode of operation and energy requirements. The average crystal size influences the method of supersaturation along with thermodynamics and crystallizer type. Purity is a major selection criterion for choosing the method of supersaturation generation. These choice parameters provide the minimal information to start the design.

8.6.2 Thermodynamic and lab-scale kinetic information

Thermodynamic information is essential for designing a separation process. Pedagogically, it can be thought of a map describing which phases will be present under what conditions. Once the solvent and solute are known, thermodynamic information such as solubility curves, melting and boiling points as well as basic nucleation and growth rates can be either measured in a MSMPR crystallizer or determined from literature. One should remember that the nucleation and growth rates determined here would be first estimates, as they do not incorporate the final geometry details, hydrodynamics, that is, production scale kinetics.

For the basic design, the needed information is the following:

(i)   Thermodynamic properties: Phase diagrams of solute/solvent systems need to be identified. If the solute and solvent show different crystal forms under operation conditions, this needs to be identified and taken into account.

(ii) Physical solvent and solute properties: Their densities, heat capacity and temperature dependence of physical properties such as viscosity need to be identified. It is also critical to know the degradation conditions for the solute to avoid chemical degradation during process.

(iii) Feed composition information: Mass flow rate, solute concentration, temperature and pressure are essential for design. For instance, for systems that have large variations in mass flow rate, auxiliary buffer tanks need to be incorporated in the design.

(iv) Information of impurities: List of all potential impurities that can be incorporated in crystal and influence the crystal quality needs to be identified. The operation conditions or method of supersaturation generation need to be altered to avoid impurity incorporation if the product specs are focused on purity.

8.6.3 Design decisions

Once the design and product specifications as well as thermodynamic information are gathered, we can make decisions on the design. The most important choices made in the decisions stage are the following:
(i) The method of supersaturation generation
(ii) The mode of operation: Batch versus Continuous
(iii) The number of stages needed for operation
(iv) The type of crystallizer to be used

We will provide detailed discussions about decisions mentioned above in upcoming section.

8.6.3.1 Selecting method of supersaturation generation
The initial decision to choose a method of supersaturation generation also referred as method of crystallization is decided on the basis of equilibrium concentration as discussed in Section 8.5.7 and illustrated in Figure 8.18. After the initial decision based on equilibrium concentration, design and product specifications need to be considered. The process-specific design requirements or limitations of methods such as tendency to form scaling can alter initial choice. Particularly, relative energy consumption is a critical one in decision between cooling and evaporative crystallization. In addition to slope of solubility curve, the relative energy consumption defined as energy required to evaporate, E_e, a given amount of feed relative to energy required to cool it, E_c.

$$\frac{E_e}{E_c} = \frac{\Delta H_{evap}}{c_p} \left(\frac{1}{c^*} - 1 \right) \frac{dc^*}{dT} \qquad (8.49)$$

In eq. (8.49), c^* is the saturation concentration in units of kg/kg solution, c_p and ΔH_{evap} are the specific heat and evaporation enthalpy, respectively. For aqueous systems, heat of crystallization can be neglected in eq. (8.49). If the relative energy consumption is larger than one, cooling crystallization is favored. Below this threshold value, evaporative crystallization is favored. As a rule of thumb, the energy demand of evaporative crystallization is much higher than cooling crystallization as the solvent evaporation is more energy-intensive than cooling the solution.

Below we will discuss special conditions where the initial decision illustrated in Figure 8.18 can be overturned for each supersaturation generation method.

Cooling crystallization: The major reason not to use cooling crystallization is scaling or encrustation. Even if cooling crystallization is the more energy favorable option, in the presence of severe scaling vacuum crystallization, direct cooling, anti-solvent or evaporative crystallization emerge as good alternatives. As a rule of thumb, the heat flux through the exchanger must stay below 500 W/m² for inorganic compounds. Use of wipers or scrappers as auxiliary equipment might enhance this critical limit and make cooling crystallization the viable option. If considerable amounts of solute are still in solution after cooing crystallization, cooling crystallization can be followed by other methods such as anti-solvent crystallization.

Evaporative crystallization: Thermally unstable compounds should not be utilized in evaporative crystallization. In addition to thermal stability, energy consumption discussed above and relatively high operating cost are the major reasons to avoid evaporative crystallization. Most evaporative crystallization processes are operated at high temperatures, which bring about high operating cost unless an energy source is readily available. Efficiency of these crystallizers depends on the crystallizer geometry, particularly liquid–vapor interfacial area. Scaling and encrustation can also be major problems in this type of crystallizers due to enhanced supersaturation generation at the liquid–vapor and vapor–liquid–solid interface.

Precipitation: This method often produces low-quality crystals in terms of size distribution due to uncontrolled supersaturations characteristically emerging precipitation. Moreover, amorphous forms are also common in this method. As a rule of thumb, if the emphasis is on crystal qualities, precipitation should be avoided or extreme care in design is required to suppress challenges in crystal quality.

Anti-solvent crystallization: Anti-solvent crystallization is the best alternative for heat-sensitive compounds that cannot tolerate evaporative crystallization. It is also commonly used in high value products such as pharmaceutical industry to recover the solute remaining in solution after cooling crystallization. However, just like precipitation, anti-solvent crystallization suffers from very high supersaturations emerging during the mixing of the solvent and anti-solvent. Unless extreme care is taken in the design, these high supersaturation ratios can lead to low crystal quality.

Melt crystallization: It is mostly applied to react with very high purities say from 90% to 99.9%. Melting crystallization cannot be applied to heat sensitive materials and it is in general more expensive than cooling crystallization. This higher cost is due to larger heat exchanger areas required to melt the target compound, the need for auxiliary heating in feed tanks to avoid solidification and the need for post-process washing to reach high purities.

8.6.3.2 Operation modes: Batch versus Continuous operation

A crystallization process can be designed to operate either in batch or in continuous mode. This choice is often referred as the *mode of operation*. In batch crystallization, the crystallizer is initially filled with undersaturated solution. Thermodynamic state of the solution is changed from undersaturated to supersaturated with a method supersaturation generation such as cooling. The supersaturated solution is kept inside the crystallizer for a preestablished batch time. At the end of the process, the suspension containing the crystals and the solution is send to a solid–liquid separation process such as to isolate crystals. In continuous mode, undersaturated mother liquor (solution or liquid) is fed to a crystallizer continuously. The product is formed as dispersed phase and the crystals are kept suspended via continuous stirring. The suspension containing product crystals and saturated solution leaves the reactor continuously, as shown in Figure 8.20. The suspension leaving the continuous crystallizer is also called slurry. The slurry is then carried out to another solid–liquid handling process such as filtration or centrifuge that separates the crystals from solution. For both batch and continuous modes, the crystals are often washed, as crystallization is often the last process before the product is shipped out.

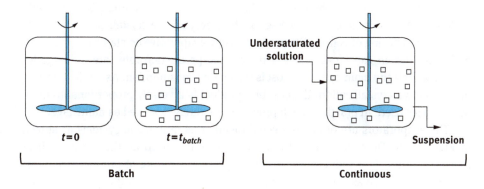

Figure 8.20: Illustration of batch and continuous processes.

Continuous operation utilizes the equipment, utilities and raw materials more efficiently compared to batch processing. This leads to lower manufacturing costs per kilogram product for continuous manufacturing. Batch manufacturing is susceptible to batch-to-batch variations leading to less uniform product quality. This lower crystal quality in batch processing propagates to downstream processing problems, increasing the manufacturing cost. Overall, continuous processing is often the preferred mode of operation. Yet, there are cases where continuous manufacturing is not feasible.

For systems resulting in severe scaling, that is, formation of unwanted crystals on confining walls, continuous operation may be constantly interrupted leading to increased down time. Such systems are produced in batch so that the scaling can be removed with repeated washes at the end of each wash. For processes with very low production capacity (less than one meter cube/day), the design of continuous crystallizer to ensuring the crystals are suspended may be challenging due to low flow rates. Important advantages of batch operation are the simplicity of equipment, ease of equipment cleaning at the end of each cycle to remove scaling and avoid contamination.

Another critical factor influencing the decision between continuous or batch operation is the production capacity. If the production capacity exceeds 50 tons per day, continuous operation mode is preferred. For processes with severe scaling tendency or small production rates, batch operation mode is preferred. As a rule of thumb, cooling and vacuum crystallization is operated batch wise, whereas for evaporative crystallization, continuous mode is the preferred mode of operation. Operating evaporative crystallization in batch mode needs large capital investment costs due to large vapor head needed, consequently continuous mode is preferred for evaporative crystallization.

8.6.3.3 Number of stages: single- and multistage operation selection

A continuous crystallization process can be designed either as a single-stage or multistage operation based on the cost calculations. Multistage operation is particularly preferred for energy-intensive processes such as evaporative crystallization. In large-scale manufacturing of commodity products using evaporative crystallization process, multistage design can be attractive if the desired production rates are too large for one vessel or if the main operative cost is the energy consumption. As a rule of thumb in evaporative crystallization, the cost of a product is inversely proportional to number of stages used. In practice, batch processes are rarely designed as multistage.

The evaporators are often connected together as multistage systems as shown in Figure 8.21. Here, the vapor from the first unit is used as the steam to heat the second unit. This reduces the steam requirements per kg of product. The pressure must be varied as shown in the figure so that the condensing vapor will be at a temperature greater than the boiling temperature in the second stage. A variety of ways connecting the different stages have been developed.

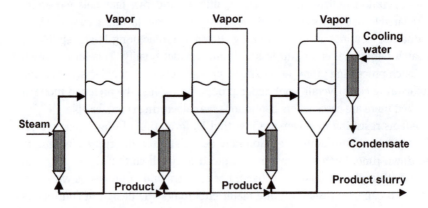

Figure 8.21: Multiple-effect evaporating crystallizer cascade for ammonium sulfate.

8.6.3.4 Selecting the crystallizer type

The last basic design decision to be made is choosing the crystallizer type. The choice of crystallizer type is done considering all the product and design specifications as well as thermodynamic and lab-scale kinetic information collected. Within this set of information, the most critical pieces determining the crystallizer type are the production capacity, method of supersaturation generation and desired crystal quality parameters, particularly, crystal size and purity. A considerable number of industrially accepted crystallizer types are available and their design will be discussed in detail. Moreover, these industrial crystallizer types can be further fitted with auxiliary equipment to match desired design specifications.

Depending on the supersaturation generation method, heat can be removed or added to the crystallizer directly or indirectly. In *direct cooling/heating,* the solution is brought in contact with an external heat transfer surface such as coils immersed in the crystallizer or the crystallizer walls can be used as heat transfer surfaces, a jacketed setup. Direct cooling/heating crystallizers have an internal heat exchange unit. In *indirect cooling/heating*, a fluid or gas is injected to cool/heat the crystallizer contents. For solute/solvent systems prone to crusting, indirect cooling/heating is preferred. In next sections, we will discuss various crystallizer types namely, stirred tank and disk crystallizers, forced circulation crystallizers, draft tube crystallizer and fluidized bed crystallizers as well as melt and vacuum crystallizers. It is important to note that this is only a limited sample out of wide range selections available to process engineers.

8.6.3.5 Stirred tank and cooling disk crystallizer

The most widely used cooling crystallizer is the *stirred tank crystallizer* with a cooling jacket due to its simple construction. It is the go-to type for small-scale manufacturing and lab experiments. It can be used with all methods of supersaturation generation. For systems showing intense crusting behavior where the crystals stick and decrease, the heat transfer efficiency is over time. To avoid this issue, crystallizers with large flat surfaces periodically cleaned with scrapper are used. As it is easier to scrap a flat surfaces, *cooling disk crystallizer* is utilized for such systems. Disk cooling crystallizers use two circular disks as heat transfer surface and constantly scrap these surfaces with wipers to counteract the crust formation. Both stirred tank and cooling disk crystallizers can be used in batch and continuous mode. Stirred tank crystallizers are utilized in cooling, evaporative and vacuum crystallization approaches. Disk crystallizers are most commonly used in cooling crystallization. The stirred tank crystallizer may be as simple as a shallow open pan heated by an open fire. Steam-heated crystallizers are widely used to produce common salt from brine and sugar refining. These applications often use an evaporative crystallizer containing *calandria* (steam chest[3], see also Figure 2.11). The magma (mixture of crystals and solution) circulates by dropping through the central downcomer and then rises as it is heated in the calandria. At the top, some of the solution evaporates increasing the supersaturation causing crystal growth.

---

3 shell-and-tube unit used as a reboiler for evaporation (or distillation)

8.6.3.6 Forced circulation crystallizers

For large-scale production, more efficient heat transfer is required to match high volume production rates. The *forced circulation crystallizer* is a simple and highly effective unit designed to provide high heat-transfer coefficients in evaporative, cooling or vacuum mode. They are commonly used with indirect heat transfer and crystallization units, so that the heat transfer rates are kept high in an external heat exchanger. A schematic diagram in which slurry is withdrawn from the crystallizer body and pumped through an external heat exchanger is shown in Figure 8.22. Depending on the heat exchanger function, the forced circulation crystallizer can be operated in cooling mode or evaporation mode. Most systems are operated with a high recirculation rate to provide good mixing inside the crystallizer and high rates of heat transfer to minimize encrustation. Moreover, the inlet to the crystallizer is often placed tangential to promote formation of swirls as shown in Figure 8.22. These swirls promote mixing consequently increase heat/mass transfer, counteract encrustation and keep the crystals suspended. The tendency of the solute to form encrustations on the cooling surface often limits their operation by restricting the temperature of the cooling liquid and the temperature decrease of the slurry flowing through the heat exchanger. High heat-transfer rates reduce the formation of encrustations considerably by minimizing the temperature difference over the heat transfer surface. The feed is commonly introduced into the circulation loop to provide rapid mixing and minimize the occurrence of regions of high supersaturation, which can lead to excessive nucleation.

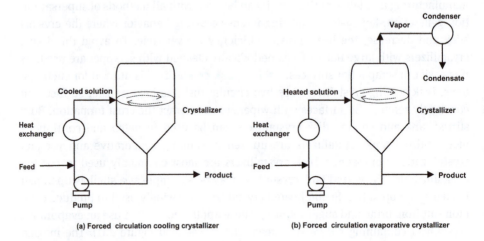

Figure 8.22: Schematic of (a) forced circulation cooling crystallizer and (b) force circulation evaporative crystallization with external heat exchanger.

The use of an external heat exchanger can be avoided by employing *direct-contact cooling* (DCC) in which the product liquor is allowed to come into contact with a cold heat-transfer medium. Other potential advantages include better heat transfer and smaller cooling load. However, problems include product contamination from the coolant and the cost of extra processing required recovering the coolant for further use. Crystallization processes employing DCC have been used successfully in recent years for dewaxing of lubricating oils, desalination of water and production of inorganic salts from aqueous solutions.

Evaporative-forced circulation crystallizers supersaturate the solution by removing solvent through evaporation as shown in Figure 8.22b. These crystallizers are used when temperature has little effect on solubility (e.g. NaCl) or with inverted solubilities (e.g. calcium acetate). On large-scale, also many types of *forced circulation evaporating crystallizers* are used. Operational problems with evaporative crystallizers can be caused by scale formation on the heat exchanger surfaces or at the vapor–liquid interface in the crystallizer. Such problems can be overcome by not allowing vaporization or excessive temperatures within the exchanger and by proper introduction of the circulating magma into the crystallizer. The latter may be accomplished by introducing the magma below the surface of the magma in the crystallizer and tangentially. This configuration introduces swirls that increase the evaporation area while keeping the crystals suspended.

8.6.3.7 Draft tube baffle crystallizer

Figure 8.23a shows a *draft-tube-baffle (DTB) crystallizer* designed to provide preferential removal of both fines and classified product. A relatively low speed propeller agitator is located in a draft tube, which extends to a few inches below the liquor level in the crystallizer. The steady movement of magma up to the surface produces a gentle, uniform boiling action over the whole cross-sectional area of the crystallizer. Between the baffle and the outside wall of the crystallizer, agitation effects are absent. This provides a *settling zone* that permits regulation of the magma density and control of the removal of excess nuclei. Flow through the annular zone can be adjusted to only remove crystals below a certain size and dissolve them in the *fines dissolution exchanger*. Feed is introduced to the fines circulation line to dissolve nuclei resulting from feed introduction.

8.6.3.8 Fluidized bed crystallizers

Another type of continuous crystallizer is the *Oslo fluidized-bed crystallizer* shown in Figure 8.23b. In units of this type, a bed of crystals is suspended in the vessel by the upward flow of supersaturated liquor in the annular region surrounding a central downcomer. The objective is to form a supersaturated solution in the upper chamber and then relieve the supersaturation through growth in the lower chamber. The use of the downflow pipe in the crystallizer provides good mixing in the growth chamber.

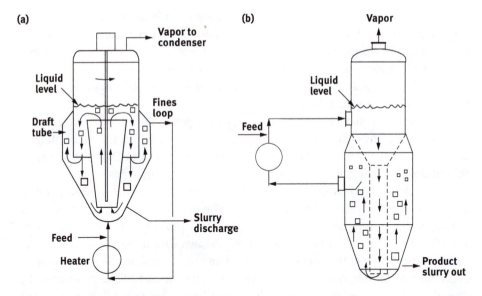

Figure 8.23: (a) Draft-tube-baffled crystallizer and (b) Oslo fluidized bed crystallizer. The arrows show the direction of fluid flow.

CSDs produced in a perfectly mixed continuous crystallizer are highly constrained. The form of the CSD in such systems is entirely determined by the residence time distribution of a perfectly mixed crystallizer. Greater flexibility can be obtained through introduction of selective removal devices that alter the residence time distribution of materials flowing from the crystallizer. *Clear-liquor advance* is simply the removal of mother liquor from the crystallizer without simultaneous removal of crystals. The primary objective of *classified-fines removal* is preferential withdrawal of crystals whose size is below some specified value. A simple method for implementation of classified-fines removal is to remove slurry from a settling zone in the crystallizer. Constructing a baffle that separates the zone from the well-mixed region of the vessel can create the settling zone. The separation of crystals in the settling zone is based on the dependence of settling velocity on crystal size. Such crystals may be redissolved and the resulting solution returned to the crystallizer. *Classified-product removal* is carried out to remove preferentially those crystals whose size is larger than some specified value.

8.6.3.9 Melt crystallizers

Melt crystallization is the process where crystalline material is separated from a melt of the crystallizing species by cooling. Melt crystallization can be conducted via multiple countercurrent contacting stages somewhat analogous to distillation. The main incentive is to attain higher purity product than can be achieved in a single stage of conventional crystallization. The concept of a *column crystallizer* is to form a crystal

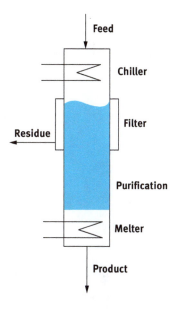

Figure 8.24: Vertical column melt crystallizer.

suspension and to force the solids to flow counter currently against a stream of enriched reflux liquid by gravity or rotating blades. At the end of the crystallizer, the crystals are melted. A portion of the melt is removed as product and the remainder is returned to the system as enriched reflux to wash the product crystals. One of the early column crystallizers, shown schematically in Figure 8.24, was developed for the separation of xylene isomers. In this unit, p-xylene crystals are formed in a scraped-surface chiller above the column and fed to the column. The crystals move downward counter currently relative to impure liquid in the upper portion of the column and melted p-xylene in the lower part of the column. Impure liquor is withdrawn from an appropriate point near the top of the column of crystals while pure product, p-xylene, is removed from the bottom of the column. An inherent limitation of column crystallization is the difficulty of controlling the solid-phase movement, because the similar phase densities make gravitational separation difficult.

The Sulzer MWB[4] system, schematically depicted in Figure 8.25, is an example of a commercial melt crystallization process that uses the gradual deposition of solids on a chilled surface. Crystal growth is on the inside of a battery of tubes through which melt is flowing. During crystallization, the front of crystals advances into the direction of the mother liquor. This built up of a solids layer requires sequential operation. Steps include partial freezing of a falling film of melt inside vertical tubes, followed by slight heating, a "sweating" operation and complete melting and recovery

---

4 Metallwerk Bruchs

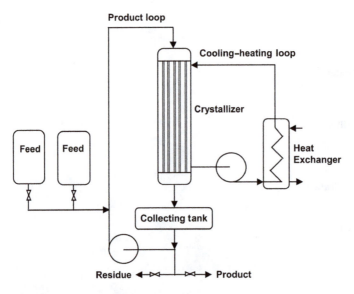

Figure 8.25: Sulzer MWB melt crystallizer system.

of the refined product. The recovered product melt can be put through the cycle again to increase purity or fresh feed can be introduced to the cycle. The process has been used on a large scale in the purification of a wide range of organic substances.

8.6.3.10 Vacuum crystallizers

Vacuum crystallizers utilize evaporation to both concentrate the solution and cool the mixture. They combine the operating principles of evaporative and cooling crystallizers. The hot liquid feed enters the crystallizer at a higher pressure and temperature, where it flashes and part of the feed evaporates. This flashing causes adiabatic cooling of the liquid. Crystals and mother liquor exit the bottom of the crystallizer. Because of the vacuum equipment required to operate at pressures of 5 to 20 mbar, the vacuum crystallizers are considerably more complex than other types of crystallizers and most common in large systems.

Beyond the aforementioned widely used crystallizers, there are several other crystallizers developed for specific purposes such as membrane-assisted crystallizers, oscillatory draft tubes and microfluidic crystallizers. Membrane-assisted crystallizers aim to control evaporation rate through a membrane consequently dictate the supersaturation in evaporative crystallization. Oscillatory crystallizers utilize structures such as draft tubes and baffles combined with oscillatory flows to enhance mixing hence enhance growth rates. Microfluidic crystallizers can provide enhanced control over heat and mass transfer within the crystallizer at the expense of production rates. With parallelization, microfluidics can improve intrinsically low manufacturing rate, however, industrial practice they have been limited to high value per weight compounds.

8.7 Crystallizer calculations

When one is purely interested in the yield of a certain crystallization process, solving the mass balances for the solvent, solutes and the solids will be sufficient. Solubility and phase relationships influence the choice of crystallizer and method of operation. Although equilibrium calculations are very important, they alone are not sufficient for a complete crystallizer design as for other equilibrium-staged separations. Equilibrium calculations can tell the total mass of crystals produced, but cannot tell the size and the number of crystals produced. However, as soon as the CSD of the product becomes of interest, nucleation and crystal growth rates must be incorporated together with population balances. Nucleation leads to the formation of crystals. Growth is the enlargement of crystals caused by deposition of solid material on an existing surface. The relative rates at which nucleation and growth occur determine the CSD. Acceptable operating conditions for the minimization of uncontrolled nucleation and encrustation of heat-exchange surfaces are defined by metastable limits. In this final section, we will cover mass and enthalpy balances as well as population balance equations mixed solids with product crystallizer.

8.7.1 Mass balance

Crystallization is a physical transformation, consequently the identity of molecular species involved do not change only their phase changes. Therefore, total and individual mass balances for each compound should consider both phases in this binary system. Similar to absorption (Chapter 3) and extraction (Chapter 5), it is common in crystallization to use *solute-free solvent flow rates and solute weight ratios* in calculations. In this section, we limit ourselves to the formation of pure crystals that contain no solvent. Mass balances concerning solvated (hydrated) crystals, though more complex; follow the same line of reasoning.

For a system wherein a pure solid component is crystallized by cooling or evaporation, the overall mass balance provides the maximum solute yield and crystal production rate. From the overall material balance, it follows that:

solute free solvent in + dissolved solute in = evaporated solvent out
+ solute free solvent out + dissolved solute out + crystals out

$$(8.50)$$

$$F' + F'X_F = V' + S' + S'X_C + C \qquad (8.51)$$

The meaning of the quantities F', S', V', C, X_F and X_C are illustrated in Figure 8.26 and defined as:

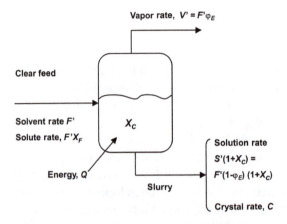

Figure 8.26: Mass balance for a binary crystallization system.

F' = mass feed rate of solute free solvent entering the crystallizer (kg solvent/hr)
V' = mass flow rate of solvent vapor leaving the crystallizer (kg solvent vapor/hr)
S' = mass flow rate of solute free solvent leaving the crystallizer (kg solvent/hr)
C = Crystal removal rate (kg crystal /hr)
X_F = weight ratio of dissolved solute in the feed per unit mass of solvent (kg solute/kg solvent)
X_C = weight ratio of dissolved solute in solvent discharged from the crystallizer per unit mass of discharged solvent (kg solute/kg solvent)

The vapor rate and solute-free solvent out are obtained from a mass balance over the solvent and become:

$$V' = F'\phi_E \qquad S' = F'(1 - \phi_E) \tag{8.52}$$

in which ϕ_E is the ratio of the solvent mass evaporated per unit mass of solvent fed.

The preceding expressions can be solved provided the composition of the exit stream is known. In many instances, the exit stream composition corresponds to saturation conditions.

8.7.2 Basic yield calculations

For crystallization, a high recovery of refined solute is generally the desired design objective. The theoretical maximum product recovery or yield from the crystallizer is then defined by:

$$Yield = \frac{F'X_F - S'X_C}{F'X_F} = \frac{C}{F'X_F} = 1 - \frac{S'X_C}{F'X_F} \tag{8.53}$$

The solute-free solvent feed rate F' and the weight ratio of dissolved solute to solvent in the feed X_F are usually fixed by upstream requirements. The weight ratio of dissolved solute in the liquid leaving the crystallizer X_C is determined by solubility, which is set by the temperature of the operation. The other adjustable variable is the solute-free solvent discharge rate S'. In evaporative systems, adding heat to the system evaporates a fraction ϕ_E of the solvent. Thus, S' decreases and recovery increases as more solvent is evaporated and removed as a separate product stream. The conclusion from the preceding discussion is that for a given feed concentration and solubility relationship, *the mode of crystallizer operation governs the maximum recovery of solute*. For a cooling crystallization system, where no solvent is removed, the only control that affects maximum solute recovery is crystallizer temperature. For evaporative systems, recovery is influenced by both the system temperature and by the quantity of solvent vaporized and removed from the system.

Introduction in eq. (8.52) provides the *maximum obtainable yield*:

$$Y_{max} = 1 - \frac{X_C}{X_F}(1 - \phi_E) \tag{8.54}$$

For specific operating modes, this simplifies to:

$$\text{Cooling crystallization only} \quad \phi_E = 0 \quad \Rightarrow \quad Y_{MAX} = 1 - \frac{X_C}{X_F} \tag{8.55}$$

$$\text{Evaporation only} \quad X_C = X_F \quad \Rightarrow \quad Y_{MAX} = \phi_E \tag{8.56}$$

Substitution of eq. (8.52) into eq. (8.51) and rewriting gives the *crystal production rate C*:

$$C = F'[X_F - X_C(1 - \phi_E)] \tag{8.57}$$

8.7.3 Enthalpy balance

In evaporative crystallization, heat needs to be introduced to the crystallizer to increase the temperature of the solution in the crystallizer to boiling point and consequently evaporate the solvent. In addition to this, crystallization is an exothermic process. Enthalpy of crystallization is released to the surroundings and heats up the crystallizer contents. In a stationary fluid, that is, not actively mixed, the heat that should be added to the system is given by an enthalpy balance:

$$Q = F\rho_L c_p \Delta T + V\rho_v \Delta H_v - C\Delta H_{cr} \tag{8.58}$$

where the ρ_L and ρ_v are the density of liquid and vapor phases. F, V and C are the mass flow rates of feed, vapor phase and crystal phase. The symbols $c_p, \Delta H_{cr}$ and ΔH_v represent the heat capacity of clear fluid, enthalpy of crystallization and evaporation enthalpy, respectively. Enthalpy and mass balances are applied for the design of all crystallizer types and operation modes. Solving these equations produce complete flow sheet including the heating and cooling duties.

8.7.4 Crystallizer dimensions

Basic design of crystallizers can be made from the suspension volume required to fulfill desired production capacity. If the volume of the suspension containing the crystals and saturated solvent, V_{susp} is known, the crystallizer diameter, D_{CR}, in a cylindrical vessel can be calculated as:

$$D_{CR} = \sqrt[3]{\frac{4V_{susp}R_{HD}}{\pi}} \tag{8.59}$$

where the R_{HD} is the height to diameter ratio dictated by the crystallizer construction materials. This value is a first estimate and it does not take into account considerations such as liquid entrainment, skit baffles and auxiliary equipment.

For evaporative crystallization, the limiting factor in design is the cross-sectional area where evaporation occurs. The cross-sectional area, $\pi D_{CR}^2/4$, can be tuned by altering the V_{susp} and R_{HD} is equation above. In addition to optimizing the cross-sectional area, the vapor flow rate has to be optimized in evaporative crystallizers to minimize entrainment of liquid droplets to the condenser. If the vapor flow rate is too large, liquid droplets will be carried to the condenser and block the condenser or decrease the efficiency of heat transfer surfaces. To avoid this, constitutive relations have been proposed. The most commonly used one is that the linear vapor velocity, v_v, should not exceed 0.025 m/s. The diameter of the crystallizer minimizing the liquid entrainment, D_v, can be calculated inserting $v_v = 0.025$ into the equation below.

$$D_v = \sqrt{\frac{4V}{\pi v_v \left(\frac{\rho_L - \rho_v}{\rho_v}\right)}} \tag{8.60}$$

where V is the vapor flow rate, ρ_L and ρ_v are the density of liquid and vapor phases. While designing an evaporative crystallizer, both D_v and D_{CR} should be evaluated and the larger one should be taken to carry out the design.

8.8 Final design procedure

The final design procedure involves the optimization of the basic design considering the economic evaluation and pilot-scale kinetic data.

8.8.1 Economic evaluation

The basic idea of economic evaluation is to maximize the profit by minimizing the
total cost with respect to design and operation costs for a given set of process and
product specifications. In this process, the total profit function is first formulated as a
function of design variables such as crystallizer diameter, mass flow rates of input
and output streams then optimized by altering the design parameters. The total profit
function also referred as economic object function consists of capital investment and
operation costs along with economic material balance. Capital investment costs include the construction cost of crystallizer and auxiliary equipment. Operational costs
are the costs associated with operating the process such as heating/cooling costs,
electricity costs for pumps, stirrers and other equipment. The economic material balance considers the input and output stream economic values.

Capital investment costs, ψ_{capital}, are a strong function of the diameter of the
crystallizer D_{CR}. Once the crystallizer diameter is calculated from eq. (8.59), the capital costs can be calculated with correlations such as the one given below.

$$\psi_{\text{capital}} = \frac{1}{T_{\text{dep}}} 20.875 D_{\text{CR}}^2 - 7.776 D_{\text{CR}} + 38.25 \qquad (8.61)$$

where T_{dep} is the depreciation value for the investment.

The most important operational costs ψ_{OP} are the impeller, pump and heat exchanger duties. The operational cost function is calculated considering the duties
of impellers W_{stirrer}, pumps W_{circ} and heat exchangers W_{HEX}, that is, how much electricity they require to operate, cost of electricity in Euros per kWh, C_{elec} and the operation time, t_{op} describing how many hours that these devices operate per year.

$$\psi_{\text{op}} = [W_{\text{stirrer}} + W_{\text{circ}} + W_{\text{HEX}}] C_{\text{elec}} t_{\text{op}} \qquad (8.62)$$

The economic material balance takes into account the economic value of the input
and output streams. The material balance profit function ψ_{mb} reflects the total
profit gained by operating the crystallizer and producing crystalline products. The
feed stream is considered a cost while the product stream is a gain. The vapor
stream produced in evaporative crystallization is considered a profit as it produces
pure solvent. Considering the monetary values of streams in euros per kilogram,
the mass flow rates in kilogram per hours and the operation time t_{on} in hours
per year, the material balance profit function can be written as

$$\psi_{mb} = [C_{product}C + C_{vapor}V - C_{feed}F]t_{on} \qquad (8.63)$$

where the monetary values of product, vapor and feed streams are denoted as C_p, C_{vapor} and C_{feed}, the mass flow rates of product, vapor and feed streams are indicated as C, V and F.

The overall profit function is written as

$$\psi_{profit} = \psi_{mb} - \psi_{capital} - \psi_{oper} \qquad (8.64)$$

Writing the overall profit function in terms of design parameters allows systematic optimization of the profit. For instance, one may think simply increasing the feed flow rate would increase the profit of a process linearly. However, the reality is more complex and overall profit function allows us to quantify how much increasing the feed flow rates would contribute to profit. As the feed flow rate increases, more products are produced so the process becomes profitable. Yet increasing the feed flow rate also increases the diameter of the crystallizer and heat duties of heat exchangers, stirrers and pumps. Both operational and capital costs are increased. Economic evaluation and an overall profit function allow us to quantify how much the profit will increase by increasing the feed flow rate. This is a critical step in making final decisions.

8.8.2 Detailed design with production scale kinetic data

In the final design, the most challenging stage is to scale up the lab-scale process to industrial scale. Fundamental challenge is product quality parameters such as CSD, morphology intimately depend on the exact crystallizer geometry, scale and operating conditions. In essence, the product quality is determined by rates at which a crystal is nucleated, attrited, grow, dissolve and agglomerate dictated by local environment in crystallizers. The local hydrodynamic environment in lab scale and industrial scale is significantly different. In other words, the kinetic information acquired from lab-scale information such as metastable, nucleation, growth, agglomeration and attrition rates does not correspond to kinetic information at industrial scale. This can be easily rationalized thinking about the hydrodynamics stresses a particle experiences. At laboratory scale, a crystal particle experiences milder shear rates due to low stirring speeds corresponding to low-to-intermediate Reynolds number flows. However, at industrial scale due to large-scale operation, a crystal particle experiences much larger stresses in turbulent flows. In industrial scale, the high shear rates created by highly turbulent flows break the particle more efficiently, leading to higher attrition and secondary nucleation rates.

In the detailed designed part, scale-independent kinetic information is coupled to detail hydrodynamic modeling of industrial crystallizers. Despite the difficulties in getting scale independent kinetic information from lab-scale and pilot-scale experiments, considerable progress has been made on this front with the development of large-scale hydrodynamic simulations and better process analytical tools (PAT). The final design stage is an iterative approach. The detailed design of crystallizer parts such as stirrer geometry and its speed is optimized to reach target-quality parameters.

Nomenclature

B^0	Homogeneous primary & secondary nucleation rate,	nuclei/unit time/unit volume
cv	coefficient of distribution	–
$c*$	Equilibrium saturation concentration	mol m^{-3}
C	Crystal production rate	kg/unit time
F'	Mass feed rate, solute free	kg/unit time
G	Crystal growth rate	m / unit time
ΔG	Excess free energy	J mol^{-1}
ΔH_{melt}	Enthalpy of melting	J mol^{-1}
L_D	(dominant) crystal size	m
m	Population density in terms of mass	kg/unit volume
M_T	Total crystal mass (eq. (8.47))	kg crystals/unit volume suspension] *or* total crystal mass (Exercise 8.10) kg crystals/unit volume solvent
n	Population density, eq. (8.37)	number of nuclei m^{-1} m^{-3}
n_0	Population density of nuclei ($L = 0$)	number of nuclei m^{-1} m^{-3}
N	Nuclei concentration, eq. (8.36)	number of nuclei/unit volume
Q	Volumetric discharge rate	m^3/unit time
r	Radius	m
S	*Supersaturation ratio*	–
σ	Relative supersaturation	–
Δc	Absolute supersaturation	mol, kg/kg solvent, kg/kg solution...
S'	Mass discharge rate of solvent, solute free	kg/unit time
$T (T_{melt})$	Temperature (at melting point)	K
V'	Vapor rate	kg/unit time
V_m	Molar volume	m^3 mol^{-1}
V_M	Volume of magma (suspension or clear liquor)	m^3
x_S	Solubility of solids, mole fraction	–
X	Weight fraction	–
Y	Yield	–
γ	Interfacial tension	N m^{-1}
γ_s	Activity coefficient	–

v	Number of moles ions in one mole of electrolyte	–
σ_D	Standard deviation of distribution	m
τ	Residence time in crystallizer	unit time
ϕ_E	Unit solvent evaporated per unit solvent fed	–
ϕ_T	Volumetric holdup of crystals	unit volume crystals / unitvolume suspension

Exercises

1 Determine the maximum yield of anhydrous sugar crystals deposited at equilibrium in the following situations. Initially, 100 kg of water is present.
   a. A saturated sugar solution is cooled from 80 °C to 20 °C.
   b. A saturated sugar solution at 80 °C has half of its water evaporated and the solution is then cooled to 20 °C

Temperature:	20 °C	80 °C	
Solubility:	2.04	3.62	(kg anhydrous sugar/kg water)

2 A solution of 80 wt% of naphthalene in benzene is cooled to 30 °C to precipitate naphthalene crystals. The solubility of naphthalene at this temperature is 45 wt% in solution. The feed rate is 5,000 kg solution/h.
   a. Calculate the maximum production rate of crystals.
   b. Calculate the maximum yield in kg crystals per unit weight of solute fed.

3 A cooling crystallizer is used to crystallize sodium acetate, $NaC_2H_3O_2$ (MW = 0.082 kg/mol), from a saturated aqueous solution. The stable form of the crystals is a hydrate with 3 moles of water. The feed is initially saturated at 40 °C, while the crystallizer is cooled to 0 °C until equilibrium conditions are established.

   If we initially dissolved the anhydrous salt in 100 kg of water/h, how many kg of crystals/h are collected?

Temperature:	0	10	20	30	40	60	(°C)
Solubility:	0.363	0.408	0.465	0.545	0.655	1.39	(kg anhydrous salt/kg water)

4 Copper sulfate is crystallized as $CuSO_4.5H_2O$ by combined evaporative/cooling crystallization. 1,000 kg/h of water is mixed with 280 kg/h of anhydrous copper sulfate (MW = 0.160 kg/mol) at 40 °C. The solution is cooled to 10 °C and 38 kg/h of water is evaporated in the process. How many kg/h of crystals can be collected theoretically?

Temperature:	10 °C	40 °C	
Solubility:	17.4	28.5	(g anhydrous salt/100 g water)

5 Calculate the maximum (theoretical) yield of pure crystals that could be obtained from a saturated solution of sodium sulfate with 0.25 kg anhydrous salt/kg free water by:

a. Isothermal evaporation of 25% of the free water.

b. Isothermal evaporation of 25% of the original water.

c. Cooling to 10 °C and assuming 2% of the original water lost by evaporation; $X_C = 0.090$ kg anhydrous salt/kg solute-free solvent.

d. Adding 0.75 kg ethanol/kg free water; $X_C = 0.030$ kg anhydrous salt/kg solute-free solvent.

e. Adding 0.75 kg ethanol/kg free water; $X_C = 0.068$ kg $Na_2SO_4 \cdot 10aq$/kg solute-free solvent $MW(Na_2SO_4.0aq) = 0.142$ kg/mol.

6 Assuming that $G \propto \Delta c$, determine whether or not the Arrhenius equation is valid for $(NH_4)_2SO_4$ crystallization. If it is valid, determine ΔE.

Note: c^* depends on T and ammonium sulfate crystallizes as an anhydrate.

Given				
Temperature:	30	60	90	°C
Solubility:	78	88	99	g anhydrous salt/100 g water
s:	0,05	0,05	0,01	–
G:	5,0	8,0	0,6	10^{-7} m/s

7 Estimate the supersaturation s of an aqueous solution of K_2SO_4 at 30 °C required to get a heterogeneous nucleation rate of 1 nucleus per second per unit volume. Equation (8.20) is applicable with an apparent interfacial surface tension y of $2 \cdot 10^{-3}$ Pa.s and a preexponential coefficient $A = 10^{25}$ nuclei per second per unit volume. The density of the K_2SO_4 crystals is 2,662 kg/m³, its molecular weight 0.174 kg/mol.

8 Potash alum was crystallized from its aqueous solution in a laboratory scale continuous MSMPR crystallizer of 10 L capacity at steady state, supersaturation being achieved by cooling. The size analysis of the produced crystals resulting from a steady-state sample taken from the crystallizer operated at means residence time of 900 s is given in the table below. Assume that the crystal density is 1,770 kg/m³ and the volume shape factor is 0.47.

Sieve Analysis Data:	Standard sieve Size (μm)	Weight on the sieve (g/kg of water)
	850	–
	710	2.3
	500	16.3
	355	20.9
	250	28.9
	180	18.0
	125	10.7
	90	3.7
	63	1.5
	45	0.6
	< 45	0.2

The steps to be taken are:
a. Calculate the population density distribution of the product crystals at mean size L over a size interval ΔL.
b. Determine the growth and nucleation rate graphically from the population density plot on a semilog scale.

9 A sodium acetate crystallizer receives a saturated solution at 60 °C. This solution is seeded with 0.2 mm average size crystals and cooled to 10 °C. Use a basis of 100 kg of entering water to which 0.25 kg seeds is added. The sodium acetate crystallizes as the trihydrate.
a. What is the mean size of the product crystals L_p?
b. If the residence time is 21 hours, what linear growth rate G is required?

Temperature: 10	60 (°C)	
Solubility:	40,8 139	(g anhydrous salt/100 g water)

10 Design a crystallizer to produce potassium sulfate crystals assuming an MSMPR unit operating continuously at steady state. The following specifications and data are available:

relative nucleation kinetics:	$B = 1 \times 10^{19} \, M_T \, G^2$ (no/s kg water)
production rate:	$P = 1,000$ kg/h
product dominant size:	$L_D = 490 \, \mu m$
magma concentration:	$M_T = 0.100$ kg/kg water
crystal density:	$\rho_c = 2,660$ kg/m^3
shape factors:	$k_v = 0.525$
outlet concentration:	$C_{out} = 0.1167$ kg/kg water
density of product solution:	$\rho_{soln} = 1,090$ kg/m^3

11 The solubilities of three commercial salts in water are given in the table below (in gram/100 gram water).

Salt	20 °C	40 °C	60 °C	100 °C
1	20	23	26	29
2	0.033	0.040	0.046	0.052
3	11	16	29	69

What is the most suitable method of supersaturation generation for each salt?

12 Molasses contains calcium which is being removed by precipitation. The feed amounts 15 m3/hr molasses with a density of 1,400 kg/m3 and contains 0.5 wt% Ca^{2+} (MW = 40 g/mol) of which 98% needs to be removed by precipitation as $CaSO_4$ (MW = 136 g/mol, density = 2,600 kg/m3) with concentrated sulfuric acid (H_2SO_4, MW = 98 g/mol). The solubility product of $CaSO_4$ is given by:

$$K_S^{CaSO_4} = [Ca^{2+}][SO_4^{2-}] = 5 \times 10^{-5} \frac{mol^2}{kg^2}$$

a. Calculate the concentration $[SO_4^{2-}]$ required to achieve the desired Ca^{2+} removal?

b. How much (kg/hr) sulfuric acid needs to be dosed?

c. Calculate the resulting slurry concentration (vol%) of $CaSO_4$ particles?

13 A crystallization plant produces 250.000 ton dry salt per year. The feed consists of saturated brine with a concentration of 300 kg/m3 NaCl (MW = 58 kg/kmol, dichtheid 2,170 kg/m3) and a density of 1,200 kg/m3. This saturation concentration can be considered as temperature independent. Salt is crystallized from the saturated brine in a three-step crystallizer in which 20% of the water from the initial feed stream is evaporated in each step. The crystallization plant operates 8,000 hr/year.

a. Calculate the yield of the crystallization plant?

b. Calculate the crystal production rate C (kg/hr)?

c. Determine the brine feed stream (kg/hr)?

We now consider the first stage of the crystallization plant, which can be considered a MSMPR-crystallizer. The nucleation rate Bo in number of nuclei per unit of time and per m^3 solution depends on the growth rate G and the Magma concentration M_T (kg/m^3 solution) as follows:

$Bo = 1.6 \cdot 10^{18} \cdot M_T G^2$

The residence time of the slightly supersaturated solution in the first stage needs to be selected such that the dominant crystal size equals 0.5 mm. Volume form factor kv = 0.56. NaCl crystallizes an anhydrate (0 aq).

d. Calculate the required residence time τ in the crystallizer?

e. Calculate the needed slurry filled volume of the first stage to obtain the required residence?

f. Time.

g. Calculate the volume of the crystals in the first-stage crystallizer?

Chapter 9
Sedimentation and Settling

9.1 Introduction

In industrial separations, sedimentation or settling is referred as the separation of solid particles dispersed in a liquid or an immiscible or partially miscible liquid phase dispersed in another liquid with action of physical forces. Solid particles suspended in a liquid are called suspensions, whereas liquid droplets dispersed in a liquid are called emulsions. The separation process of two phases (solid–liquid or liquid–liquid) is designed according to which phase(s) are considered valuable as a product or intermediate for future processing.

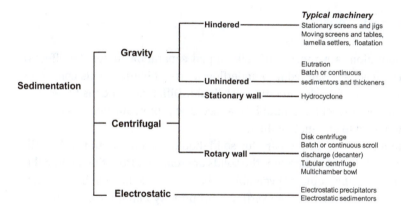

Figure 9.1: General classification of sedimentation equipment with respect to driving force and typical machinery used corresponding to each classification category.

Sedimentation exploits *the density difference* between two phases for the partial separation or concentration leveraging gravity or centrifugal forces. Electrostatic separation leverages *the induced charge to separate particles in air* with electrostatic forces. Sedimentation can be classified into the physical driving force in design, namely gravity, centrifugal or electrostatic. These groups of sedimentation technology are further subdivided into the operations and equipment listed in Figure 9.1. Additionally sedimentation may be divided into the functional operations *thickening* and *clarification*. The primary purpose of thickening is to increase the concentration of suspended solids in a feed stream, while that of clarification is to remove a relatively small quantity of suspended particles and produce a clear effluent. Separation of two liquid phases, immiscible or partially miscible liquids, is a common requirement in the process industries. The simplest

https://doi.org/10.1515/9783110654806-009

form of equipment used to separate liquid phases is the gravity-settling tank. Their design is largely analogous to the design of gravity-settling tanks for solid–liquid separation. For difficult to separate liquid–liquid mixtures such as emulsions, centrifugal separators are used. In addition to solid–liquid separation, it is often desirable to remove either the coarse or the fine particles from the product. This process is referred to as *classification* or solid–solid separation and can be achieved in many types of solid–liquid separation equipment.

A perfect solid–liquid separation would result in a stream of liquid going one way and dry solids going another. Unfortunately, none of the separation devices works perfectly. Typically, there may be some fine solids leaving in the liquid stream, and some of the liquid may leave with the bulk of the solids. This imperfection is characterized by the mass fraction of the solids recovered and the residual moisture content of the solids.

9.2 Gravity sedimentation

Gravity sedimentation is a process of solid–liquid separation under the effect of gravity. A slurry feed is separated into underflow slurry of higher solids concentration and an *overflow* of predominantly clear liquid. Difference in density between the solids and the suspending liquid is a necessary prerequisite. Flocculation agents are often used to enhance settling.

Sedimentation is used in industry for solid–liquid separation and solid–solid separation. In solid–liquid separation, the solids are removed from the liquid either because the solids or the liquid are valuable or because these have to be separated before disposal. If the clarity of the overflow is of primary importance, the process is called *clarification* and the feed slurry is usually dilute. If the primary purpose is the production of concentrated slurry, the process is called *thickening* and the feed slurry is usually more concentrated. In solid–solid separation, the solids are separated into fractions according to size, density, shape or other particle property. Sedimentation is also used for size-based separation of solids. One of the simplest ways to remove the coarse or dense solids from a suspension feed is by sedimentation. Sedimentation speed depends on the particle size and density difference between the particle and the surrounding liquid in a predictable manner for low volume fractions (Section 9.2.2). Successive decantation steps can easily separate a particle mixture with significant difference in size and density. In all cases, the assessment of the sedimentation behavior of the solids within the fluids will allow the correct size of vessel to be determined. It is therefore vital to know how solids behave during sedimentation.

9.2.1 Sedimentation mechanisms

Gravity sedimentation is the separation of particles from liquids under gravity. The particle sedimentation rates are dependent upon particle properties such as size distribution, shape and density as well as the surrounding medium density and viscosity. Physical analysis of sedimentation behavior can explain the interdependence of these parameters for dilute suspensions (Section 9.2.2) and concentrated suspension (Section 9.2.3). Particles with particle diameters of the order of a few microns settle to slow for most practical operations due to Brownian motion (Section 9.2.6). Wherever possible, such particles are agglomerated or flocculated into relatively large clumps called flocs that settle out more rapidly. Spherical or near-spherical particles and agglomerates settle considerably more rapidly than plate or needle-like particles of similar weights.

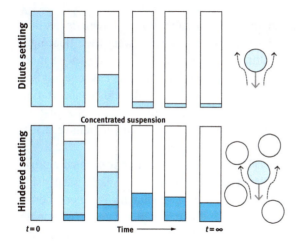

Figure 9.2: Schematic of a batch sedimentation experiment with macroscopic particles dispersed in a liquid as a function of time for dilute and concentrated suspensions. The darker color indicates denser collection of particles.

The main factors that determine the settling behavior are the concentration of the particulate solids and particle size distribution. The volume fraction, that is, how much of the total volume is occupied by the dispersed phase relative to total volume of dispersed and continuous phase is commonly used in sedimentation literature to indicate concentration. The volume fraction effect on the characteristics of sedimentation is best understood through analyzing a batch settling experiment, as in Figure 9.2. Solid particles without the tendency to cohere with each other generally settle at a steady rate provided that they are far away from other particles. At low volume fractions, they can be described as discrete particles. At low solid concentrations, the individual particles settle as individuals in the return fluid, which is displaced upward as illustrated in Figure 9.2. Regardless of their properties, the particles are sufficiently far apart to settle freely in this *dilute sedimentation* region. Increasing the particle concentration in a fluid decreases the settling rate of each individual particle. This phenomenon, known as *hindered settling*, is readily appreciated if it is considered that the settling of each particle is accompanied by a displacement of solvent in the opposite direction illustrated by dotted arrows in Figure 9.2. Since the fluid is unable to pass through the particles, it needs to displace surrounding particle to settle. Displacing neighboring particles costs energy and momentum, consequently particles slow down. At intermediate concentrations, the particles in close vicinity, which are in loose mutual contact, settle by *channeling*. The channels are of the same order of magnitude as the particles and are developed during an induction period in which an increasing quantity of return fluid forces its way through the mass.

The other phenomenon influencing the sedimentation behavior is particle size and the aggregation state of the particles. Contact between two flocculent particles may result in cohesion, resulting in an increase in size and hence a more rapidly sedimenting particle. Particle interference by collision and coagulation are other factors. For flocculent particles, the effect of initial solids concentration on sedimentation behavior has been observed to exhibit three quite distinct modes. In dilute suspensions, the individual particles or flocs again behave as discrete particles.

Particles near the bottom of the cylinder pile up into a concentrated sludge, whose height increases as more particles settle. This continues until the suspension zone disappears and all the solids are contained in the sediment. This condition is known as the *critical sedimentation point*. As illustrated in Figure 9.3 with letter C, the solid–liquid interface follows an approximately linear relationship with time until this critical sedimentation point is reached. In such concentrated suspensions, however, fluid flow is only possible through the minute voids between the primary particles. The hydrodynamic resistance of the touching particles below drastically reduces the sedimentation rate to a relatively low compaction rate. In this *compression regime*, the rate of sedimentation is a function of both the solid concentration and the depth of settled material in the tank. Particles closer to the base will be compressed by the mass of solids above, resulting in more concentrated sediment by slowly expelling the liquid, which turns the flocs into the deposit. This continues until equilibrium is established between the weight of the flocs and their mechanical strength of sediment at the bottom of the tank. The discussed effects of particle coherence and concentration of the settling characteristics of a feed suspension is summarized in Figure 9.4. It is important to realize that although the feed stream may start in one regime, it may pass through all of these regimes during clarification or thickening.

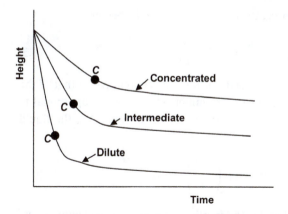

Figure 9.3: Effect of concentration on sedimentation.

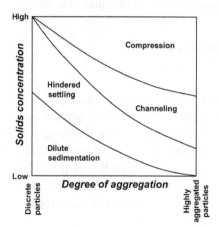

Figure 9.4: Effect of particle aggregation and solids concentration on the settling characteristics of a suspension.

9.2.2 Dilute sedimentation

Force balance around a settling sphere sedimenting in a viscous liquid under the influence of gravity is given in eq. (9.1). The net acceleration of a *free falling sphere* of mass m_p is caused by the gravitational, F_G, buoyant, F_B, and drag forces, F_D, acting on this sphere:

$$\text{acceleration force} = \text{gravitational force} - \text{buoyant force} - \text{drag force} \quad (9.1)$$

$$F_A = F_G - F_B - F_D \quad (9.2)$$

For a spherical particle of diameter d_p, the acceleration force F_A becomes

$$F_A = m_p \frac{dv}{dt} = \frac{\pi d_p^3}{6} \rho_S \frac{dv}{dt} \tag{9.3}$$

where ρ_S is the solid density, v is the velocity of the particle relative to that of the liquid and t is the time. The drag forces on an immersed body are proportional to the $A_\perp$ is the projected area of a particle in the direction of flow, $A_\perp$, the liquid density, ρ, and a drag coefficient, C_D:

$$F_D = C_D A_\perp \frac{1}{2} \rho v^2 = C_D \left(\frac{\pi d_p^2}{4} \right) \left(\frac{1}{2} \rho v^2 \right) \tag{9.4}$$

In fluid mechanics ρv^2 is the scaling for the inertial pressure. Consequently, $A_\perp \rho v^2$ becomes inertial force acting on the particle and C_D is a drag coefficient, a prefactor accounting for the particle shape and different flow regimes. C_D is a function of the *particle Reynolds number, Re,* that characterizes the nature of liquid flow around the particle where η is the fluid viscosity:

$$Re = \frac{\rho v d_p}{\eta} = \frac{\rho v^2}{\frac{\eta v}{d_p}} = \frac{\text{Inertial pressure}}{\text{Viscous pressure}} \tag{9.5}$$

At low Reynolds numbers, the particle does not accelerate meaning that the viscous pressure dominates over the inertial pressure responsible from fluid acceleration. This can be plainly seen if the parameters in Reynolds number are reorganized as shown in eq. (9.5). At low Reynolds numbers, the fluid acts as thin lamina of viscous materials sliding past the sedimenting object; hence the low Reynolds number regime is called laminar flow. At high Reynolds numbers, inertial forces acting on the particle dominate the viscous forces creating turbulence, another flow regime, with different C_D.

In gravity sedimentation, the dominant force is due to the *gravitational acceleration* acting on the buoyant mass of the particle:

$$F_G - F_B = m_p g - \frac{m_p}{\rho_S} \rho g = \frac{\pi d_p^3}{6} (\rho_S - \rho) g \tag{9.6}$$

Thus, the force balance equation in eq. (9.2) becomes

$$\frac{\pi d_p^3}{6} \rho_S \frac{dv}{dt} = \frac{\pi d_p^3}{6} (\rho_S - \rho) g - C_D \left(\frac{\pi d_p^2}{4} \right) \left(\frac{1}{2} \rho v^2 \right) \tag{9.7}$$

After a short initial acceleration period, the particle will cease accelerating and attain a *constant velocity.* This velocity calculated by setting the acceleration term in eq. (9.2) to zero and equilibrating buoyancy, gravitational and drag forces. This gives for the so-called *terminal settling velocity* v_∞ of a single sphere in a viscous liquid:

$$\frac{\pi d_p^3}{6}(\rho_S - \rho)g = C_D\left(\frac{\pi d_p^2}{4}\right)\left(\frac{1}{2}\rho v_\infty^2\right) \qquad (9.8)$$

hence

$$v_\infty = \left(\frac{4(\rho_S - \rho)g\,d_p}{3\rho\,C_D}\right)^{1/2} \qquad (9.9)$$

which can be solved if the value of C_D as a function of Re is known. In *dilute sedimentation* the particle Reynolds numbers are usually low (<1) and the relationship between C_D and Re is described by

$$C_D = \frac{24}{Re} = \frac{24\eta}{\rho v_\infty d_p} \qquad (Re < 1) \qquad (9.10)$$

modifying the terminal settling velocity equation to *Stokes' law* for spherical particles.

$$v_\infty = \frac{d_p^2}{18\eta}(\rho_S - \rho)g \qquad (Re < 1) \qquad (9.11)$$

The above expression shows that the terminal sedimentation speed of a particle is determined by the physical characteristics of the particle and the continuous phase. Increased sedimentation rates are obtained for larger particle diameters, greater density difference between the particle and the continuous phase and lower viscosity of the continuous phase. *There are two restrictions on the applicability of Stokes' law.* One concerns the effect of concentration and the other is that the particle Reynolds number should be less than 1.0. At higher values of Re, the loss of flow symmetry results in an increased C_D value, which can be estimated from the following correlations:

$$C_D = \frac{24}{Re}\left(1 + \frac{3}{16}Re\right) \qquad (1 < Re < 5) \text{ and} \qquad (9.12)$$

$$C_D = 18.5\,Re^{-0.6} \qquad (5 < Re < 500) \qquad (9.13)$$

Thus, for Reynolds numbers above 1 only a numerical solution is possible for which the force balance is written as follows:

$$C_D Re^2 = \frac{4}{3}\frac{d_p^3\rho(\rho_p - \rho)g}{\eta^2} \qquad (9.14)$$

The right-hand term of this equation is a known constant for a given problem and can be used to find the relevant Re number, from which the stationary settling velocity is calculated according to

$$v_\infty = \frac{\eta Re}{\rho d_p} \qquad (9.15)$$

9.2.3 Hindered settling

As the concentration of the suspension increases, the particles get closer together and no longer settle as individuals. For a particle to settle the surrounding fluid must be displaced in the opposite direction. At these high concentrations, a settling fluid needs to displace not fluid but other particles in its vicinity. Moreover, it needs to flow through a smaller space as the particle concentration is increased. Consequently, the rate of sedimentation in a swarm of particles is reduced compared to that of a single particle. This *hindered rate of sedimentation* appears to depend only on volume fraction of the particles, ε:

$$\frac{v_s}{v_\infty} = \varepsilon^2 f(\varepsilon) \qquad (9.16)$$

where v_s is the hindered settling velocity of a particle and v_∞ is the terminal settling velocity. The void fraction ε is the volume fraction of the fluid defined by

$$\varepsilon = \frac{V_L}{V_L + V_S} \qquad (9.17)$$

The voidage function $f(\varepsilon)$ has different forms depending on the theoretical approach adopted. A commonly used empirical relation is that of Richardson and Zaki ($n = 2.65$) and of Brinkman ($n = 2.5$):

$$f(\varepsilon) = \varepsilon^n \qquad (9.18)$$

Another important form is the Carman–Kozeny equation:

$$f(\varepsilon) = \frac{\varepsilon}{10(1-\varepsilon)} \qquad (9.19)$$

These correlations apply only to the cases where aggregation and flocculation are absent. Suspensions of fine particles often flocculate and therefore their effective sizes increase; consequently, they exhibit different behavior than the one captured by eq. (9.16).

9.2.4 Continuous sedimentation tank (gravity-settling tank)

Figure 9.5 illustrates a tank for the continuous removal of solid particles from a process liquid. The liquid is introduced at one end of the tank and flows toward the

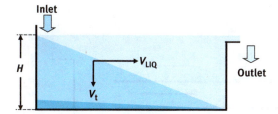

Figure 9.5: Tank for continuous removal of solid particles from a process liquid.

outlet at the other end. The dispersed particles are separated out and fall to the bottom with their terminal settling velocity v_∞ or hindered settling velocity v_s depending on the volume fraction. These processes are described most simply by the ideal continuous sedimentation tank model, which equates the required settling time of the particle and the residence time derived from the horizontal flow of the liquid. The residence time τ_{liq} of the liquid in the tank is obtained by dividing the volume V of the tank by the volumetric flow rate of liquid Q:

$$\tau_{liq} = \frac{V}{Q} = \frac{AH}{Q} \qquad (9.20)$$

where H is the tank depth and A its area. During the *residence time* of the liquid in the tank τ_{liq}, the particle must have time to fall to the bottom of the tank, thus:

$$\tau_{liq} = \frac{H}{v_\infty} \qquad (9.21)$$

The minimal required area A to allow particles with a stationary settling velocity v_∞ sufficient time to settle is now found by equating these two expressions, thereby eliminating the depth of the tank from the equations:

$$A \geq \frac{Q}{v_\infty} \qquad (9.22)$$

This equation defines the *maximum permissible throughput* to separate particles of a given diameter with their corresponding terminal settling velocity v_∞. Since the terminal settling velocity is a function of the particle diameter, see eq. (9.11), the collection efficiencies for a given tank area A can be estimated for a range of particle size values giving the *grade efficiency curve*

$$\xi(d_p) = \frac{v_\infty A}{Q} \qquad (9.23)$$

Two important conclusions may be drawn from eq. (9.22). The first is that the height H of the tank does not influence the throughput but is mainly taken between 0.7 and 1 m to allow volume for collection and compaction of the solids. The second is

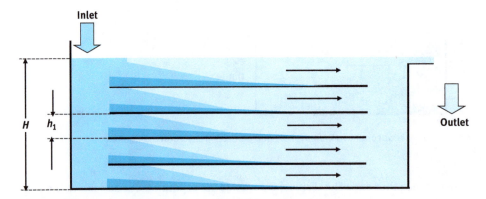

Figure 9.6: Tank with horizontal plates.

that the throughput of this type of tank is directly proportional to the area that can be utilized for separation. Accordingly an increased throughput Q is obtained when the tank area A is increased through fitting in a number of horizontal plates as illustrated in Figure 9.6. This increases the *number of separation channels* to N, each with a throughput:

$$Q = v_\infty A \tag{9.24}$$

Giving for the total throughput of the tank:

$$Q = v_\infty N A \tag{9.25}$$

It is therefore the total area $N{\cdot}A$ that determines the throughput. In the case of continuous separation, horizontal channels will eventually become clogged with sediment and separation will cease. If inclined plates are used as shown in Figure 9.7, the sediment slides down the plates under the influence of gravity and collects at

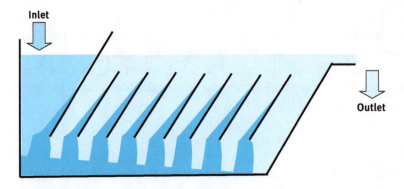

Figure 9.7: Tank with inclined plates.

the bottom of the tank. For the calculation of the throughput of this type of tank the *projected plate area* is used:

$$Q = v_\infty N A \cos \alpha \qquad (9.26)$$

In case of concentrated suspensions, the hindered sedimentation speeds have to be taken into account by replacing the terminal settling velocity, v_∞ with hindered settling velocity, v_s. However, it is challenging to exactly know the volume fraction of a sedimenting fluid in the tank. Consequently, the initial volume fraction can be taken as a maximum value keeping in mind the volume fraction will change as the suspension enters and sediments in the tank.

9.2.5 Gravity sedimentation equipment

Two distinct forms of sedimentation vessels are in common usage. The *clarifier* is used for the clarification of a dilute suspension to obtain an overflow containing minimal suspended solids. In a *thickener* the suspension is concentrated to obtain an underflow with high solid content while also producing a clarified overflow. Sedimentation equipment can be divided into batch settling tanks and continuous thickeners or clarifiers. Most commercial equipment is built for continuous sedimentation in relatively simple settling tanks.

The largest user of clarifiers is probably in the mineral processing and wastewater treatment. The most common thickener is the *circular basin type* shown in Figure 9.8. After treatment with flocculant, the feed stream enters the *central feed well*, which dissipates the stream's kinetic energy and disperses it gently into the thickener. In an operating thickener, the downward increasing solid concentration gives stability to the process. The settling solids and some liquid move downward. Most of the liquid flows upward and into the *overflow*,

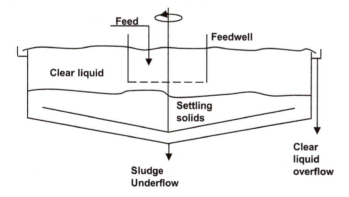

Figure 9.8: The circular basin continuous thickener.

which is collected in a trough around the periphery of the basin. The bottom gently slopes to the center and the settled solids are pushed down by a number of *scraper blades*. The conventional one-pass clarifier is designed for the lowest specific overflow rate, which is typically 1–3 m h^{-1} depending on the degree of flocculation. Typical heights range from 0.7–1 m to allow sufficient volume for collection and compaction of the solids. Stacking of sedimentation units in vertical arrangements increases the capacity per unit area.

9.2.6 Brownian motion and its influence on sedimentation

If we systematically decrease particle size in a suspension starting from millimeter to nanometer, their sedimentation behavior changes below a critical size. Simply put, the expected time sedimentation times increase dramatically and deviate from the predicted values in eq. (9.11) below this critical size. This deviation is due to a physical phenomenon that goes unnoticed in our daily experience with macroscopic objects, Brownian motion, kicks in altering how such small particles settle and behave. The easiest reporting of this phenomenon is attributed to a Botanist Robert Brown in 1827. Robert Brown observed that once he grinded plant seeds below a critical size and suspended them in water. He realized that they exhibit random motion. He first suspected that this was a biological effect. To test his hypothesis, he repeated the same procedure with all the materials he could get his hands on, including fossils. Finally, he concluded that this random motion occurs regardless of the materials they are made of. Later, Albert Einstein's theoretical contributions in 1905 and conclusive experiments of Jean Perrin in 1908 allowed us to understand this phenomenon. Through the efforts of several scientists, we now have a decent understanding of Brownian motion. Any suspension containing particles exhibiting Brownian motion are called colloidal suspensions, a term coined by Michael Faraday.

Brownian motion occurs in any object immersed in a fluid. It arises from random collisions of surrounding solvent molecules with the object. At any given time, solvent molecules bombard colloidal particles unequally as shown in Figure 9.9. If we count the number of collisions on the right and left half of a colloidal particle, they will not be equal. This collision imbalance leads to random motion driving the colloid particle to move randomly across the fluid. This indicates that as you read these lines you are exhibiting Brownian motion. However, the gravitational energy attracting you to the Earth's gravitational center is significantly larger than the kinetic energy you receive from random collisions of air molecules. Consequently, the Brownian motion you exhibit is extremely small. This can be quantified by calculating the Peclet number, *Pe*, comparing the gravitational potential energy required to move a colloid its own diameter with the average kinetic energy solvent molecules transfer to the colloid. The gravitational energy gain is mgd_p where m, g, d_p are the mass of the colloid dispersed, the gravitational acceleration and colloid diameter,

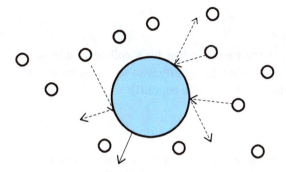

Figure 9.9: Illustration of a colloid exhibiting Brownian motion. The brown and white spheres illustrate colloidal particle and solvent molecules, respectively.

respectively. The average kinetic energy is kT from kinetic theory where k_b is the Boltzmann constant (1.380649×10^{-23} J K^{-1}) and T is the temperature in Kelvin, K.

$$Pe = \frac{mgd_p}{kT} \qquad (9.27)$$

If Pe is equal and smaller than 1, this indicates the system is colloidal and it exhibits Brownian motion and diffuses in a solvent. As Pe scales with d_p^4 for spherical particles as shown below, altering particle size has a significant effect on Pe number:

$$Pe = \frac{\rho g d_p^4}{6kT} \qquad (9.28)$$

For particles dispersed in a fluid, Brownian motion is quantified by Einstein–Stokes–Sutherland equation. Due to the aforementioned collisions, colloids exhibiting random motion are quantified by diffusion coefficient, D:

$$D = \frac{k_b T}{3\pi \eta d_p} \qquad (9.29)$$

where η is the viscosity and d_p is the diameter of a spherical particle.

It is informative to predict the critical size when the Brownian motion will be relevant in designing a sedimentation process. For colloidal suspension, gravity sedimentation is not a reasonable choice as Brownian motion randomizes the particles counteracting sedimentation. To sediment a colloidal suspension, a sedimentation agent first aggregates particles and the aggregated particles with larger particle size are sedimented by gravity. Any additions compromising colloidal stability are called aggregation agents, for instance, multivalent salts. To quantify the relative strength of diffusion over sedimentation, the characteristic timescale for Brownian motion, t_{diff}, and sedimentation, t_{sed}, should be calculated as follows:

$$t_{\text{diff}} = \frac{d_p^2}{D} \tag{9.30}$$

The physical interpretation of t_{diff} is the time required for a colloid to diffuse its own diameter. t_{sed} indicates the time required for a particle to sediment its own diameter, where v_∞ is the sedimentation speed given in eq. (9.11):

$$t_{\text{sed}} = \frac{d_p}{v_\infty} \tag{9.31}$$

Their ratio gives us a quantitative measure of how relevant is Brownian motion in sedimentation. If this ratio is equal or larger than 1, the particle will diffuse at least as fast as it sediments:

$$\frac{t_{\text{diff}}}{t_{\text{sed}}} = \frac{\frac{d_p^2}{D}}{\frac{d_p}{v_\infty}} = \frac{v_\infty d_p}{D} \tag{9.32}$$

We can argue that if this ratio is larger than 1, the diffusion will dominate and make gravity sedimentation futile.

9.2.7 Colloidal stability

Colloidal suspensions can be extremely stable making their separation challenging by gravity sedimentation. Particles are pulled to the bottom of sedimentation tank due to density difference between the particle and the surrounding medium. However, Brownian motion disperses them counteracting the gravitational pull. In essence, the colloids are using the kinetic energy transferred to them by solvent molecule collisions to counteract the gravitational pull. To sediment colloids, we need to understand what keeps them stable.

Colloids can be stable over centuries. They can stay suspended in a fluid without aggregating or sedimenting as shown in Figure 9.10. The first person to investigate colloidal stability was Michael Faraday in mid-1850s. Some of his original colloidal suspensions are still stable more than 150 years later. In his early work, Faraday realized that he could destabilize colloids by adding salts. Later research has also shown that colloids move in electric fields indicating that they are charged. These evidence provided clues on why colloids are stable. Once solid colloidal particles are immersed in a liquid, the surface groups on the colloid surface get charged. This brings a net charge to the colloids, yet the solution needs to be charge neutral as the charged neutral state is the least energy state, hence thermodynamically favorable state compared to charged state. This is achieved by bringing the ions in the solution close to the colloid in what we call a Debye layer or double layer. The Debye layer acts as a "charge shield" counteracting the attractive forces

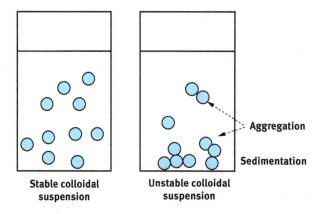

Figure 9.10: Illustration of colloidal stability stable and unstable colloidal suspensions.

such as van der Walls forces. van der Walls forces are attractive forces emerging from mutual repulsion between the atoms' electron cloud. Two teams of scientists developed this understanding independently. Boris Derjaguin and Lev Landau in Soviet Union, Evert Verwey and Theodoor Overbeek in the Netherlands conceived the idea that colloids are stable due to balance of attractive Van der Waals forces and repulsive electrostatic double layer force in early 1940s.

According to their theory, one can think of the destabilization of colloids by adding salt analogous to charging a capacitor. As more salt is added, the double layer capacitor is charged. The Debye layer becomes smaller consequently less effective as a "charge shield." With decreasing effect of Debye layer, attractive Van der Walls forces take over and aggregate subsequently sediment. Building on this fundamental understanding, more effective sedimentation agents have been developed such as multivalent salts that can charge the "double layer" capacitor more effectively per mole of salt added. These sedimentation agents increase the effective particle size consequently increasing Pe and the gravitation pull relative to Brownian motion.

9.3 Centrifugal sedimentation

Centrifugal sedimentation increases the force on particles over that provided by gravity and extends sedimentation to finer particle sizes and to emulsions that are normally stable in the gravity field. Centrifugation equipment is divided into *rotating wall* (sedimentating centrifuges) and *fixed-wall* (hydrocyclones) devices.

9.3.1 Particle velocity in a centrifugal field

In a liquid-filled rotating vessel, the generated *centrifugal acceleration* increases with the distance of the particle from the axis of rotation and the angular velocity. Centrifugal acceleration is described by

$$a = r\omega^2 \tag{9.33}$$

which acts toward the center of a circle for a restrained object. Particles in suspension are unrestrained and free to move tangentially outward from the center of rotation with a centrifugal acceleration equal and opposite to the centripetal force. Replacing the gravitational acceleration in eq. (9.9) by the acceleration in a centrifugal field provides the sedimentation velocity of particles in a rotating vessel:

$$v_\infty = \left(\frac{4(\rho_S - \rho)d_p}{3\rho C_D} r\omega^2\right)^{1/2} \tag{9.34}$$

Comparison of eq. (9.34) with eq. (9.9) provides the commonly quoted *g-factor*:

$$g-\text{factor} = \frac{r\omega^2}{g} \tag{9.35}$$

However, unlike in gravity settling where the acceleration is constant, the particle velocity in a centrifugal field depends on the radial distance from the center because the radius r appears in the acceleration expression. Hence, to derive the design equations for centrifugal sedimenting machines, the velocity must be written in the differential form:

$$v_\infty = \frac{dr}{dt} = \left(\frac{4(\rho_S - \rho)d_p}{3\rho C_D} r\omega^2\right)^{1/2} \tag{9.36}$$

which can be arranged by using Stokes' law ($C_D = 24/Re_p$) for the drag force:

$$v_\infty = \frac{dr}{dt} = \frac{d_p^2}{18\eta}(\rho_S - \rho) r\omega^2 \qquad (Re < 1) \tag{9.37}$$

9.3.2 Sedimenting centrifuges

Industrial centrifuges are commonly divided into batch, continuous and semicontinuous. The common laboratory centrifuge is a simple batch bottle centrifuge designed to handle small batches of material for laboratory separations. The basic structure is usually a motor-driven vertical spindle supporting various heads or rotors. There are three types of rotors: swinging bucket, fixed-angle head or small perforate or imperforate baskets for larger quantities of material. The *bottle centrifuge* is often used for preliminary testing to provide the basis of the design of commercial centrifuges.

Process centrifuges are more complex and available in a variety of sizes and types. *Tubular centrifuges* are used to separate liquid–liquid mixtures or to clarify liquid–solid mixtures with less than 1% solid content and fine particles. The solids are collected at the bowl wall and removed manually when sufficient bowl cake has accumulated. Liquid is discharged continuously. As the name suggests, these machines (Figure 9.11) have long tubular bowls that rotate around their vertical axis. Feed material is introduced at the base of the rotor. The longer the feed material spends in the bowl, the longer the centrifugal force is allowed to act on the particles, resulting in a progressively clarified cleaned feed stream as it flows up the length of the tubular bowl. *Multichamber centrifuges*, Figure 9.12, utilize a closed bowl that is subdivided into a number of concentric vertical cylindrical compartments through which the suspension flows in series. Their efficiency is high because of the reduced traveling distance to the collecting surface. Cleaning of multichamber centrifuges is more difficult and takes longer than for the tubular type.

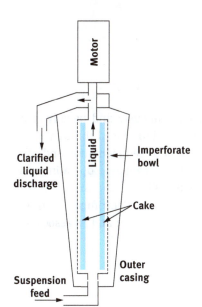

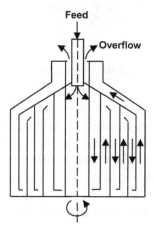

Figure 9.12: Schematic of a multichamber bowl.

Figure 9.11: A tubular centrifuge.

Centrifuges that channel feed through a large number of conical disks combine high flow rates with high theoretical capacity factors. The basic idea of increasing the settling capacity by using a number of layers in parallel is the same as the lamella principle in gravity sedimentation. Both liquid–liquid and liquid–solid separations are performed for slurries with solid concentrations below 15% and small particle sizes. The general flow patterns in a *disk stack centrifuge* are illustrated in Figure 9.13. Feed enters near the center of the bowl from either the top or the bottom. The clarified medium is discharged at a relatively small radius, generally at the top of the bowl. Solids are collected at the underside of the disks, slide outward along the surfaces and finally move from the outer edges of the disks to the bowl wall by free settling. Continuous solid discharge is achieved by sloping the inner walls of the bowl toward the discharge point. Generally, disk centrifuges have the best ability to collect fine particles at a high rate.

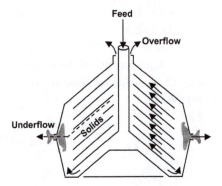

Figure 9.13: Clarifying disk stack centrifuge.

Scroll centrifuges or *decanters* discharge solids continuously and usually drier than disk and imperforate bowl centrifuges. The feed is introduced through a stationary axial tube. Solids are collected on the bowl wall by sedimentation and continuously moved up a sloping beach by a *helical screw conveyor* operating at a differential speed with respect to the bowl. As shown in Figure 9.14, the solids discharge is usually at a radius smaller than that of the liquid. Centrifugal fields are lower than in disk or tubular centrifuges because of the conveyor and its associated

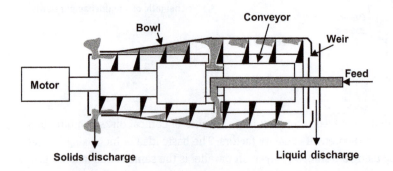

Figure 9.14: Continuous-scroll discharge decanter.

mechanism. For clarification, this type of centrifuge recovers medium and coarse particles from feeds at high or low solids concentration.

9.3.3 Bowl centrifuge separation capability

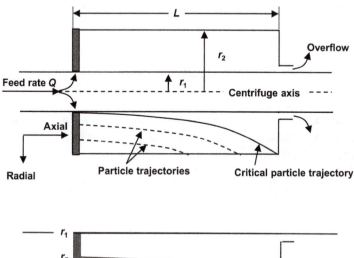

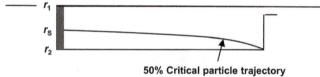

Figure 9.15: Bowl centrifuge: critical particle trajectories for 100% and for 50% capture efficiency.

The separation capability of cylindrical bowl centrifuges can be analyzed by equating the time required for settling of a spherical particle to the bowl wall to the time required for the feed liquid to travel to the discharge. In the model derivation plug flow is assumed to apply for the liquid flowing down the machine axis. The particles are considered to settle under Stokes conditions and reach their terminal settling velocity the moment they enter the centrifuge pond. The particles that reach the wall of the centrifuge are removed from the system while particles that do not reach the wall will be swept out of the machine in the overflow. Under these conditions, the particle that is just captured within the centrifuge will travel along the *critical particle trajectory* shown in Figure 9.15. The critical particle trajectory reflects the particle diameter for which the trajectory goes from the top surface (inner radius) of the centrifuge bowl to the bottom surface (outer radius) in the residence time within the machine. Particles of a similar diameter entering the machine between r_1 and r_2 will not present a problem as they will follow a parallel trajectory and intercept the wall before the end of the machine. The *residence time* of the particle in the axial direction equals:

$$\tau_{\text{liq}} = \frac{V_C}{Q} \tag{9.38}$$

where V_C is the volume of the centrifuge and Q is the volume flow rate of material fed to the machine. The radial velocity of the sedimenting particles is given by eq. (9.37). Integrating this equation using the limits of $r = r_1$ at $t = 0$ and $r = r_2$ at $t = \tau_{\text{liq}}$, and rearranging for the *sedimentation time* give

$$\tau_{\text{liq}} = \frac{18\eta \, \ln(r_2/r_1)}{d_p^2 \, (\rho_S - \rho) \, \omega^2} \tag{9.39}$$

Equating the equation for the sedimentation time to that for the residence time, multiplying numerator and denominator by the acceleration of gravity, g, and solving for the volumetric flow rate Q give

$$Q = \left(\frac{d_p^2}{18\eta} \, (\rho_S - \rho) \, g \right) \frac{V_C \, \omega^2}{g \, \ln(r_2/r_1)} \tag{9.40}$$

where the first bracketed term equals the Stokes terminal settling velocity v_∞. Substituting this term gives

$$\frac{Q(100\%)}{v_\infty} = \frac{V_C \, \omega^2}{g \, \ln(r_2/r_1)} \tag{9.41}$$

9.3.4 The sigma concept

The left-hand side of eq. (9.41), commonly indicated by the symbol Σ, has the dimensions of area and is equivalent to the surface area in m^2 of a gravitational settling tank with the same performance (100% efficiency, all particles captured) as a bowl centrifuge with an angular velocity ω. Its right-hand side is called the *machine sigma parameter*, which comprised solely of the dimensional parameters of the centrifuge and represents the clarification ability of the machine. The foregoing sigma theory is derived on the assumption of 100% collection of a particle of a critical diameter. It is more common to characterize and compare machines on the basis of *50% collection efficiency*. For this reason, the sigma theory is modified by assuming that only 50% of the particles with a given diameter need to be removed from the entering homogeneous suspension. The critical particle trajectory is therefore altered to that shown in Figure 9.15. The critical particle enters the machine and starts its journey at a radial position somewhere between r_1 and r_2. The residence time of the particle in axial direction remains:

$$\tau_{liq} = \frac{V_C}{Q} \tag{9.42}$$

Instead of r_1 the start radius r_S will now enter into the radial residence time:

$$\tau_{liq} = \frac{18\eta \ln(r_2/r_S)}{d_p^2(\rho_S - \rho)\omega^2} \tag{9.43}$$

The knowledge that equal volumes of the machine must exist between the sections r_1-r_S and r_S-r_2 can now be used to eliminate r_S in $\ln(r_2/r_S)$:

$$r_S^2 - r_1^2 = r_2^2 - r_S^2 \quad \text{or} \quad r_S^2 = \frac{r_2^2 + r_1^2}{2} \quad \text{and} \quad \ln\frac{r_2}{r_S} = \frac{1}{2}\ln\left(\frac{r_2}{r_S}\right)^2 = \frac{1}{2}\ln\frac{2r_2^2}{r_2^2 + r_1^2} \tag{9.44}$$

After substituting this result into the expression for the radial residence time and combining the residence time, equations provide the **machine parameter** at 50% collection efficiency:

$$\frac{Q(50\%)}{V_\infty} = \frac{2V_C\omega^2}{g\ln(2r_2^2/(r_2^2 + r_1^2))} = 2\Sigma_{Machine} \tag{9.45}$$

in which the effective volume for tubular centrifuges will be

$$V_C = \pi L(r_2^2 - r_1^2) \tag{9.46}$$

The assumptions and conditions for deriving the machine parameter equations impose several limitations on the application of the Σ-concept. The first concerns the particulate material that is assumed to be spherical in shape and uniform in size. Particles are assumed to settle as individual particles without interaction with the Stokes terminal settling velocity. The second group of assumptions concerns flow conditions. Fresh feed is introduced uniformly into the full space available for its flow and not disturbed by the layer of deposited material on the centrifuge wall. Although in practice only few of the assumed conditions are fully satisfied, the Σ-concept is a valuable tool that allows a comparison of different centrifuges operating on the same feed material. The Σ-value of several machines can be measured, which is the most realistic means of comparing. According to the derived machine parameter equations, the sedimentation performance of any two similar centrifuges operating on the same feed suspension is equal if the quantity Q/Σ is equal for both. This leads to

$$Q_2 = Q_1 \frac{\Sigma_2}{\Sigma_1} \tag{9.47}$$

This relation requires identical sedimentation characteristics when operating on the same material. The Σ-concept permits scale-up between similar centrifuges solely on the basis of sedimentation performance.

9.3.5 Capacity of disk centrifuges

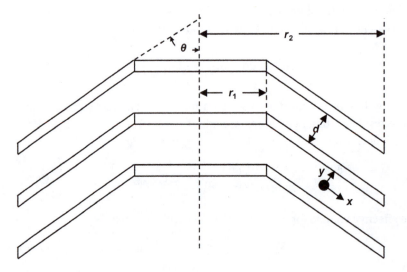

Figure 9.16: Schematic of separation in a disk stack.

The separation of particles inside a disk centrifuge is illustrated in Figures 9.13 and 9.16. The continuous liquid phase, containing solid or liquid particles, flows from the outside of the disk stack, having radius r_2, to the inside discharge opening, having radius r_1. When the liquid phase is considered evenly divided between the spaces d between the disks, the flow in each disk space is

$$Q_n = \frac{Q}{n} \tag{9.48}$$

where Q is the total flow through the entire disk stack containing n disks. The flow of the continuous liquid phase is also assumed to be in a radial plane and parallel to the surfaces of the disks. Here again an equation can be established to describe the trajectory of a particle under the combined effect of liquid transport velocity acting in the x-direction and the centrifugal settling velocity in the y-direction. The liquid flows between closely spaced conical disks from the outside to the axis. The settled solids slide down the underside of the disks and into the chamber outside the disk. For a stack of n disks with a distance d between the disks we obtain

$$\frac{dx}{dt} = \frac{dr}{dt \sin \theta} = \frac{Q/n}{2\pi r d} \tag{9.49}$$

and

$$\frac{dy}{dt} = \frac{dr}{dt} \cos \theta = \frac{d^2}{18\eta} (\rho_s - \rho) r\omega^2 \cos \theta \tag{9.50}$$

After integration, these two expressions for residence time are equated in a critical particle trajectory analysis as described earlier. The condition that defines the throughput Q for which 50% of the entering particles are collected becomes

$$\frac{Q}{v_\infty} = 2 \left(\frac{2\pi n \omega^2}{3g \tan \theta} (r_2^3 - r_1^3) \right) \tag{9.51}$$

where θ is half the included angle of the disks. The resulting expression for the machine parameter Σ corresponds again to the area of a gravity-settling tank capable of the same separation performance as the disk stack:

$$\Sigma = \frac{2\pi n \omega^2}{3g \tan \theta} (r_2^3 - r_1^3) \tag{9.52}$$

9.3.6 Hydrocyclones

Hydrocyclones offer one of the least expensive means of solid–liquid separation from both an operating and an investment viewpoint. They are cheap, compact, versatile and similar in operation to a centrifuge, but with much larger values of g-force. This force is, however, applied over a much shorter residence time. The most significant difference with a centrifuge is that centrifugal forces are generated without the need for mechanically moving part other than a pump. The velocity of that liquid delivers the energy needed for the rotation of the liquid. Cyclones have been employed to remove solids and liquids from gases and solids from liquids and are operated at temperatures as high as 1,000 °C and pressures up to 500 bars.

The principle features and flow patterns are shown schematically in Figure 9.17. In a hydrocyclone, the liquid path involves a double vortex with the liquid spiraling downward at the outside and upward at the inside. The *primary vortex* at the outside carries suspended material down the axis of the hydrocyclone. In the inside, material is carried up the axis and into the overflow vortex finder by the *secondary vortex*. The *overflow* usually consists of a dilute suspension of fine solids, while the underflow is a concentrated suspension of more coarse solids. Depending on their design, either the thickening or classifying action of the hydrocyclone is enhanced. The long cone shown in Figure 9.18 provides thicker underflow concentrations, but poorer sharpness of separation than the long cylinder. The *vortex finder* is important in reducing the loss of unclassified material.

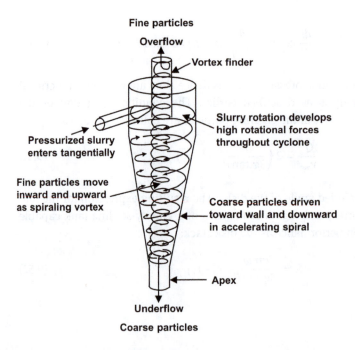

Figure 9.17: Principal features and flows inside a hydrocyclone. Reproduced with permission from [17] .

As a first approximation, a hydrocyclone can be considered as a rolled-up settling chamber in which gravitational acceleration is replaced by centrifugal acceleration. A good separation is obtained when the settler criterion is fulfilled:

$$A \geq \frac{Q}{v_\infty} \tag{9.53}$$

For small spherical particles, the terminal settling velocity toward the cyclone wall under conditions where Stokes law applies is given by

$$v_\infty = \frac{a(\rho_S - \rho)d_p^2}{18\eta} \tag{9.54}$$

where a is the centrifugal acceleration of the particle in the hydrocyclone. In the calculation of the centrifugal acceleration, the *tangential liquid velocity* inside the hydrocyclone is very important. It is the means by which a suspended particle following the liquid flow path will experience the centrifugal force. At the outer radius of the hydrocyclone, the tangential liquid velocity is approximately equal to inlet velocity of the feed v_f, which is equal to the ratio of the volume throughput Q, and the surface area A_I of the cyclone inlet. Although the inlet surface area may differ for different cyclone designs, often a value around $\frac{1}{6}D^2$ is used, giving

$$v_f = \frac{Q}{A_I} = \frac{6Q}{D^2} \quad \text{or} \quad Q = v_f \frac{D^2}{6} \tag{9.55}$$

where D is the cyclone diameter as indicated in Figure 9.18. The centrifugal acceleration of a particle in terms of the tangential velocity at the outside of the hydrocyclone now becomes

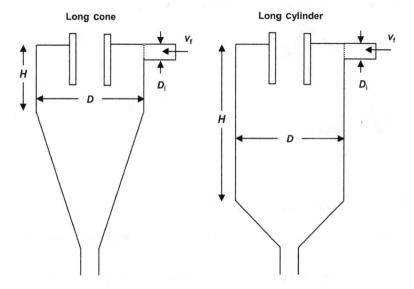

Figure 9.18: Basic hydrocyclone designs.

$$a = r\omega^2 = \frac{v_f^2}{r} = \frac{2v_f^2}{D} \tag{9.56}$$

giving for the terminal settling velocity:

$$v_\infty = \frac{2v_f^2}{D} \frac{(\rho_s - \rho)d_p^2}{18\eta} \tag{9.57}$$

When it is considered that the majority of the settling process will take part in the cylindrical part of the cyclone, the surface area of the rolled up settler is equal to

$$A = \pi DH \tag{9.58}$$

where H is the height of the cylindrical part as indicated in Figure 9.16. Introducing the derived expressions in eqs. (9.58), (9.57) and (9.55) in eq. (9.52) to eliminate A, Q and v_∞ gives, after some rewriting, the *theoretical minimum particle diameter* d_{100} for which *100% collecting efficiency* should be obtained as a function of the tangential velocity v_f and the cyclone throughput Q:

$$d_{100} \geq d_p = \left[\frac{3}{2\pi} \frac{\eta D^2}{v_f (\rho_s - \rho) H} \right]^{1/2} = \left[\frac{1}{4\pi} \frac{\eta D^4}{Q (\rho_s - \rho) H} \right]^{1/2} \qquad (9.59)$$

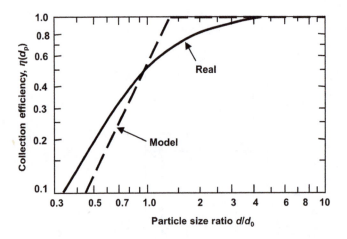

Figure 9.19: Comparison between theory and reality for the separation efficiency of a hydrocyclone.

In reality, the distinction between particles that are and are not separated by hydrocyclones is not that sharp. Although it is shown in Figure 9.19 that eq. (9.59) gives a reasonable indication where the separation can be expected, it is more common to describe the efficiency of hydrocyclone performance in terms of the diameter where 50% of the particles are collected. For a collection efficiency of 50%, a particle must travel from the center of the inlet pipe to the wall of the hydrocyclone within the residence time provided by the settling area. The concept that only half of the uniformly suspended solids in the feed is processed giving the particle diameter at 50% collection efficiency was used before the sigma analysis and results in

$$\frac{Q}{v_\infty} = 2A \frac{r\omega^2}{g} = 2\pi H \frac{2v_f^2}{g} = 2\Sigma \qquad (9.60)$$

It can be seen that the machine parameter for a hydrocyclone is a function of the tangential liquid velocity, which is responsible for the magnitude of the centrifugal acceleration. The poor sharpness of separation can be overcome by employing hydrocyclones in series, as illustrated in Figure 9.20. High retention efficiency of

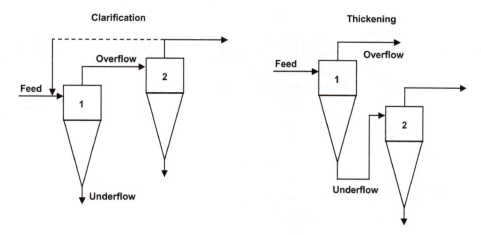

Figure 9.20: Two hydrocyclones in series for clarification with a possible partial recycle and thickening.

solids is obtained in hydrocyclones of low diameter. This imposes limitations on the throughput of a single device. Therefore, it is common to install packages with multiple hydrocyclone units operating in parallel.

9.4 Electrostatic precipitation

9.4.1 Principles

In electrostatic precipitation, *electrostatic forces* are used to remove the solid or liquid particles suspended in a gas stream. An efficient separation is achieved by creating an *electrostatic charge* on the particles through the generation of ions in the gas, which unload their charge on the particles by collision or diffusion. In electrostatic precipitators, ions are created in the gas by exposure to *corona discharge* in high-density electrostatic fields. This is accomplished by applying high voltages (30–100 kV) to an assembly of negatively charged wires and positively grounded collection plates as shown in Figure 9.21. At the wires, the corona discharge occurs when positive ions in the gas impinge on the negatively charged wires and release electrons that ionize the gas into positive ions and electrons. While the positive ions produce further ionization, the produced electrons associate with gas molecules to form negative ions traveling across the gas flow to the positively charged collecting electrodes. This continuous flow of negative ions effectively fills the space between the electrodes with negative ions that charge the particles passing through the space with the gas by two mechanisms, bombardment field charging and diffusion charging. *Field charging* occurs by ions that move toward the positive electrode with high

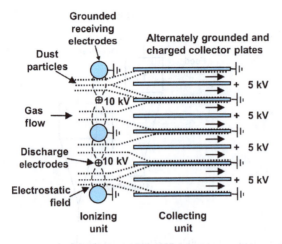

Figure 9.21: Schematic of a two-stage electrostatic precipitating unit.

velocity and impact on the particles. In *diffusion charging* the thermal movement of the ions in the gas leads to particle–ion collisions. The charging continues until the so-called *equilibrium charge* on the particle repels further ions.

As soon as the dust particles acquire some negative charge, they will migrate toward the positive collector electrodes away from the negative discharge electrode. Although the overall picture is very complex, the calculation of the particle drift velocity is usually based on a relatively simple model with the following assumptions:

1. The particle is considered fully charged during the whole of its residence in the precipitating field.
2. The distribution of particles through the precipitator cross section is uniform.
3. Particles moving toward the electrode normal to the gas stream encounter fluid resistance in the viscous flow regime and Stokes law can be applied.
4. Repulsion effects between the particles are neglected.
5. There are no hindered settling effects in the concentrated dust near the wall.
6. The effect of the movement of the gas ions, sometimes called the electric wind, is neglected.
7. The gas velocity through the precipitator does not affect the migration velocity of the ions.
8. The dust particles move at their terminal velocity.

Although these assumptions are rather simplistic, the migration velocity based on these eight assumptions has been found to give reasonable estimates of the **cross-stream drift velocity** measured in experiments. For a particle with saturation charge q moving in a field of constant intensity E, the terminal migration velocity v_m toward the collector electrode may be determined from equilibrium between the electrostatic force and the drag force:

$$E.q = 3\pi\eta d_p v_m \tag{9.61}$$

Rewriting gives the *migration velocity* of particle:

$$v_m = \frac{E.q}{3\pi\eta d_p} \tag{9.62}$$

On arrival at the earthed collector, the particles adhere and are discharged to earth potential. The strong electrostatic field and adhesive properties of the particulate inhibit re-entrainment. When a layer of particles has formed, the layer is broken into individual particle by mechanical vibrations. Finally, the particles fall into a hopper for collection.

9.4.2 Equipment and collecting efficiency

Precipitators can be classified as single- or two-stage equipment. In the two-stage designs, used for air purification, air conditioning or ventilation, particle charging and collection are carried out in two separate stages. As illustrated in Figure 9.21, the dust particles are first charged in a separate charging section, providing only a fraction of a second residence time to avoid collection. Particle collection follows in the second stage, which consists of alternately charged parallel plates. As laminar flow conditions usually prevail in two-stage precipitators, the *collecting efficiency* $\xi(d_p)$ can be expressed in direct analogy with sedimentation tanks:

$$\xi(d_p) = \frac{v_m A}{Q} \tag{9.63}$$

where v_m is the effective migration velocity of the dust particles, A is the collecting area and Q is the gas flow rate. Equation (9.57) can be rewritten in terms of the gas velocity v_g, the electrode spacing W and the length of the collector electrodes L:

$$\eta(d_p) = \frac{v_m L}{v_g W} \tag{9.64}$$

Application of this equation is limited to the two-stage precipitator. The two-stage precipitators are built for only laminar flow cases. In most industrial applications, single-stage units are used that are built for turbulent flow causing turbulent diffusion and re-entrainment. It was found experimentally that the efficiency of a precipitator was an exponential function of the gas stream residence time in the precipitator field. Under the conditions that the dust is uniformly distributed in the beginning, the uncollected dust remains uniformly distributed and the migration velocity is effectively constant. Deutsch derived the following equation for the collecting efficiency:

$$\eta(d_p) = 1 - \exp\left(-\frac{v_m A}{Q}\right) = 1 - \exp\left(-\frac{v_m L}{v_g W}\right) \tag{9.65}$$

In single-stage units, the particles are charged and collected in the same electrical field, thus making the design simpler. For a round wire axially suspended in a tube, a radial distributed electrostatic field is obtained. To accommodate larger gas flows, banks of tubes are nested together vertically to facilitate the removal of the deposited dust by gravity (Figure 9.22). *Wire-in-tube separators* are employed for small quantities of gas. The wire in tube precipitator was the earliest type used but has now been almost entirely superseded by the plate type. *Wire-and-plate or simply plate precipitators* consist of vertical parallel plates with vertical wires arranged at intervals of 0.1–0.2 m along the centerline between the plates (Figure 9.23). The wires have negative charge while the plates are normally earthed. A plate precipitator may have 10–15 m tall parallel flat plates with 0.2–0.4 m horizontal spacing and discharge electrodes that are suspended midway between the plates.

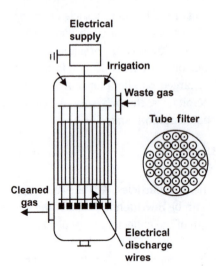

Figure 9.22: Diagram of a tube-type single-stage electrostatic precipitator plate separator.

The dust particles collected on both the discharge and collector electrodes must be removed at preset time intervals. This process is called rapping. It is vital for precipitator performance as clean electrodes are essential for the generation of an effective corona discharge and electrostatic field. Particles are removed by mechanically vibrating the collection plates, thereby dislodging the particles, which drop into collection hoppers. Swing hammers or camshaft, scrapping and coil springs are most commonly used for

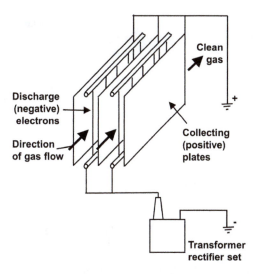

Figure 9.23: Schematic of a single-stage electrostatic.

mechanical rapping. Electromagnetic or pneumatic devices allow easier adjustment of the timing and intensity of the rapping than the mechanical devices. In the wet two-stage tubular precipitator, the deposited dust is removed from the collector electrodes by a flowing water film. This device consists of a short ionizing section followed by a relatively long collection system. The discharge electrode is in the form of a rod or a tube with a section of sharp discharge points at the end, centered in the collection tube.

Electrostatic precipitators are applied wherever very large volumes of gases have to be cleaned with high collection efficiency on fine particles. It should be noted that electrostatic precipitators utilize high energy arcs, consequently they should only be employed when there is no explosion risk. The plants are invariably used for fly ash collection in large (800–1,200 MW) coal-fired electric power stations and for the collection of dusts in the cement industry. Precipitators are also employed for large-scale fume collection systems in the metallurgical industry.

Nomenclature

A	Area	m^2
a	Centrifugal acceleration	$m\,s^{-2}$
C_D	Drag coefficient	–
d, D	Diameter	m

E	Electric field strength	$V\ m^{-1} = N\ C^{-1} = m\ kg$ $s^{-3}\ A^{-1}$
F	Force	N
g	Gravitational acceleration	$m\ s^{-2}$
H	Height	m
L	Length	m
m	Mass	kg
Re	Reynolds number	–
t	Time	s
q	Charge	$C = A.s$
Q	Volume flow rate	$m^3\ s^{-1}$
$Q(50\%), Q$	50% and 100% collection efficiency, respectively, eqs.	$m^3\ s^{-1}$
(100%)	(9.41) and (9.45)	
r	Radius	m
v	Velocity	$m\ s^{-1}$
V	Volume	m^3
W	Electrode spacing	m
ε	Fluid volume fraction, eq. (9.16)	–
η	Viscosity	$kg\ s^{-1}$ $m^{-1} = N{\cdot}s\ m^{-2} = Pa{\cdot}s$
$\xi(d_p)$	Efficiency, eqs. (9.23) and (9.63)	–
ρ, ρ_s	Density, solid density	$kg\ m^{-3}$
Σ	Machine parameter, eq. (9.45)	
τ	Residence time	s
ω	Angular velocity	$rad\ s^{-1}$

Exercises

1   Calculate the terminal settling velocity of a 70 µm diameter sphere (density 2,600 kg/m³) in water (density 1,000 kg m⁻³, viscosity 1×10^{-3} Ns m⁻²).

2   Design a gravity-settling tank separator to handle 15 m³ h⁻¹ of an aqueous suspension containing 0.118 vol% of copper oxide. Laboratory tests have shown that the solids settling rate in a vertical vessel is 3.54×10^{-4} m s⁻¹. Density of copper oxide is 5,600 kg m⁻³.

3   A settling tank is used to separate ion exchange particles ($\rho_S = 1,200$ kg m⁻³, $d_p = 1$ mm) from a treated effluent stream ($\rho = 1,000$ kg m⁻³, $\eta = 0.001$ Pa s, $Q = 0.3$ m³ s⁻¹). Calculate the required area of this settling tank operating under laminar flow conditions.

4   Calculate the diameter of a settling tank to separate light oil from water. The oil is the dispersed phase. Oil flow rate 1,000 kg h⁻¹, density 900 kg m⁻³,

viscosity 0.003 Pa s. Water flow rate 5,000 kg h^{-1}, density 1,000 kg m^{-3}, viscosity 0.001 Pa s. The droplet diameter amounts to 150 μm.

5   This exercise concerns the separation of minerals by density difference. A water stream Q_f (0.05 m^3 s^{-1}) contains a mixture of spherical lead sulfide particles and spherical quartz (sand) particles. The quartz and lead sulfide particles should be separated to the best extent. However, this separation is complicated by the fact that smaller heavy particles can have the same settling velocity as larger lighter particles. As shown in the following illustration, the whole settler consists of three settling chambers, each with its own length L. It can be assumed that a uniform horizontal flow exists in which the particles can settle unhindered. The values of L have been chosen in such a way that no quartz is collected in the first chamber and no lead sulfide in the last chamber. The width of all three settling chambers is 5.0 m.

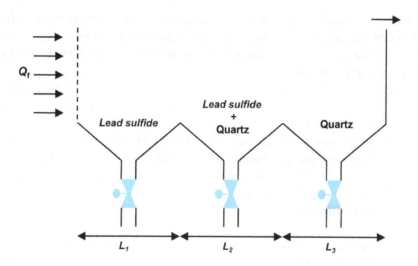

The density ρ of lead sulfide amounts to 7,500 kg m^{-3} and that of quartz 2,650 kg m^{-3}. Both minerals are present in particle sizes ranging from 10.0 to 25.0 μm. The density ρ_f of water at 20 °C equals 998 kg m^{-3} and the viscosity $\eta_f = 1.005 \times 10^{-3}$ kg m^{-1} s^{-1}. The gravitational acceleration g amounts to 9.81 m s^{-2}.
a.   Calculate the smallest particle diameter of lead sulfide in the first chamber.
b.   What is the minimal value of $L_1 + L_2$ to ensure that no lead sulfide particles settle in the last chamber?

6   Compare the gravitational and centrifugal sedimentation rate at 5,000 rev min^{-1} of yeast cells ($d_p = 8$ μm, $\rho_s = 1,050$ kg m^{-3}) in water. ($\rho = 1,000$ kg m^{-3}, $\eta = 10^{-3}$ Pa s). The centrifuge bowl has an inner radius of 0.2 m.

7   If a centrifuge is 0.9 m diameter and rotates at 20 Hz, at what speed should a labo-
     ratory centrifuge of 150 mm diameter be run if it is to duplicate plant conditions.

8   A continuous tube centrifuge with a bowl of 1.5 m long and 0.75 m diameter
     operating with a pool depth of 0.1 m at 1,800 rpm is clarifying an aqueous sus-
     pension at a rate of 5.4 m^3 min^{-1}. All particles of diameter greater than 10 μm
     are being removed. Calculate the efficiency of this machine. The solid and liq-
     uid densities are 2,800 and 1,000 kg m^{-3}, and the liquid viscosity is 0.001 Pa s.

9   A centrifuge basket of 600 mm long and 100 mm internal diameter has a dis-
     charge weir of 25 mm diameter. What is the maximum volumetric flow of liquid
     through the centrifuge such that when the basket is rotated at 200 Hz all particles
     of diameter greater than 1 μm are retained on the centrifuge wall. Solid density
     2,000 kg m^{-3}, liquid density 1,000 kg m^{-3}, liquid viscosity 10^{-3} Pa s.

10   A low-concentration suspension of clay in water is to be separated by centrifu-
     gal sedimentation. Pilot runs on a laboratory tubular bowl centrifuge operating
     at 20,000 rev min^{-1} indicate that satisfactory overflow clarity is obtained at a
     throughput of 8 × 10^{-6} m^3 s^{-1}.
         The centrifuge bowl is 0.2 m long, has an internal radius of 0.022 m and the
     radius of the liquid surface is 0.011 m. Solid density 2,640 kg m^{-3}, liquid density
     1,000 kg m^{-3}, liquid viscosity 10^{-3} Pa s.
     a.   If the separation is to be carried out in the plant using a tubular centrifuge
          of 0.734 m long with an internal radius R = 0.0521 m and $R - r_1$ = 0.0295 m,
          operating at 15,000 rev min^{-1} with the same overflow clarity, what produc-
          tion flow rate could be expected?
     b.   What is the effective cut size?

11   Estimate the diameter of hydrocyclones (H = 4D) required for the separation of
     polypropylene particles (d = 10^{-4} m, ρ = 900 kg m^{-3}) from butane (η = 10^{-4} Pa s,
     ρ = 600 kg m^{-3}). The inlet velocity is limited to 3 m s^{-1} to limit the pressure
     drop. How many cyclones should be arranged in parallel to handle a stream of
     600 kg s^{-1} slurry?

12   As a last step produced, fruit juices need to be clarified by removing the very
     fine particles (10 kg m^{-3} fruit juice). Assume that all particles have the same
     size. Before the final filtration, the particles are concentrated by sedimentation.
     In the laboratory, the following sedimentation rates have to be measured:

Time (h)	0	1	2	3	6	10
h (cm)	35	29	23	17	13	11

where h is the height of the interface between clear juice and the slurry layer.

a. Calculate the size of the particle in the slurry.

b. How large should be the area of a settling tank when 0.01 m³ s⁻¹ fruit juice needs to be processed?

As an alternative, a centrifuge is used that is constructed in such a way the clear juice stream leaves in axial direction and while a part of the juice with the particles is discharged radial. The two radii r_1 and r_2 equal 0.2, respectively, 0.4 m, while L = 1.4 m (compare Figure 9.13). This centrifugal sedimentation yields a slurry with 100 kg solids per m³ slurry.

$\rho_S = 1{,}050$ kg m⁻³ $\rho_L = 1{,}000$ kg m⁻³ $\eta = 10^{-3}$ Pa s $g = 9.81$ m s⁻²

---

c. Which rpm is required to separate all particles with this centrifuge?

d. Calculate the volume flow (m³ s⁻¹) of the resulting slurry.

Chapter 10
Filtration

10.1 The filtration process

In this chapter we focus on the industrially important filtration process where solid particles dispersed in liquid and gas are separated by a *porous filtration medium*. The suspension with targeted particles enters the filtration unit housing the filtration medium (Figure 10.1). The solid particles are not permitted to pass through the filter whereas the solvent can pass through. The resulting stream carrying significantly lower amounts of particles is called filtrate. Flow through the porous filtration medium, also referred as filter, is usually achieved by applying a static pressure difference as the driving force.

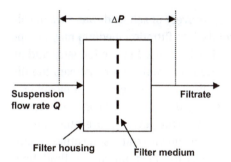

Figure 10.1: Schematic of a filtration system.

The most widely used filtration types are cake, or surface filtration, and deep bed filtration. In *cake filtration* (Figure 10.2a), separation is to be achieved upstream with a relatively thin filter medium. The targeted particles for separation must be larger than the pores of the medium or they should form bridges to cover the pores as illustrated in Figure 10.2a. In the latter case the initial breakthrough of particles will stop as soon as the bridges are formed. In bridge formation, particles jam around a pore blocking the pore collectively while they can pass through the pore individually. On top of this first particle layer, successive layers of solids deposit and form a cake. Bridging over the pores and porosity of the cake enables filtration of particles smaller than the pore size. Performance of cake filters is usually improved with relatively high feed solids concentration. The main disadvantage of *cake filtration* is the declining flow rate due to the increasing pressure drop across the growing cake. This is circumvented in *cross-flow filtration* systems (Figure 10.2b), where a sweeping flow continuously shears off the cake and counteracts cake formation. In this approach,

https://doi.org/10.1515/9783110654806-010

little or no cake is allowed to form on the medium consequently the pressure drop due to cake formation is avoided.

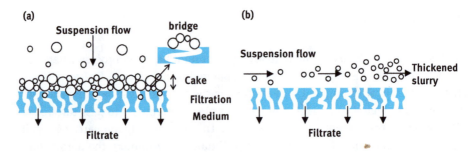

Figure 10.2: Schematic of the (a) cake filtration with an illustration of the bridge and cake formation and (b) cross-flow filtration mechanisms.

Cake filters are used in clarification of liquids, recovery of solids, dewatering of solids, thickening of slurries and washing of solids. The filtration medium may be operated in various modes. *Vacuum operation* is widely used in the industry and at the laboratory scale. In the filters a low pressure is maintained downstream the filter medium by vacuum pumps. The pressure difference between the atmospheric pressure in upstream and the created low pressure downstream forces the suspension though filter. *Pressure filters* operate at pressure levels above atmospheric and the high pressure is created upstream while the downstream is the atmospheric pressure. The pressure differential created across the medium causes fluid flow through the filter. In *centrifugal filtration*, fluid flow is created by the centrifugal force resulting from spinning the suspension. Such centrifugal filters are found in many applications in the food, beverage and pharmaceutical industries. The different operation modes and typical machinery used are shown in Figure 10.3. The diversity in the machinery used reflects the complexity involved in the processing of solids, particularly those in small particle size ranges.

An important first step in the rationalization of solid-liquid separation problems is to choose between sedimentation, filtration or their combination. Filtration is typically used when (i) it is desired to obtain relatively dry solids or (ii) when the sedimentation techniques become impractical due to small density differences or particle diameters or low-solids concentrations. Combination with sedimentation may be advantageous for prethickening of suspension. As with other unit operations, filtration cannot achieve 100% separation efficiency. Some of the solids may be left in the liquid stream, and some liquid will be entrained with the separated cake. Separation of solids is measured by the *fraction recovered* while the separation of liquid is usually characterized through the *slurry moisture content* or *solids concentration*. Although filter media such as a woven wire mesh easily retain coarse materials, "screens" with smaller

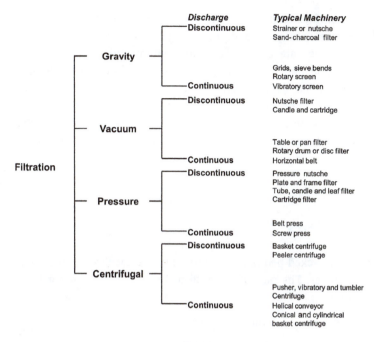

Figure 10.3: General classification of filtration equipment.

openings or pores such as nonwoven cloths or membranes are required as the size of the particulates decreases. When the particulates are small compared to the medium pores, particle deposition inside the pore network in the depths of the filter medium may occur. This so-called *deep bed filtration* is fundamentally different from cake filtration in principle and application. In deep bed filtration, the filter medium is a deep bed filter with extended network and pore size much greater than the targeted particles it is meant to remove. Particles penetrate into the medium where they sediment due to gravity settling, diffusion and inertial forces. The sediment particles form a thin filtration layer inside the pore network of the filter. This thin filtration can be thought of as an internal cake that achieves filtration of particles smaller than the pore size. Deep bed filters were developed for potable water treatment as the final polishing process. They also find increasing application in industrial wastewater treatment.

Filtration is frequently combined with other processes to enhance its performance. For instance, *washing* is applied to replace the mother liquor in the solids stream with a clean wash liquid. Especially for high purity products, washing may represent a dominant portion of the total filtration cost. The three most common washing techniques are displacement, reslurrying and successive dilution. *Dewatering* is used to reduce the moisture content of filter cakes either by mechanical compression of the filter cake or by displacement of the liquid with air. It is enhanced by addition of dewatering aids to the suspension in the form of surfactants that reduce surface tension. Another important aspect to improve the performance of a filtration step may be conditioning or pretreatment of the feed suspension. Common techniques are coagulation, flocculation and the addition of inert filter aids. Coagulation and flocculation increase the effective particle size with the accompanying benefits of higher settling rates, higher permeability of filter cakes and better particle retention in deep bed filters. Coagulation brings particles into contact to form agglomerates. Flocculation agents such as natural or synthetic polyelectrolytes interconnect colloidal particles into giant flocs up to 10 mm in size.

10.2 Filtration fundamentals

10.2.1 Flow through packed beds

During its flow through a packed bed of solids, the liquid passes through the open spaces between the particles. The part of the total bed volume available for fluid flow is called the porosity:

$$\varepsilon = \frac{\text{volume of voids}}{\text{total bed volume}} \tag{10.1}$$

As the liquid flows through the bed, hydrodynamic friction due to solid packing leads to a pressure drop. With experiments, Henry Darcy discovered that during filtration the superficial filtrate velocity, v_F, is proportional to the imposed pressure difference, ΔP, and inversely proportional to the viscosity of the flowing fluid, η. The proportionality constant, R_{tot}, is determined by the resistances of the cake, R_C, and that of the filter medium, R_M. This is schematically shown in Figure 10.4 and may be written as:

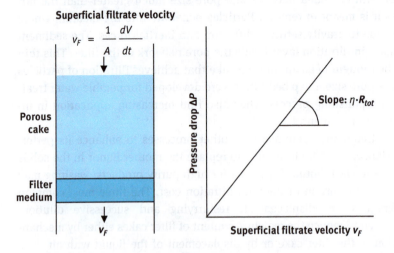

Figure 10.4: Schematic of flow through a porous medium.

$$v_F = \frac{\Delta P}{\eta\, R_{tot}} \tag{10.2}$$

The total resistance characterizes the ease of liquid flow through the filter cake and, subsequently, the porous filter medium. Because these are two resistances in series, the sum of the cake resistance, R_C, and the resistance of the filter medium, R_M, can replace R_{tot}:

$$v_F = \frac{\Delta P}{\eta\,(R_M + R_C)} \tag{10.3}$$

Implicitly, it is assumed that the resistance of the filter medium does not change during the process. Although true for the medium, filter cake resistances can vary considerably depending on porosity, particle shape and packing, particle size and distribution, cake formation rate and slurry concentration. This is the reason that the relations derived for the resistance of idealized particle beds should only be used as guidelines and are not applicable to practical filtration operations. More elaborate relations considering caking behavior are also available and discussed later in the chapter. In filtration, it is common practice to deduce an empirical permeability from simple laboratory tests or existing operating data as shown in Figure 10.4. Finally, the filtrate velocity is replaced by the amount of filtrate dV collected in period dt divided by the filter cross-sectional area A:

$$v_F = \frac{1}{A} \frac{dV}{dt} \tag{10.4}$$

10.2.2 Cake filtration

The mathematical description of the cake filtration process, already shown in Figure 10.2a, starts by combining both relations of the filtrate velocity (eqs. (10.3) and (10.4)) into *Darcy's law* to relate the filtrate flow rate and total pressure drop:

$$\Delta P = \Delta P_C + \Delta P_M = \eta \frac{R_C + R_M}{A} \frac{dV}{dt} \tag{10.5}$$

During filtration the solid deposition increases the cake height, which leads to an increase in the *total cake resistance*. A material that exhibits the linear increase in cake resistance with the cake height shown in Figure 10.5 is known as an *incompressible cake*. In an incompressible cake filtration process, the solids concentration or porosity of the filter cake remains constant and the cake volume increases by a

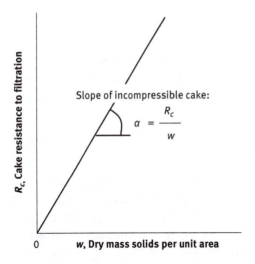

Figure 10.5: Specific resistance of an incompressible cake.

constant amount for each unit volume of suspension. However, when filtering at constant pressure, the rate of filtration and solids deposition declines as shown in Figure 10.6, because each new element of filter cake increases the total resistance to filtrate flow through the deposited cake. The total cake resistance can be represented through the proportionality constant, α, and the deposited mass of dry solids per unit area of the filter, w:

$$R_C = \alpha w \tag{10.6}$$

The proportionality constant, α, is known as the *specific cake resistance* and has the units of meter per kilogram (m/kg). The amount of deposited dry solids per unit area can be obtained from the dry solids concentration, c, in the suspension and the total amount of suspension,[1] V, divided by the filter area, A:

$$w = \frac{cV}{A} \tag{10.7}$$

Introduction of the specific cake resistance into eq. (10.5) provides the two resistances in series form of Darcy's law which we will use in the further analyzes of filtration processes:

$$\Delta P = \frac{\eta c \alpha}{A^2} V \frac{dV}{dt} + \frac{\eta R_M}{A} \frac{dV}{dt} \tag{10.8}$$

---

1 c in weight/unit volume suspension (or liquid), V refers to suspension (or liquid) volume.

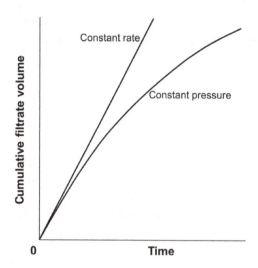

Figure 10.6: Constant pressure and constant rate filtration.

10.2.3 Constant pressure and constant rate filtration

For an incompressible cake, this modified Darcy's law equation contains three variables and five constants. The equation can be solved analytically only if one of the three variables (time t, filtrate volume V and pressure difference ΔP) is held constant. This reflects the two main operation modes of industrial filters. Vacuum filtration tends to be under constant pressure and pressure filtration is often under constant rate dV/dt. Under constant pressure, eq. (10.8) can be rearranged and integrated using the boundary condition of zero filtrate at zero time:

$$\int_0^t dt = \frac{\eta c \alpha}{A^2 \Delta P} \int_0^V V dV + \frac{\eta R_M}{A \Delta P} \int_0^V dV \tag{10.9}$$

After integration, the following equation known as the *linearized parabolic rate law* is obtained:

$$\frac{t}{V} = \frac{\eta c \alpha}{2 A^2 \Delta P} V + \frac{\eta R_M}{A \Delta P} \quad \textit{constant pressure filtration} \tag{10.10}$$

Thus, a graph of the experimental data points of t/V against the independent variable V permits determination of the slope and the intercept from which the specific cake resistance and resistance of the filter medium can be calculated.

Constant rate filtration is encountered when a positive displacement pump feeds a pressure filter. Due to the increasing cake resistance, the pressure delivered by the pump must increase during the filtration process to maintain a constant filtration rate. Constant rate filtration is easily observed on a plot of filtrate volume against time, as illustrated in Figure 10.6. Because under these circumstances:

$$\frac{dV}{dt} = \frac{V}{t} = \text{constant} \tag{10.11}$$

and eq. (10.8) can be rearranged to give:

$$\Delta P = \left(\frac{\eta c \alpha}{A^2} \frac{V}{t}\right) V + \left(\frac{\eta R_M}{A} \frac{V}{t}\right) \qquad \textit{constant rate filtration} \tag{10.12}$$

which is a straight line when the filtration pressure is plotted against the filtrate volume. In the same manner as with constant pressure filtration the slope and the intercept taken from the graph provides the values for the specific cake and filter medium resistance.

10.2.4 Compressible cakes

In practice, all filter cakes display some form of compressibility. Increasing the filtration pressure results in an increased cake solids concentration, accompanied by a higher cake resistance. A useful check on the compressible characteristics of a material under investigation follows from constant pressure filtrations carried out at different pressures. Rearranging eq. (10.10):

$$\Delta P \frac{t}{V} = \left(\frac{\eta c \alpha}{2A^2}\right) V + \left(\frac{\eta R_M}{A}\right) \tag{10.13}$$

and plotting the left-hand side against the cumulative filtrate volume for different pressure levels provides a direct indication whether the specific cake resistance is a function of the operating pressure and hence whether or not the cake is compressible. Fortunately, many materials give a cake of roughly constant concentration under constant pressure conditions. Increasing the filtration pressure results in a cake with another uniform though higher solids concentration. This observation has led to the *average cake concentration and resistance concept* in which the specific resistance and pressure drop over the cake ΔP_c are related through the following empirical relation:

$$\alpha = \alpha_0 \, \Delta P_C^n \tag{10.14}$$

in which α_0 and n are empirical constants. The ratio of pressure drop over the full filter cake to the average specific resistance is equal to the integral of their differential amounts, yielding after substitution of eq. (10.14):

$$\frac{\Delta P_C}{\alpha_{av}} = \int_0^{\Delta P_C} \frac{d\Delta P}{\alpha} = \frac{1}{\alpha_0} \int_0^{\Delta P_C} \frac{d\Delta P_C}{\Delta P_C^n} \tag{10.15}$$

which integrates and rearranges to give:

$$\alpha_{av} = \alpha_0 (1-n) \Delta P_C^n \tag{10.16}$$

For *constant pressure* filtrations, this can be substituted into eq. (10.10) to give the general constant pressure filtration equation for compressible cakes:

$$\frac{t}{V} = \frac{\eta c \alpha_0 (1-n) \Delta P_C^n}{2A^2 \Delta P} V + \frac{\eta R_M}{A \Delta P} \tag{10.17}$$

For *constant rate* filtrations, eq. (10.12) can be rearranged to

$$\Delta P = \left(\frac{\eta c \alpha_0 (1-n) \Delta P_C^n}{A^2} \frac{V}{t} \right) V + \left(\frac{\eta R_M}{A} \frac{V}{t} \right) \tag{10.18}$$

Neglecting the resistance of the filter medium ($R_M << \frac{c \alpha_{av}}{A}$ hence $\Delta P \approx \Delta P_C$) gives:

$$\Delta P^{1-n} = \frac{\eta c \alpha_0 (1-n)}{A^2} \left(\frac{V}{t} \right)^2 t \tag{10.19}$$

Thus at constant rate filtration (V/t = constant) a logarithmic plot of pressure drop over the filter cake against time yields a straight line if average values of specific resistances and cake concentration exist.

10.3 Filtration equipment

An important factor in the optimization of filtration processes is the thickness of the filter cake. A too thick filter cake leads to an uneconomic lengthening of the filter cycle due to low filtering, dewatering and washing rates. On the other hand, thin filter cakes may be difficult to remove from the equipment, again increasing the filter cycle time. The *productivity* of all filters (kg dry solids/s) is related to the time required to complete a full filtration cycle and may be described as:

$$\text{Productivity} = \frac{\text{Mass of dry solids per cycle}}{\text{Cycle time}} = \frac{cV}{t_C} \tag{10.20}$$

where V is the volume of filtrate produced per cycle (m³) and c the mass of dry solids deposited per unit of filtrate volume (kg/m³). Besides filtration, t_F, additional time is required for dewatering, t_D, cake washing, t_W, and finally discharge of the filter cake and cleaning/reassembly/filling of the filter. It is usual to lump the latter two and other not mentioned operations into a total downtime period, t_{DW}. The *overall cycle time* is then given by:

$$t_C = t_F + t_D + t_W + t_{DW} \tag{10.21}$$

10.3.1 Continuous large-scale vacuum filters

Vacuum filters are the only truly continuous filters that can provide for washing, drying and other process requirements on a large scale. Examples include rotary drum, rotary disc, horizontal tilting pan and horizontal belt filters. Many of them use a *horizontal filtering surface* with the cake forming on top. The suspension is delivered to the filter at atmospheric pressure and vacuum is applied on the filtrate side of the medium to create the driving force for filtration. The *rotary vacuum drum filter* is the most popular vacuum filter. As depicted in Figure 10.7, the filtering surface of rotating drum filters is usually situated on the outer face of a cylindrical drum that rotates slowly about its horizontal axis and is partially submerged in a slurry reservoir. The drum surface is covered with a cloth filter medium and divided into independent longitudinal sections that are connected to the vacuum source by a circular connector or rotary valve. Filtration takes place when a section is submerged in the feed slurry. After cake formation dewatering and washing by vacuum displacement takes place, followed by cake discharge at the end of the rotational cycle. In some applications, compression rolls or belts are used to close

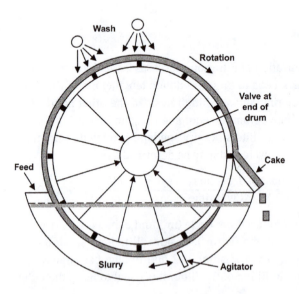

Figure 10.7: Schematic of a rotating vacuum filter.

possible cracks or to further dewater the cake by mechanical compression. Cake discharge can be achieved by knife, belt or string and roller discharge.

The theory for incompressible constant pressure (vacuum) difference cake filtration may be used to calculate the capacity of a rotary vacuum filter with a total surface, A, and an effective filter area, A_F, for a filtration time, t_F. Application of the constant pressure filtration equation, eq. (10.10), gives:

$$\frac{t_F}{V^2} = \frac{\eta \, \alpha \, c}{2 A_F^2 \, \Delta P} V + \frac{\eta \, R_m}{A_F \, \Delta P} \tag{10.22}$$

In the stationary state, the term with the medium resistance, R_M, often can be neglected and eq. (10.22) simplifies to:

$$\frac{t_F}{V^2} \approx \frac{\eta \, \alpha \, c}{2 A_F^2 \, \Delta P} \tag{10.23}$$

In that case, the filtrate volume, V, follows from:

$$V \approx A_F \sqrt{\frac{2 \, t_F \, \Delta P}{\eta \, \alpha \, c}} \tag{10.24}$$

The obtained filtrate volume can then be used to calculate the *mass of dry cake deposited* (kg) on the drum filter:

$$m = cV \tag{10.25}$$

These are the filtrate volume (m³) and mass of dry cake deposited (kg) per cycle of the drum. For a rotating drum filter the total cycle time, t_C, is the time needed for one drum evolution. The time for filtration, t_F, is determined by the fraction φ of the cycle time the filter surface is immersed in the slurry:

$$t_F = \varphi \, t_C \quad \text{and also} \quad A_F = \varphi \, A \tag{10.26}$$

Finally, the filtrate and dry solids production rate per unit of filter area are obtained by dividing both by the total drum filter area, A, and the total cycle time, t_C:

Drum filter filtrate production rate

$$V_r = \frac{V}{t_C \, A} \quad (\mathrm{m^3 \, m^{-2} \, s^{-1}}) \tag{10.27}$$

Drum filter dry solids production rate

$$w_r = \frac{cV}{t_C \, A} \quad (\mathrm{kg \, m^{-2} \, s^{-1}}) \tag{10.28}$$

Compared to drum filters, significant savings in required floor space and costs is possible with *vertical disc filters*. Rotary disc filters use a number of discs mounted vertically on a horizontal shaft and suspended in a slurry reservoir. The feed suspension is supplied continuously into troughs, in which the liquor flow is arranged in the same direction as the rotating discs. A particular disadvantage of disc filters is their inefficiency in cake washing and the difficulty they present for cloth washing.

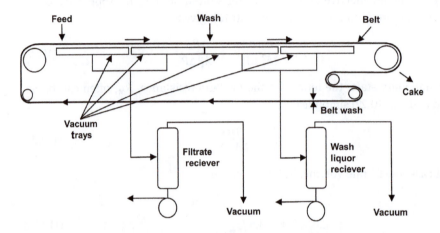

Figure 10.8: Schematic of a horizontal belt filter (adapted from [17]).

Horizontal filters largely circumvent the geometric constraints inherent in the design of rotary drum and vertical disc units. They have the advantage that gravity settling can take place before the vacuum is applied, and are ideal for cake washing, dewatering and other process operations such as leaching. Of the available industrial units in this category the *horizontal belt filter* is the most popular. A schematic diagram of a typical unit is shown in Figure 10.8. The feed slurry is supplied to the upper surface of the horizontal filter cloth that is supported by an endless belt situated above the vacuum box filtrate receivers. The top strand of the endless belt is used for filtration, cake washing and drying. There is appreciable flexibility in the relative areas allocated to each cycle step. Efficient cake discharge can be affected by separation of the belt from the cloth and directing the latter over a set of discharge rollers. Here, the produced cake is cracked and discharged by a sharp turn in the cloth over a small diameter roller. In module type horizontal belt filters, stainless steel or plastic trays replace the endless rubber belt. Horizontal belt filters are well suited to either fast or slowly draining solids, especially where washing requirements are critical. The primary advantages of this filter are its simple design and low maintenance costs. The main disadvantage is the difficulty of handling very fast filtering materials on a large scale. Cake formation on a belt filter is also described by the parabolic equation for vacuum filtration (eqs. (10.22)–(10.25)), similar to a rotating drum vacuum filter. The difference is that for a horizontal belt filter the fractional length of the belt devoted to filtration, z_F, and the belt speed, v_B, determine the *time for filtration*, t_F, and the *belt filtration area, A*:

$$t_F = \frac{z_F}{v_B} \quad \text{and} \quad A = z_F\, h_F \tag{10.29}$$

where h_F is the belt width and t_F is an incremental value which varies from zero to a maximum corresponding to the end of cake formation. After calculating the filtrate volume, V, and mass of dry cake deposited, m, from eqs. (10.24) and (10.25); the filtrate and dry solids production rate per unit of filter area are again obtained by dividing V and m by the belt filtration area, A, and time for filtration, t_F, respectively:

Belt filter filtrate production rate

$$V_r = \frac{V}{t_F\, A} \quad (\text{m}^3\,\text{m}^{-2}\,\text{s}^{-1}) \tag{10.30}$$

Belt filter dry solids production rate

$$w_r = \frac{cV}{t_F\, A} \quad (\text{kg}\,\text{m}^{-2}\,\text{s}^{-1}) \tag{10.31}$$

10.3.2 Batch vacuum filters

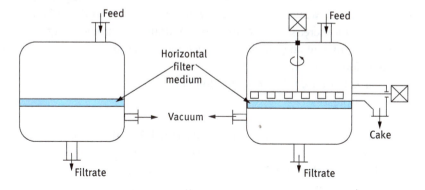

Figure 10.9: Nutsche filter.

Figure 10.10: Mechanized Nutsche filter.

The best-known batch vacuum filter units include Nutsche, horizontal table and vertical leaf filters. The *Nutsche filter* (Figure 10.9) is a scaled-up version of the simple Buchner funnel consisting of a tank divided into two compartments by a horizontal filter medium supported by a filter plate. Vacuum is applied to the lower compartment where the filtrate is collected. These filters are particularly advantageous for separations where it is necessary to keep batches separate and when rigorous washing is required. They are simple in design but laborious and prone to high wear in cake discharge. This problem is circumvented with the *mechanized*

Nutsche shown in Figure 10.10, which is provided with a shaft carrying a stirrer passing through the cover. The stirrer can sweep the whole filter area and can be lowered or raised vertically as required. The agitators are fitted to facilitate slurry agitation, cake smoothening prior to washing and cake removal. A discharge door is provided at the edge and the rotor moves the cake towards the door. Enclosed agitated filters are useful when volatile solvents are in use or when the solvent gives off toxic vapor or fume. Another advantage is that their operation does not require any manual labor. The horizontal table filter has overall features similar to the other horizontal vacuum filters, except that during filtration, washing and dewatering, the filter element is stationary. This facilitates optimization of filter cake thickness, wash times etc. The totally closed system is opened to allow band movement for cake discharge.

10.3.3 Pressure filters

In circumstances where large separating areas may be necessary because of slow settling characteristics, poor filterability, high solids content or other factors, the use of pressure filtration is beneficial. Pressure filters can be operated at constant pressure differential or at constant flow rate. At constant pressure, the liquid flow rate will decrease with time whilst at constant flow rate the pressure differential increases to compensate for increasing cake resistance. Within the extremely large variety of pressure filters three main groups may be distinguished:

1. Plate-and-frame filter presses
2. Pressure vessels containing tubular or flat filter elements
3. Variable-chamber presses

Various types of *plate-and-frame filter presses*, designed for cake formation and squeezing, are available for use within the chemical industry. The conventional plate-and-frame filter press shown in Figure 10.11 contains a sequence of perforated square, or rectangular, plates mounted on suitable supports alternating with hollow frames and pressed together with hydraulic screw-driven rams. The plates are covered with a filter cloth, which also forms the sealing gasket. Most units are batch operated. After filling frames with slurry, the filtrate is drained through the plates and the machine is disassembled or opened for cake discharge. Washing is performed by introducing the wash liquid either through the main feed port or through a separate port behind the filter cloth. Plate-and-frame filter presses are most versatile since their effective area is easily varied by blanking off some of the plates and cake holding capacity is altered by changing the frame thickness. In deciding the overall economics of the process, the time taken in discharging the cake and refitting the filter is of great importance.

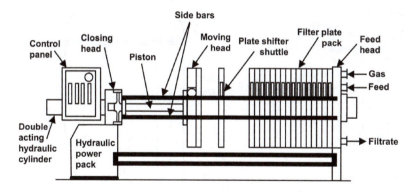

Figure 10.11: Typical arrangement of a plate-and-frame filter press.

All pressure vessel filter units consist of a multitude of leaf, candle or cartridge filter elements mounted horizontally or vertically in a pressure vessels housing (Figure 10.12). Vertical vessels with vertical leaf or candle filters are the cheapest of the *pressure vessel filters* and have the lowest volume-to-area ratio. In order to avoid filter cake bridging between the elements, serious attention has to be given to candle or leaf spacing. Deposition on the outer surface is advantageous in view of the increase in area with cake growth. Horizontal leaf filters consist of a stack of rectangular horizontal trays mounted inside the vessel that can be withdrawn for cake discharge. They have the disadvantage that half the filtration area is lost because the underside of the leaf cannot be used for filtration. Cartridge filtration is limited to *liquid polishing* or *clarification* in order to keep the frequency of cartridge replacements down.

Most continuous pressure filters have their roots in vacuum filtration technology. They have been adapted to pressure by enclosing in a pressure cover. As such one find continuous disc, drum and belt pressure filters. Special designs of continuous pressure filters are *belt presses* and *screw presses*. Belt presses combine gravity drainage with mechanical squeezing of the cake between two running belts. In screw presses (Figure 10.13), a screw mounted inside a perforated cage conveys the material along the barrel. The available volume for transport diminishes continuously along the length of the screw in order to compress the filter cake. Washing liquid can be injected at points along the length of the cage. Screw presses are only suitable for the dewatering of a high solids containing paste or sludge, because no filtration stage is included.

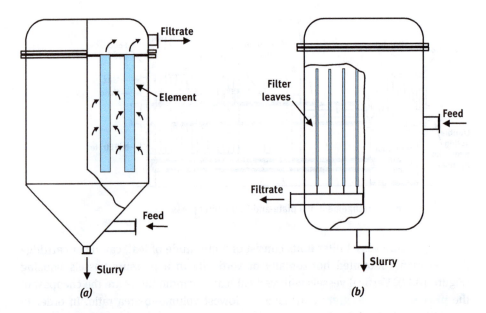

Figure 10.12: A vertical candle (a) and vertical leaf (b) pressure vessel filter.

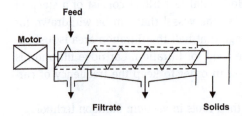

Figure 10.13: Schematic of a screw press.

10.4 Filter media

The filter medium is the heart of any filtration step and the importance of a careful selection cannot be overstated. Difficulties in various industrial processes relate to the interaction between impinging particles and the pores of the filter medium. Selection of the filter medium, however, depends largely on experience in small- and large-scale operations. The resistance to filtrate flow, clarity of filtrate and durability may characterize the performance of the filter medium. Resistance to flow depends, to a great extend, on the porosity or free area of the medium which in turn depends on the material used and the type of manufacturing. Clarity of the filtrate depends on the pore size distribution of the filter medium and the particle size as well as the ability of the particles to form bridges to cover the pores. The ideal circumstance, where all particles are retained on the surface of the filter is often not realized. Particle penetration into cloth or membrane pores leads to an increase in resistance of the medium to the flow of filtrate.

A wide variety of filter media are available to the user. Table 10.1 gives an overview of relevant filter media choices and their characteristics. For most applications the medium of particular interest will be the one that is readily installed in the filter for the process. Thus, woven and nonwoven fabrics, constructed from natural (cotton, silk, wool) or synthetic fibers, are probably the most common industrial filter medium in pressure, vacuum and centrifugal filters. Wire cloths and meshes, produced by weaving monofilaments of ferrous or nonferrous metals, are also widely used in industrial filtrations. At the small scale, filter papers are common. The same materials and also rigid porous media (porous ceramics, sintered metals, woven wires) can be incorporated into cartridge and candle filters. These filter elements are usually constructed in the form of a cylinder. Cartridge filters are either depth or surface type and widely used throughout the whole process industry for the clarification of liquids. They are particularly useful at low-solid contaminant concentration and particle sizes smaller than 40 μm.

Table 10.1: Overview of filter media.

Filter medium	Examples	% Free area	Minimum trapped particle size (μm)
Woven fabrics	Cloths, natural and synthetic fibers	20–40	10
Rigid porous media	Ceramics,	50–70	1
	sintered metal, glass		3
Metal sheets	Perforated plates	5–20	100
	woven wire		5
Nonwoven sheets	Felts,	60–80	10
	paper, mats		5
Loose solids	Sand, diatomaceous earth, perlite, carbon	60–95	

If the specific cake resistance is too high or the cake is too compressible, the addition of filter aids can improve the filtration rate considerably. Filter aids are rigid, porous and highly permeable powders that are applied as a *precoat* which then acts as a filter medium on a coarse support, or mixed with the feed suspension as *body feed* to increase the permeability of the resulting cake. A precoat of filter aids allows filtration of very fine or compressible solids from suspensions of 5% or lower solids concentration. In fact, we speak here about deep bed filtration. In body feed filtration the filter aid serves as a body builder to obtain high cake porosities in order to maintain high flow rates. Materials suitable as filter aids include diatomaceous earth, expanded perilitic rock, cellulose, nonactivated carbon, ashes, ground chalk or mixtures of those materials.

10.5 Centrifugal filtration

10.5.1 Centrifugal filters

In a centrifugal filter the suspension is directed on the inner surface of a perforated rotating bowl to induce the separation of liquid from the solids. Centrifugal filters essentially consist of a rotating basket equipped with an appropriate filter medium (Figure 10.14). The driving force for filtration is the centrifugal force acting on the fluid. During cake formation, the filtrate passes radially outwards through the filter medium and the bowl. Because the centrifugal forces tend to pull out the liquid from the cake, filtering centrifuges are excellent for *dewatering* applications. Solids removal may take place continuously or batch-wise. In comparison to other filtration equipment, centrifuges show good separation performance at the expense of relatively high costs. This is due to required special foundations to absorb vibrations combined with high manufacturing and maintenance costs associated with the parts rotating at high speed. Centrifuges are generally only applicable to the coarser particles, typically 10 μm to 10 mm.

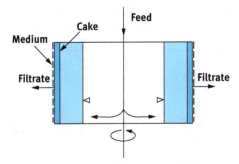

Figure 10.14: Perforated basket centrifuge.

The simplest fixed-bed centrifuge design is the *perforated basket centrifuge*, which has a vertical axis, a closed bottom and an overflow restriction at the top end. The slurry is fed through a pipe or rotating feed cone into the basket. The cake is discharged manually or by a scraper that moves into the cake after the basket slows down to a few revolutions per minute. The plough directs the solids toward a discharge opening provided at the bottom of the basket. Some residual cake remains since the plough cannot be allowed to reach too close to the cloth. The basket centrifuge has a wide range of application in the filtration of slow draining products that require long feed, rinse and draining times. It can be applied to the finest suspensions of all filtering centrifuges because filter cloths may be of pore size down to 1 μm. Making the axis horizontal may eliminate the nonuniformity due to gravity with a vertical basket. This is known as the *peeler centrifuge* shown in Figure 10.15, which is designed to operate at constant speed to eliminate nonproductive periods. The cake is also discharged at full speed by means of a sturdy knife, which peels off the cake into a screw conveyor or a chute in the center of the basket. The peeler centrifuge is particularly attractive where filtration and dewatering times are short such as high output duties with nonfragile crystalline materials.

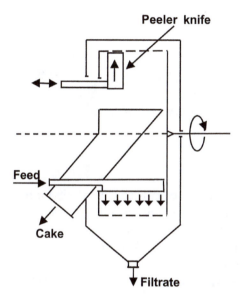

Figure 10.15: Peeler centrifuge.

Continuously operating, moving bed centrifuges use either conical or cylindrical screens. The *conical centrifuge* in Figure 10.16 has a conical basket rotating either on a vertical or horizontal axis. The feed suspension is fed into the narrow end of the cone. If the cone angle is sufficiently large for the cake to overcome its friction on the screen, the centrifuge is *self-discharging*. *Pusher-type centrifuges* have a cylindrical basket with a horizontal axis. The feed is introduced through a distribution cone at the closed end of the basket (Figure 10.17) and the cake is pushed along the basket by means of a reciprocating piston that rotates with the basket. Pusher centrifuges can be made with multistage screens consisting of several steps of increasing diameter.

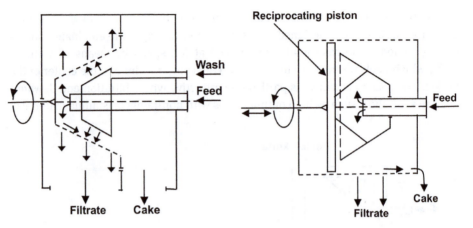

Figure 10.16: Conical centrifuge.   Figure 10.17: Pusher centrifuge.

10.5.2 Filtration rates in Centrifuges

Theory of fluid flow through porous media has not been used extensively for the design of filtering centrifuges. Nonetheless, the effect of the major process variables on the separation performance can be readily assessed through a mathematical model derived for simple centrifuge geometries. The situation obtained during slurry filtration in a basket centrifuge is illustrated in Figure 10.18. Consider a differential cylindrical element of filter cake between radii r and $r+dr$. The mass of solids contained in this element is:

$$dw = \rho_S(1-\varepsilon)\,2\pi hr dr \tag{10.32}$$

and the area available for filtrate flow equals:

$$A = 2\pi hr \tag{10.33}$$

The *actual pressure differential* over the element of the cake is given by the difference between the centrifugal body force and the frictional force.

$$-dP = dP_F - dP_R \tag{10.34}$$

Darcy's law gives the pressure drop in the fluid caused by frictional effects from the volumetric flow of liquid Q through the element:

$$\frac{dP_F}{dr} = \left(\frac{\eta\alpha(1-\varepsilon)\rho_S}{2\pi rh}\right)Q \tag{10.35}$$

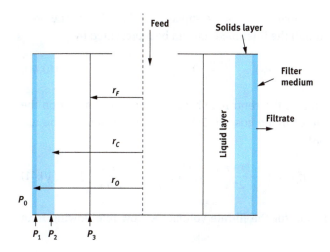

Figure 10.18: Schematic of the centrifugal filtering process.

The *centrifugal body force* acting on the liquid flowing through the cake is:

$$dP_R = \rho\omega^2 r dr \tag{10.36}$$

Introduction of Darcy's law for the pressure drop due to the friction and the centrifugal pressure effect in eq. (10.34) gives for the total effect from radius r_C to r_0:

$$-\int_{P_2}^{P_1} dp = \left(\frac{\eta\alpha(1-\varepsilon)\rho_S}{2\pi h}Q\right)\int_{r_C}^{r_0}\frac{dr}{r} - \rho\omega^2\int_{r_C}^{r_0}r dr \tag{10.37}$$

where P_2 is the pressure at the cake-supernatant liquid interface. Here, it has been assumed that both α and ε are invariant with radius, which should be reasonable true for crystalline inorganic precipitates and crystals. After integration, we obtain:

$$-(P_1 - P_2) = \left(\frac{\eta\alpha(1-\varepsilon)\rho_S}{2\pi h}Q\right)\ln\left(\frac{r_0}{r_C}\right) - \frac{\rho\omega^2}{2}(r_0^2 - r_C^2) \tag{10.38}$$

The *hydraulic pressure* resulting from the supernatant slurry layer on top of the produced filter cake is given by:

$$(P_2 - P_3) = \rho\omega^2\int_{r_F}^{r_C} r dr = \frac{\rho\omega^2}{2}(r_C^2 - r_F^2) \tag{10.39}$$

where P_3 is the pressure at the inner radius of the supernatant fluid layer. The pressure drop caused by flow through the filter medium can be represented by:

$$(P_1 - P_0) = \frac{\eta R_M}{2\pi r_0 h} Q \tag{10.40}$$

Assuming that the pressure inside the centrifuge basket is the same as that on the outside ($P_3 = P_0$) allows us to combine eqs. (10.39) and (10.40) to replace the unknown $-(P_1 - P_2)$ in eq. (10.38):

$$\frac{\rho_L \omega^2}{2}(r_C^2 - r_F^2) + \frac{\rho_L \omega^2}{2}(r_0^2 - r_C^2) = \left(\frac{\eta \alpha (1-\varepsilon)\rho_S}{2\pi h} Q\right) \ln\left(\frac{r_0}{r_C}\right) + \frac{\eta R_M}{2\pi r_0 h} Q \tag{10.41}$$

which reduces to the expression for the filtrate volumetric flow rate flowing from the centrifuge:

$$Q = \frac{\frac{\rho_L \omega^2}{2}(r_0^2 - r_F^2)}{\left(\frac{\eta \alpha (1-\varepsilon)\rho_S}{2\pi h}\right)\ln\left(\frac{r_0}{r_C}\right) + \frac{\eta R_M}{2\pi r_0 h}} \tag{10.42}$$

At this stage, it is often convenient to write this expression in terms of the *mass of solids filtered*, M_S, by introducing:

$$M_S = \rho_S(1-\varepsilon)\,\pi h(r_0^2 - r_C^2) \tag{10.43}$$

to substitute $\rho_S(1 - \varepsilon)$:

$$Q = \frac{\frac{\rho_L \omega^2}{2}(r_0^2 - r_F^2)}{\left(\frac{\eta \alpha M_S}{2\pi h[\pi h(r_0^2 - r_C^2)]}\right)\ln\left(\frac{r_0}{r_C}\right) + \frac{\eta R_M}{2\pi r_0 h}} \tag{10.44}$$

The denominator of this expression can be rewritten in terms of an *average flow area* in the cake, A_{av}, a *logarithmic mean area of flow*, A_{lm} and the bowl or medium area, A_0, by noting:

$$\frac{2\pi h[\pi h(r_0^2 - r_C^2)]}{\ln(r_0/r_C)} = \frac{2\pi h(r_0 + r_C)}{2} \frac{2\pi h(r_0 - r_C)}{\ln(r_0/r_C)} = A_{av} A_{lm} \tag{10.45}$$

Hence, the *filtrate volumetric flow rate* can be written as:

$$Q = \frac{\frac{\rho_L \omega^2}{2}(r_0^2 - r_F^2)}{\frac{\eta \alpha M_S}{A_{av} A_{lm}} + \frac{\eta R_M}{A_0}} = \frac{\Delta P}{\frac{\eta \alpha M_S}{A_{av} A_{lm}} + \frac{\eta R_M}{A_0}} \tag{10.46}$$

This equation suggests that plots of t/V versus M_S (or V) can be used to estimate cake and medium resistances during centrifugal filtration, as is the case in vacuum and pressure filtration.

10.6 Interceptive filtration

10.6.1 Deep bed filtration

Deep-bed filtration is typically used for the removal of submicrometer suspended solid particles with very low concentrations that do not settle or filter readily. Their relatively high efficiency in removing fine particles present in low concentration is used extensively for drinking water filtration and the final polishing of effluents before discharge or process liquids prior to further processing. The high capacity available in modern deep bed filters offers the implementation opportunities where conventional equipment could not produce an economical separation.

Commercial deep-bed filters are simple in construction with a holding tank for the granular bed of solids through which the suspension to be filtered is passed. Flow may be directed downwards or upwards, with typical liquid velocities of 40 and 15 m/h. The granular bed is usually made up of sand, gravel, anthracite or a variety of other materials having a particle size of 0.5–5 mm. Although the diameter of the pores is 100–10,000 times larger than the diameter of the suspended particles, suspended particles are deposited inside the filtration medium. Particle deposition occurs via a collection of mechanisms in mechanical and physiochemical in nature. In addition to mechanical interception; particles deposit inside the filter medium to electrostatic and London dispersion forces also known as Van der Waals attractive interactions. The deposited particles cause blockage and an increased pressure drop of the fluid flowing through the bed. A common design to improve the effectiveness of the deep-bed filter is to use multiple solids layers on top of each other. This is illustrated in Figure 10.19, where a dual layer of anthracite/sand is contained in a vessel. The anthracite particles made of coal, being coarser than the sand, serve to prevent the formation of surface deposits on the sand surface. During filtration the bed is contaminated with the particles from the filtered liquid. At fixed times the filter is *back-washed* with clean liquid in counter flow configuration to rinse the bed clean. The filter bed is fluidized, quickly releasing the particles deposited during the filtration stage. This is a regeneration step extending the life-time of these filters.

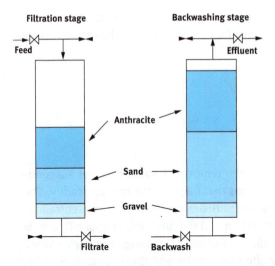

Figure 10.19: Dual layer deep bed filter.

10.6.2 Impingement filtration of gases

Baffle bundles, also known as *impingement separators*, are widely employed in industry for removal of droplets and particles from gas flows. In plate separators, rows of baffles can be arranged in series to increase the overall efficiency of impingement. In this manner, the rather modest collection efficiencies accessible in a single row configuration higher industrially practical collection efficiencies are achieved with a multiple row unit configuration. A series of parallel impact plates is commonly named a *wave-plate separator*. Multiple body contacting is also used in *fibrous filters*, which are especially important in modern particle collection technology when highly efficient collection in the finest particle size range is required. Of all collectors, these have the widest spectrum of application and thus a largest share of the market. With regard to application, design and operation, filtration devices for particle collection can be classified as deep-bed filters or surface filters.

Deep-bed filters consist of a relatively loose fiber mat with a pore volume fraction often >99%. The term filter denotes packing consisting of wires or fibers, which obstructs the cross-section of a gas flow. The filter can be arranged either horizontally or vertically and can consist of uniformly or randomly arranged wires or fibers. Particle collection takes place in the interior of the layer where the dust accumulates. As the gas flows through the fibrous layer, it must flow past a large number of cylindrical elements. Hereby, the particles in the gas continue their path until they strike a wire or fiber element by inertial or flow-line interception. For submicron particles, Brownian diffusion influences the performance of these filters. Smaller the particles diffuse, more decreasing efficiency of adsorption on fibers. An important concept for the collecting efficiency of a filter is its *overall on-flow area*. This follows from the total length of all the fibers from which the filter is made up. Based on the filter cross-sectional area, A, the overall relative on-flow area, p, is given by:

$$p = \frac{LD}{A} = f \frac{4}{\pi} \frac{G}{\rho_F} \frac{H}{D} \tag{10.47}$$

where L is the total wire length of the filter, D is the wire diameter, G is the mass of the filter, H is the thickness of the packing and ρ_F is the density of the wire material. The fact that the fibers are not at right angles to the direction of flow is accounted for through the correction factor f. With this relative on-flow area concept it is possible to transfer the collecting efficiency of a single wire to the wire filter. For a single layer of wire, the collection efficiency with relative on-flow area p_1 and single wire collection efficiency $\eta_F(0)$ becomes:

$$\eta_F(1) = \eta_F(0) p_1 \tag{10.48}$$

Extension to n consecutive layers provides the *collection efficiency* of a filter as the product of the single layer probabilities:

$$\eta_F(n) = 1 - (1 - \eta_F(0) p_1)^n \approx 1 - \exp(-\eta_F(0) p) \tag{10.49}$$

with

$$p = p_1 n \tag{10.50}$$

With eq. (10.49) and a known value of the relative on-flow area p of a wire or a fiber filter, its collection efficiency can be calculated from that of a single cylinder.

For the collection of liquid droplets, fogs and mists in-depth fiber bed filters made out of horizontal pads of knitted metal wire are used (Figure 10.20). Collection from the upflowing gas is mainly by inertial interception. Thus, efficiency will be low at low superficial velocities and for fine particles. For collection of fine mist particles, the use of randomly oriented fiber beds is preferred. Fine particle removal by filtration through a bed or granular solids has appeal because of corrosion and temperature resistance. Several types of aggregate-bed filters are available which provide in-depth filtration. Both gravel and particle-bed filters have been developed for removal of dry particulates. Important parameters for the collection efficiency in granular beds are bed thickness, gravel size and air velocity.

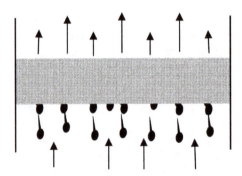

Figure 10.20: Schematic of a wire mesh separator.

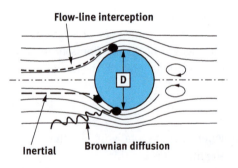

Figure 10.21: From top to bottom inertial interception, flow-line interception and Brownian diffusion particle separation mechanisms.

10.6.3 Interception mechanisms

Impingement separators require the transport of particles to a surface on which they are deposited by collision. Deposition requires a body for impact to be placed in the gas stream. This can be a wire, a fiber or a droplet of washing liquid around which the gas can flow on all sides. However, as in the case of wave plate separators, it can also be an impact surface which forces the gas flow to change its direction. If the particle contacts the surface of a fiber, it can be regarded as collected. The three important mechanisms for particle collection schematically drawn in Figure 10.21 are: inertial interception, Brownian diffusion and flow-line interception. *Inertial interception* occurs when particles approach targets such as a baffle, impaction element, fiber or droplet with sufficient velocity to cause a collision with the target by inertia of the particle. As a result of inertia, particle trajectories deviate from the streamline of the gas so that the particles can strike the collecting surface of the target. This applies to all particles of large enough diameters present in the central filament of flow in the oncoming gas. Because sufficiently large particles will always move in a straight line, their collection efficiency will be close to unity. *Brownian motion* causes the particles to diffuse at random across flow streamlines to be captured by a target. Only very small particles (<1 μm) are subject to Brownian motion, which is therefore of secondary importance in most gas-solid separations focusing in separating larger particles. *Flow-line interception* is the striking of a target by a particle that passes the target in a streamline within one particle radius. While inertial interception separation increases with droplet diameter, diffusional deposition increases with decreasing droplet size. This results in a separation minimum at droplet sizes between 0.2 μm and 0.7 μm. It is mainly this minimum range where flow-line interception can dominate. The resulting fractional collection efficiency of a single fiber is shown in Figure 10.22.

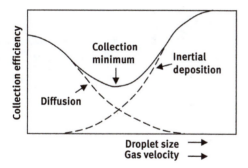

Figure 10.22: Collection efficiency of a single fiber.

For interception separators, the collection efficiency is usually estimated in terms of the *target efficiency* of the single baffle. The target or collection efficiency is described in terms of the separation number defined as:

$$N_S = \frac{V_\infty V_0}{g D_b} \qquad (10.51)$$

where g is the gravitational acceleration, D_b is the characteristic baffle dimension, V_∞ is the terminal settling velocity and v_0 is the approach gas velocity. The use of Figure 10.23 assumes that Stokes' law is applicable. The efficiency increases with increasing approach gas velocity because of the particle inertia. Likewise, efficiency increases with increasing terminal velocity. Somewhat counterintuitive is the decrease in efficiency with larger characteristic dimension. This is a result of the increased distance that the gas must be deflected.

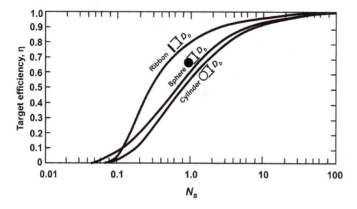

Figure 10.23: Target efficiency of single spheres, cylinders and ribbons.

10.6.4 Lamellar plate separators

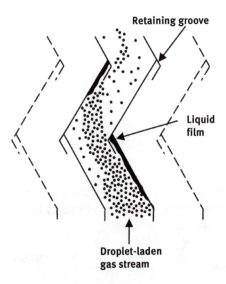

Figure 10.24: Operating principle of a wave-plate separator.

Lamellar plate separators are often used because of their simple construction and low-pressure drop. These devices consist of parallel channels with multiple deflections and retaining grooves (Figure 10.24). The droplet-laden gas stream undergoes several direction changes in the channels. When the gas is diverted, entrained particles that are sufficiently large will continue their motion due to inertia until they strike the collecting surface. On the other hand, particles small enough to be carried with flow behave almost as gas molecules and follow the gas flow. Particles intermediate in size can follow the curvatures of the gas flow only partially. Depending on their initial position and the corresponding stream line, these droplets will hit or miss the impact body. Hence, only a partial separation is possible for these droplets, separation efficiencies vary dramatically depending on the size distribution. It follows that for impingement separators, the collection efficiency varies from nil for sufficiently small particles to 100% for large particles. Like all inertial collectors, *lamellar separators* become more efficient as flow velocity increases. On account of their simple and compact construction, wave plate separators are relatively cost effective and space saving.

Nomenclature

A	Area	m^2
c	Mass of dry solid per unit volume suspension	$kg\ m^{-3}$
D	Diameter	m
g	Gravitational acceleration	$m\ s^{-2}$
h, H	Height	m
L	Length	m
m, M	mass	kg
N_S	Separation number	–
t	Time	s
Q	Volume flow rate	$m^3\ s^{-1}$
p	Relative on-NBSflow/NBS area, eq. (10.47)	m^2
P	Pressure	$N\ m^{-2}$
r	Radius	m
R_M, R_C	Resistance of filter medium and cake, resp.	m^{-1}
v	Velocity	$m\ s^{-1}$
v_r	Filtrate production rate	$m\ s^{-1}$
V	Volume	m^3
w	Mass of dry solids per unit area	$kg\ m^{-2}$
w_r	Dry solid production rate	$kg\ m^{-2}\ s^{-1}$
α	Constant in eqs. (10.6), (10.14), (10.16)	$m\ kg^{-1}$
ε	Fluid volume fraction, eq. (10.1)	–
η	Viscosity	$kg\ s^{-1}\ m^{-1} = N{\cdot}s\ m^{-2} = Pa{\cdot}s$
$\eta_F(n)$	Collection efficiency of n layers, eq. (10.49)	–
ρ	Density	$kg\ m^{-3}$
φ	Fraction of cycle time, eq. (10.26)	–
ω	Angular velocity	$rad\ s^{-1}$

Excercises

1   A slurry is filtered with a laboratory leaf filter with a filtering surface area of $0.05\ m^2$ to determine the specific cake and cloth resistance using a vacuum giving a pressure difference of 0.7 bar. The volume of filtrated collected in the first 5 min was $250\ cm^3$ and after a further 5 min, an additional $150\ cm^3$ was collected. Calculate the specific cake and the cloth resistance given that the filtrate viscosity is 10^{-3} Pa.s and that the slurry contains 5 vol% of solids (density $3{,}000\ kg\ m^{-3}$).

2   The data given in the table below were obtained from the constant pressure period of a pilot scale plate and frame filter press. The mass of dry cake per unit volume of filtrate amounts $125\ kg/m^3$. Calculate the cake resistance given: filter area $2.72\ m^2$, viscosity 10^{-3} Pa.s, filtrate pressure3 bar.

time:	92	160	232	327	418	472	538	(s)
filtrate volume:	0.024	0.039	0.054	0.071	0.088	0.096	0.106	(m^3)

3 Calculate the specific cake and medium resistance when the same slurry data apply from the following constant rate data obtained on the same pilot scale plate and frame filter press:

filtrate volume:	0.016	0.032	0.040	0.056	0.064	0.072	0.088	0.096	0.114	(m³)
pressure:		0.9	1.2	1.35	1.7	1.8	1.85	2.3	2.4	2.7 (bar)

4 Laboratory filtrations conducted at constant pressure drop on a slurry of $CaCO_3$ in water gave the data shown in the following table:
The filter area was 440 cm², the mass of solid per unit volume of filtrate was 23.5 g/L and the temperature 25°C. Evaluate the quantities α and R_m as a function of pressure drop and fit an empirical equation to the results for α.

Filtrate volume	Test I (0.45 bar)	Test 2 (1.10 bar)	Test 3 (1.95 bar)	Test 4 (2.50 bar)	Test 5 (3.40 bar)
V(l)	t(s)	t(s)	t(s)	t(s)	t(s)
0.5	17	7	6	5	4
1.0	41	19	14	12	9
1.5	72	35	24	20	16
2.0	108	53	37	30	25
2.5	152	76	52	43	35
3.0	202	102	69	57	46
3.5		131	89	73	59
4.0		163	110	91	74
4.5			134	111	89
5.0			160	133	107
5.5				157	
6.0				183	

5 Calculate the relationship between the average specific resistance and the filtration pressure from the following data, obtained from a series of constant pressure filtration experiments:

Filtration pressure:	70	104	140	210	400	800	(kPa)
Specific resistance:	1.4	1.8	2.1	2.7	4.0	5.6	(×10¹¹ m/kg)

6 A slurry, containing 0.1 kg of solid (specific gravity 2.5) per kilogram of water, is fed to a rotary drum filter 0.6 m long and 0.6 m diameter. The drum rotates at one revolution in 6 min and 20 percent of the filtering surface is in contact with the slurry at any instant. The specific cake resistance is 2.8×10^{10} m/kg and medium resistance 3.0×10^{9} m^{-1}.

a. Determine the filtrate and dry solids production rate when filtering with a pressure difference of 65 kN/m^2.

b. Calculate the thickness of the cake produced when it has a porosity of 0.5.

7 A rotary drum filter with 30 percent submergence is to be used to filter an aqueous slurry of $CaCO_3$ containing 230 kg of solids per cubic meter of water. The pressure drop is to be 0.45 bar. Calculate the filter area required to filter 40 ltr/min of slurry when the filter cycle time is 5 min. The specific cake resistance is 1.1×10^{11} m/kg and medium resistance 6.0×10^{9} m^{-1}.

8 Calculate the dry solids production from a 10 m^2 rotating vacuum filter operating at 68 kPa vacuum and the following conditions: Cake resistance 1×10^{10} m/kg, medium resistance 1×10^{10} 1/m, drum speed 0.105 rad/s (1 rpm), fraction submerged 0.3, solids concentration 0.1 kg solids/kg slurry, cake moisture 3.5 kg wet cake/kg dry cake, liquid density 1000 kg/m^3, liquid viscosity 0.001 Pa.s.

9 Calculate the filtration time, required area and operational speed and cake thickness of a 1 m wide horizontal belt filter operating with a vacuum of 60 kPa required to produce 5 m^3/h of filtrate. The slurry has the following properties: specific cake resistance 5×10^{9} m/kg, slurry solids concentration 350 kg/m^3, solids density 2,000 kg/m^3, liquid density 1,000 kg/m^3 and viscosity 0.001 Pa.s, cake porosity 0.43

Filter data:	width 2 m, linear velocity 0.1 m/s, medium resistance 2×10^{9} m^{-1}
Cake properties:	$\alpha_{av} = 7.1 \times 10^{8} \Delta P^{0.46}$ m/kg, $\varepsilon_{av} = 0.84 \Delta P^{-0.054}$, solid density 2350 kg/m^3
Filtrate:	density 1390 kg/m^3, viscosity 0.001 Pa.s

10 It is proposed to use an existing horizontal belt filter to separate phosphoric acid from a slurry containing gypsum at 30% w/w. Cake formation at a constant vacuum of 50 kPa is to be followed by displacement washing and deliquoring. Of the total 9 m belt length 1.5 m is available for the filtration stage. Calculate the solids production rate (kg/s).

11 A tank filter is operated at a constant rate of 25 ltr/min from the start of the run until the pressure drop reaches 3.5 bar and then at a constant pressure drop of 3.5 bar until a total of 5 m^3 of filtrate is obtained. What is the total filtration time required given: specific cake resistance 1.8×10^{11} m/kg, medium resistance 1.0×10^{10} m^{-1} slurry solids concentration 150 kg/m^3, viscosity 0.001 Pa.s

12 Calculate the washing rate of a filter cake 0.025 m thick deposited on a centrifuge basket (0.635 m diameter, 0.254 m height) rotating at 20 rps. Assume the
cake is incompressible, has a porosity of 0.53 and a specific resistance of
6×10^9 m/kg. A medium with a resistance of $R_m = 1 \times 10^8$ m^{-1} is used to line the
perforate basket. There is almost no supernatant liquid layer over the cake.
Solids density 2,000 kg/m^3, liquid density 1,000 kg/m^3 and viscosity 0.001 Pa.s.
   a. What time is required for the passage of two void volumes of wash if there
      is almost no supernatant liquid layer over the cake?
   b. How does the washing rate change when a 5 cm thick supernatant liquid
      layer is present?

13 To evaluate the feasibility of filtration a titanium dioxide slurry is filtered in a
laboratory setup with an area of 0.01 m^2 and a constant pressure difference of
1 bar. The solid concentration equals 50 kg dry titanium dioxide per m^3 filtrate.
The liquid viscosity is 0.001 Pa.s. A typical measuring series yields:

Time(s):	60	300	540	900
Filtrate volume $\times 10^3$ m^3:	0.25	0.59	0.79	1.03

   a. Determine the specific cake resistance α.
      The same titanium dioxide slurry is filtered with a horizontal belt filter at a
      constant pressure difference $\Delta P = 60$ kPa and a belt velocity of 0.1 m s^{-1}.
      The filter area of 2 m^2 uses the same filter cloth as used in the laboratory
      trials. The medium resistance R_M can be neglected compared to the specific
      cake resistance. The belt width equals $w = 0.5$ m
   b. Calculate the production capacity of this belt filter in kg dry titanium diox-
      ide per hour.

14 A slurry that contains 100 kg dry solid particles per m^3 of slurry has been evaluated in a laboratory filter with an area of 0.050 m^2. These tests were performed
with a constant pressure difference of 0.6 bar. After each half liter of filtrate volume the time was recorded. A representative series is:

Time (s)	45	101	180	265
Filtrate volume (liter)	0.5	1.0	1.5	2.0

a. Determine the specific cake resistance α (m kg^{-1}) and the filter resistance
   R_M (m^{-1})

$\rho_s = 1050$ kg m^{-3}	$\rho_L = 1000$ kg m^{-3}	$\eta = 10^{-3}$ Pa.s	$g = 9.81$ m s^{-2}

The large scale filtration of this slurry is executed under constant rate conditions in a pressure filter. The maximum overpressure of the filter housing amounts 2 bar. The filtration starts with a pressure difference of 0.6 bar on a clean filter medium with the same properties as used in the laboratory tests. The filter area A = 4 m².

b. Calculate the initial filtration rate?

c. Calculate the filtrate amount at the moment the overpressure equals 2 bar.

d. How long does it take to reach this 2 bar overpressure?

Chapter 11
Membrane Filtration

11.1 Introduction

Landfill gas, comprising approximately equal parts of methane and carbon dioxide, is formed by anaerobic degradation of organic materials in municipal and/or agricultural waste. *Membrane processes* are frequently used to bring the caloric value of landfill gas to the specifications of domestic natural gas by removing the small amounts of contaminants and a considerable part of carbon dioxide. Membrane gas permeation installations, operating generally at 1,000 to 100,000 m^3/day, are widespread.

With conventional filter particles with sizes of about 1 µm and up can be separated, as discussed in Chapter 10. With membranes, particles down to molecular sizes can be removed from gas and liquid streams. That is why a variety of membrane separation processes are used by so many industries. This technology is applied to important practical separations such as

- filtration of particles from liquids or gases (*micro-* and *ultrafiltration*)
- removal of low molecular weight solutes from a solvent (e.g., desalination of seawater or decoloring of wastewater by *reverse osmosis* or *nanofiltration*)
- removal of low molecular weight materials from macromolecular mixtures (e.g., separation of urea from proteins by *dialysis* as in artificial kidneys)
- *gas separation* (production of pure nitrogen or oxygen from air, hydrogen purification)
- separation of liquids (e.g., dehydration of organic liquids by *pervaporation* or *vapor permeation*)

These many types of membrane separations differ in type of membrane, driving force and module design (combination of module and housing). Membrane processes extend from the separation of particles as large as 5 µm from a liquid or a gas, to molecules and ions as small as 0.5 nm. It goes without saying that in a single chapter like this, all aspects of the broad field of membrane technology cannot be covered in sufficient detail. This chapter will be limited to *pressure-driven* membrane separations, which covers major part of the membrane market worldwide. Special membrane separations such as electrodialysis, membrane distillation, electrofiltration and separation by liquid membranes are not covered here. For an in-depth treatment we refer dedicated literature (see references and further reading).

A thorough discussion about membrane selection, transport mechanisms, module design and flow patterns are beyond the scope of this book. In this chapter these topics will just be touched upon, only to understand the formulation of the applicable rate expressions, membrane fouling mechanisms and the consequences for basic module design.

https://doi.org/10.1515/9783110654806-011

11.2 Membrane selection

A formal definition of a membrane would be a *semipermeable phase, which restricts the motion of certain species in a different way*. Therefore, the relative rates of transport of various species through this barrier are controlled by its properties. The differences in *rejection* (relative low velocity) and *permeation* (relative high velocity) through the membrane are the fundamental base of membrane separation. Any pressure driven membrane separation process can be depicted as in Figure 11.1. The membrane splits a feed stream into a *permeate* stream, enriched in particles capable of permeating the membrane and a *retentate* stream that leaves the module without passing through the membrane. The permeate is sometimes called filtrate or penetrant, the retentate is also known as concentrate, residue, reject or raffinate.

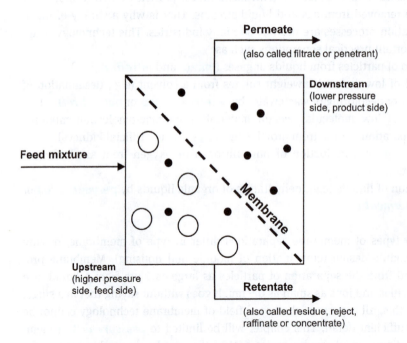

Permeate

(also called filtrate or penetrant)

Downstream
(lower pressure
side, product side)

Feed mixture

Membrane

Upstream
(higher pressure
side, feed side)

Retentate

(also called residue, reject,
raffinate or concentrate)

Figure 11.1: Outline and terminology of a membrane filtration process.

A membrane can be polymeric, inorganic or even metallic, as long it is semipermeable to at least one component in the mixture to be separated. The structure of a membrane determines the mechanism by which it performs separation:

- By *selective retardation*

  The membrane contains very narrow pores (of molecular dimensions) in which transport takes place by Knudsen diffusion. The rate of diffusion varies with the molecular weight, retarding heavier molecules.
- By *size exclusion*

  The membrane contains pores of such dimensions that smaller molecules can pass through while others, with larger molecular cross sectional areas, cannot.
- By *solution-diffusion*

  Dense membranes can separate by a mechanism known as solution-diffusion. The upstream penetrant molecules dissolve in the membrane material, move to the downstream permeate side by molecular diffusion, where they are desorbed.

Basically, two families of membranes can be distinguished: symmetrical and asymmetrical, see Figure 11.2. *Symmetrical membranes*, usually with a thickness of several microns, can be either dense (nonporous) or porous. Porous membranes are used in microfiltration and dialysis, whereas dense membranes find applications in gas separation. A major improvement has been the development of *asymmetric membranes*, comprising a very thin selective toplayer, either dense or (micro)pores fitted on a much thicker, porous support. The very thin toplayer, usually a few nanometers thick, allowed much higher fluxes compared to the much thicker symmetric membranes (see 11.4.1).

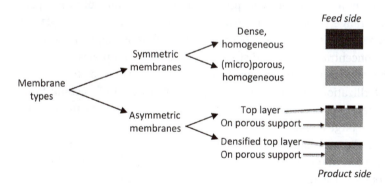

Figure 11.2: Type of membranes.

An example of a ceramic asymmetric membrane is given in Figure 11.3. A thin silica membrane of approximately 30 nm thick – where the actual separation takes place – is supported on a porous alumina layer of a few microns thickness. The two layers are carried by a 2 mm thick disk which provides mechanical strength. In the category of asymmetric membranes, we find applications such as ultrafiltration, gas separation, nanofiltration, reverse osmosis and others.

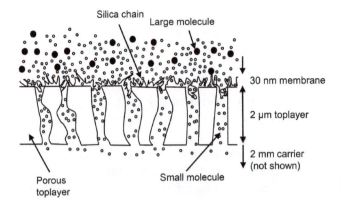

Figure 11.3: Asymmetric membrane: silica membrane supported on a porous alumina toplayer.

11.3 Membrane filtration processes

Membranes are used for a variety of separations. Membranes can be classified not only by membrane structure, but also by the separation process they are applied in or according to size of particles to be separated. Important membrane separations used in industry are microfiltration (MF), ultrafiltration (UF), nanofiltration (NF), reverse osmosis (RO) and gas separation (GS). Figure 11.4 illustrates the nomenclature relating separation process and particle size.

Microfiltration is a pressure-driven process used to separate ultrafine particles – in the size range of 0.1–10 μm – from liquid or gas streams. Examples of such particles are colloids, dyes, micro organisms and large macromolecules. Removal from or concentration in liquid streams is an important application of microfiltration. Other applications include purification of a variety of feed streams, liquid or gas and sterilization (by filtering bacteria), as in removal of bacteria from potable water. In fact, MF was developed to separate microorganisms from water and is the oldest and largest application of membrane separation.

Ultrafiltration, like microfiltration, is a pressure-driven process used to concentrate particles. However, in ultrafiltration the particle size being rejected from the permeate stream is orders of magnitude smaller: approximately 2–100 nm. Examples of such small particles are proteins, viruses and latex particle emulsions. Solvents, ions and other small soluble species will pass through UF-membranes. Ultrafiltration is used to recover latex particles from wastewater, to concentrate proteins, to separate oil-water emulsions and macromolecular solutions. Another example is the clarification of fruit juices. Also, microfiltration as well as ultrafiltration are employed as a pretreatment step for reverse osmosis and nanofiltration. UF is an essential tool in the manufacturing of sterile vaccines by removing viruses from solution. With UF, whey, a by-product of cheese manufacturing, is separated into an aqueous solution of lactose and salts (permeate), retaining casein, butterfat, bacteria and proteins in the retentate.

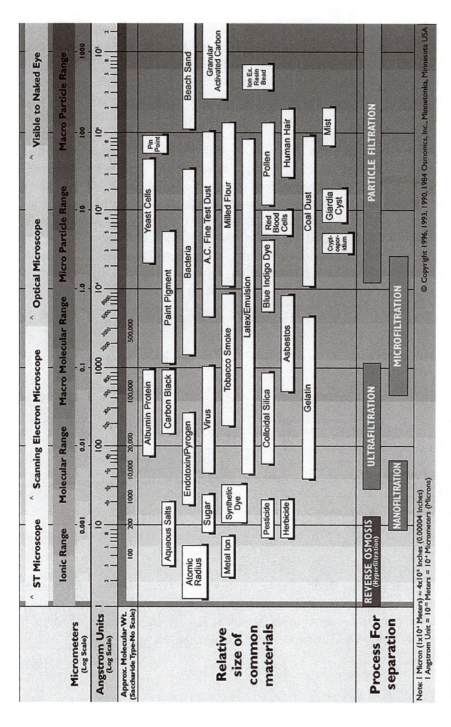

Figure 11.4: Classification of membrane filtration processes (Courtesy of Osmonics).

Nanofiltration, also a pressure-driven process, separates the solvent from particles and dissolved molecules smaller than about 2 nm. NF is characterized by a membrane that selectively restricts flow of solutes, while permitting flow of the solvent (almost always water). Nanofiltration membranes are usually charged, permitting the permeation of salts, while low MW organic compounds are being retained.

While *Reverse osmosis* and nanofiltration are closely related in terms of applications, their separation mechanisms are quite different. Where nanofiltration can be considered a submicron filtration process, though for very small (nanometers) particles, reverse osmosis is based on reversing the phenomena of osmosis. The principle is explained in Figure 11.5. A system comprising an aqueous solution and pure water, separated by a membrane, permeable to water but not permeable to the solutes, is not in equilibrium. The thermodynamic potential of water in the aqueous solution is lower, offering pure water a so-called *osmotic pressure*, π_{osm}, as the driving force for it to move through the membrane into the solution. The process continues till equilibrium is reached. This is the case if build up *hydrostatic pressure*, $\rho g \Delta h$ (see Figure 11.5), balances the osmotic pressure related to the final concentration[1]:

$$\Pi_{osm} \approx cRT = \rho g \Delta h \qquad (11.1)$$

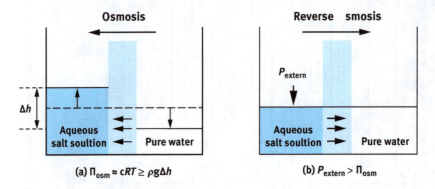

(a) $\Pi_{osm} \approx cRT \geq \rho g \Delta h$ (b) $P_{extern} > \Pi_{osm}$

Figure 11.5: Principle of (a) osmosis and (b) reverse osmosis.

This process is reversed for desalination of seawater in reverse osmosis. An external hydrostatic pressure, larger than the existing osmotic pressure, increases the thermodynamic potential of water in the seawater, forcing water into the direction of the product clean water. This process continues as long as:

$$P_{extern} > \Pi_{osm}(c) \qquad (11.2)$$

---

1 The Van't Hoff relation $\pi_{osm} = cRT$ applies only to dilute solutions; c is the total amount of solute per unit pure solution, see Table 11.1.

A complication may be the increasing salt concentration in the remainder of the solution, causing the osmotic pressure to increase to such an extent that the required value of P_{extern} gets too high for practical purposes. As an example, Table 11.1 shows the osmotic pressure of various aqueous solutions.

Table 11.1: Osmotic pressures of various aqueous solutions [1].

Sodium chloride solutions			Sea salt solutions		Sucrose solutions	
g mol NaCl / kg H_2O	Density (kg/m^3)	Osmotic pressure (atm)	Wt.% salts	Osmotic pressure (atm)	Solute mol frac x 10^3	Osmotic pressure (atm)
0	997.0	0	0	0	0	0
0.01	997.4	0.47	1.00	7.10	1.798	2.48
0.10	1001.1	4.56	3.45[a]	25.02	5.375	7.48
0.50	1017.2	22.55	7.50	58.43	10.69	15.31
1.00	1036.2	45.80	10.00	82.12	17.70	26.33
2.00	1072.3	96.2				

[a]Value for standard seawatter.

The production of potable water, which contains less than 1,000 ppm of total dissolved solids,[2] requires lots of effort. Reverse osmosis plays an important role in providing an alternative to meet such a requirement, especially in places where conventional processes are not feasible. Small RO units find applications in remote areas like ships or islands, but RO-plants can be very large, producing up to 1 m^3 s^{-1} of fresh water.

11.4 Flux equations and selectivity

The performance of a membrane is determined by experimental conditions (such as pressure difference across a membrane), the geometry of a membrane module and, last but not least, by membrane properties. In this section, we focus on two important membrane properties:
- the *flux*, how much of the component to be removed is passing through the membrane per unit time and per unit membrane area, and
- the *selectivity*,[3] a measure of the relative permeation rates of different components through the membrane.

---

2 As defined by the World Health Organization.
3 Also expressed is the fraction of solute in the feed retained by the membrane, known as the retention.

11.4.1 Flux definitions

In this section, both properties will be addressed. To start with the former one, the flux J_i of a component i through a membrane is related to the driving force for transport. In case of a pressure difference across a membrane, the flux of component i, J_i, is defined through the following flux equation:

$$J_i \equiv P_{M,i} \frac{p_0 - p_\delta}{\delta_M} = P_{M,i} \frac{\Delta p_{\text{extern}}}{\delta_M} \tag{11.3a}$$

The flux, expressed in amount per unit time per unit surface area, is proportional to the absolute value of the – positive – pressure difference p_0–p_δ over the membrane with thickness, δ_M. The proportionality constant is called the permeability $P_{M,i}$ of component i through the chosen membrane, which is a unique value for a particular combination of membrane and component.

In case the osmotic pressure cannot be neglected, the *effective pressure difference* for transport is decreased with the difference in osmotic pressures between feed and permeate solutions:

$$J_i = P_{M,i} \frac{\Delta p_{\text{extern}} - \Delta \Pi_{\text{osm}}}{\delta_M} \tag{11.3b}$$

where $\Delta \Pi_{\text{osm}} = \pi_0 - \pi_\delta$.

In absence of a pressure difference across a membrane, the driving force for transport through the membrane may be the concentration difference. Then the related flux equation becomes

$$J_i \equiv P_{M,i} \frac{c_0 - c_\delta}{\delta_M} \tag{11.3c}$$

Permeability should be expressed in consistent units. Comparing the permeability with the coefficient of diffusion in Fick's equation and realizing that the dimensions of pressure, p, and concentration, c, differ by a factor RT, one may conclude that the dimension of $P_{M,i}$ is $m^2\ s^{-1}$ divided by $J\ mol^{-1}$ (the dimension of RT). However, for historical reasons and practical purposes, *permeability*, especially for gas separations, usually is expressed in *barrer*, defined as

$$1\ \text{barrer} \equiv \frac{10^{-10}\text{cm}^3(\text{STP})\ \text{cm}}{\text{cm}^2\text{s cmHg}} \tag{11.4}$$

In other words, the permeability is a transport flux through a membrane per unit transmembrane driving force per unit membrane thickness. The factor 10^{-10} is added just for convenience. The IUPAC applies the same definition, but recommends expressing permeability in SI units ($kmol\ m\ m^{-2}\ s^{-1}\ kPa^{-1}$) rather than in cgs units.

An alternative way to express how fast a certain species can flow through a membrane is the *permeance*, which is defined as the ratio between permeability and membrane thickness or as a transport flux per unit transmembrane driving force. Thus, permeance is a proportionality constant just like a mass transfer coefficient. If the driving force in the flux equation is expressed as concentration difference (see eq. (11.3c)), permeance is given in ms^{-1}. The values of permeability are closely connected to molecular and/or membrane properties, depending on the type of mass transport through a membrane, which can be porous or dense. In porous membranes with relatively wide pores, viscous flow takes place with a pressure difference as driving force. In small pores (small compared to the mean free path) *Knudsen flow* is predominant. In some cases the diameter of a molecule exceeds that of the pore, resulting in so-called *size exclusion*. In dense membranes a *solution-diffusion* mechanism takes place. The different mechanisms are depicted in Figure 11.6. Viscous flow does not offer any selectivity, whereas size exclusion only depends on molecular size relative to pore diameter. Therefore, in the following sections relations for permeability and selectivity will be derived just for diffusion through porous membranes and for solution-diffusion through dense membranes.

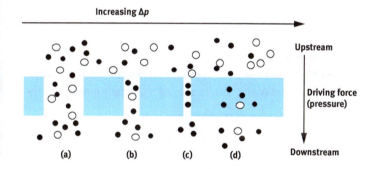

Figure 11.6: Transport mechanisms in membranes. (a) Viscous flow through macro- and mesopores; (b) diffusion through micropores; (c) size exclusion; (d) solution through a dense membrane.

11.4.2 Permeability for diffusion in porous membranes

Mass transport in pores not only occurs by viscous flow. *Molecular diffusion* will contribute when a difference in concentration across the membrane exists. The flux equation for combined convection and diffusion in a binary mixture with components i and j is

$$J_i = - c_{tot} D_{im} \frac{\partial x_i}{\partial z} + \frac{c_i}{c_{tot}} (J_i + J_j) \tag{11.5}$$

where D_{im} is the coefficient of diffusion of species i in the mixture. In case of equimolar counter diffusion the second term of the right-hand side becomes zero and eq. (11.5) transfers into the well-known *Fick's law*:

$$J_i = -C_{tot} D_{im} \frac{\partial x_i}{\partial z} \approx D_{im} \frac{c_{i,0} - c_{i,\delta}}{\delta M} \tag{11.6}$$

This equation only applies to diffusional transport through membranes if pressure on either side of a membrane is kept equal. However, eq. (11.6) is still useful as an approximation for unimolecular diffusion in *dilute systems* ($c_i/c_{tot} \ll 1$). For ideal gas mixtures eq. (11.6) transfers into

$$J_i = -\frac{D_{im}}{RT} \frac{\partial p_i}{\partial z} \approx \frac{D_{im}}{RT} \frac{p_{i,0} - p_{i,\delta}}{\delta M} \tag{11.7}$$

By analogy with eq. (11.3a), the permeability appears to be proportional with molecular diffusivity and the selectivity for diffusive transport through membranes is determined by the ratio of the selectivities.

For ideal gases the diffusivity can easily be related to molecular properties by simple kinetic theory:

$$D_i = \frac{1}{3} \lambda_i \bar{v}_i \tag{11.8}$$

where $\lambda i = \frac{RT}{\sqrt{2}\sigma_i p_{tot} N_{Av}}$, the mean free path and $\bar{v}_i = \sqrt{\frac{8RT}{\pi M_i}}$, the mean speed.

σ_i is cross-sectional area of species i and N_{av} the Avogadro number. Thus, diffusivity is proportional to the reciprocal of the square root of molecular mass[4] and the selectivity is given by

$$\text{Selectivity} = \frac{D_i}{D_j} = \sqrt{\frac{M_j}{M_i}} \tag{11.9}$$

When the pore diameter is of the same magnitude as the mean free path or smaller, Knudsen diffusion is the main transport mechanism. In this case, collisions of molecules with the pore wall are more frequent than intermolecular collisions. Equation (11.7) is still valid as long as the Knudsen diffusivity D_{iK} replaces the molecular diffusivity D_{im}. The *Knudsen diffusivity* is calculated from

$$D_{iK} = \frac{d_p}{3} \bar{v}_i \tag{11.10}$$

---

4 Assuming that the ratio of cross-sectional areas is approximately unity.

Unlike the value of D_{im}, the value of D_{iK} is not just a molecular property, but also depends on the pore diameter d_p. The selectivity, however, still obeys eq. (11.9). In case of size exclusion, the larger molecule cannot permeate at all and the selectivity goes to infinity.

11.4.3 Permeability for solution-diffusion in dense membranes

The transport mechanism in dense membranes is quite different from that determining transport through (micro)porous membranes and, actually, cannot be considered a special kind of filtration. Nevertheless, dense membranes are often applied and a basic knowledge is valuable.

Dense membranes let permeating species dissolve in its matrix, creating a concentration gradient across the membrane thickness. The concentration gradient offers the driving force for dissolved molecules to travel to the other side of the membrane by molecular diffusion. Once at the other side, the dissolved molecules are freed and transferred into the permeate phase. Thus, permeability in this case will depend on both solubility and diffusivity, as will be quantified in the following.

Unlike porous membranes, where no jump in concentration at the membrane surface exists, dense membranes show different concentrations at both sides of the membrane surface. Mass transport through the membrane takes place, the driving force being the concentration difference $c_{i0}-c_{i\delta}$. Both concentrations are defined at either membrane surface just inside the membrane. In Figure 11.7 two possible concentration gradients are drawn. The usual situation is given by the solid line. Upstream, at the feed side, the feed concentration in the boundary layer just outside the membrane is assumed to be in equilibrium with the concentration of solute i, c_{i0}, at the same surface just inside the membrane, see Figure 11.7. The same assumption applies to the product side: the product concentration in the boundary

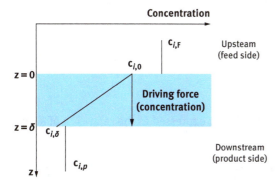

Figure 11.7: Solution-diffusion mechanism in a dense membrane.

layer just outside the membrane is supposed to be in equilibrium with the solute concentration just inside the downstream membrane surface.

In the special case where no mass transfer limitation exists outside the membrane, the boundary values just outside the membrane equal the bulk feed concentration, c_{iF}, respectively the bulk product concentration, c_{iP}. Thermodynamics control the concentration ratio at both surfaces:

$$K_i = \frac{c_{i,0}}{c_{i,F}} = \frac{c_{i,\delta}}{c_{i,P}} \qquad (11.11)$$

Elimination of the concentration difference inside the membrane from eqs. (11.6) and (11.11) and, again, assuming ideal gas behavior, leads to the following *flux equation for dense membranes*:

$$J_i = K_i D_{im} \frac{c_{iF} - c_{iP}}{\delta_M} = \frac{K_i D_{im}}{RT} \frac{p_{iF} - p_{iP}}{\delta_M} \qquad (11.12)$$

Comparison with the original equation, eq. (11.3a), shows that now the permeability in a dense membrane is proportional to the product of thermodynamic equilibrium partition constant, K_i, and the diffusivity through that membrane, D_{im}. Equation (11.12) is still valid in the usual case that mass transfer limitations outside the membrane do have an influence on the overall transport rate, as long as the upstream and downstream boundary concentrations are applied instead of the bulk values.

Comparison of eqs. (11.9) and (11.12) shows that the selectivity found in separations with dense membranes will be much higher than those obtained with porous membranes:

$$\text{Selectivity} = \frac{K_i\, D_{im}}{K_j\, D_{jm}} = \frac{K_i}{K_j} \sqrt{\frac{M_j}{M_i}} \qquad (11.13)$$

The reason is that the *ratio of solubilities*, K_i/K_j, usually overrules the importance of molecular weights. A limiting situation arises when a certain species in a mixture cannot dissolve in the membrane material at all: the related selectivity goes up to infinity.

11.4.4 Selectivity and retention

In membrane filtration, the selectivity is usually expressed as the *retention* of a certain species, by the formula:

$$R_i = \frac{c_{Ri} - c_{Pi}}{c_{Ri}} \qquad (11.14)$$

where R_i is the retention of species i, C_{Ri} is the concentration of i in the retentate and C_{Pi} is the concentration of i in the permeate. It is obvious that, if C_{Ri} equals C_{Pi}, there is no separation and $R_i = 0$. However, if $C_{Pi} = 0$, the retention of i is complete. Membranes are characterized by a *molecular weight cut-off* (MWCO) of, for instance, 60,000 Da. The definition of an MWCO of 60,000 Da is that the membrane retains 90% of a compound with an MW of 60,000 Da.

In Figure 11.8, the MWCO of two membranes are shown. Membrane A has a very shallow profile and Membrane B a sharp one. With Membrane B, insulin can easily be separated from human serum albumin, because the retention of insulin is 0%, while the retention of human serum albumin is about 96%. With Membrane A, the insulin retention is 32%. This means that only 68% of the insulin ends up in the permeate. With Membrane A, the retentate must be purified in a second step for a more complete separation. In *gas separations*, the use of retention is not practical, but selectivity is used:

$$S_{ij} = \frac{\left(c_i/c_j\right)_{permeate}}{\left(c_i/c_j\right)_{retentate}} \tag{11.15}$$

where S_{ij} is the selectivity of species i over species j and c_i and c_j are the concentrations of i and j in the permeate and retentate.

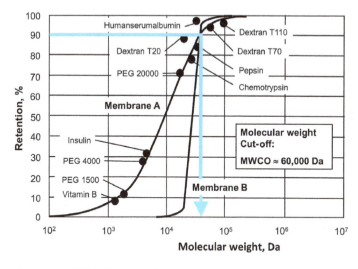

Figure 11.8: MWCO of two membranes.

11.5 Concentration polarization

The concentration gradients shown in Figure 11.7 represent a special situation: here the flux is sufficiently small to prevent any mass transfer limitation. This is usually the case in dense, symmetrical membranes, for example, gas separation, which are relatively thick. Concentration gradients observed in liquids flowing through porous membranes are different in two respects:

- no concentration jump at the liquid-membrane interface because there is no phase transition
- concentration gradients near the interface may exist because the fluxes can be relatively high.

A general picture is given in Figure 11.9. The retained solutes can accumulate at the membrane surface, where their concentration will gradually increase. Such a concentration build-up will generate a diffusive flow back to the bulk of the solution. A steady state condition is reached when the convective transport of the solute towards the membrane is equal to the sum of the permeate flow plus the diffusive back-transport of the solute:

$$J * c + D \frac{dc}{dx} = J * c_P \tag{11.16}$$

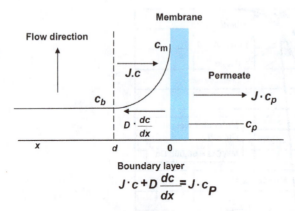

Figure 11.9: Concentration polarization near a membrane.

where J is the flux through the membrane, m³/m².hr, c is the concentration in the boundary layer, kg/m³, c_P is the concentration in the permeate, kg/m³, D is the diffusion coefficient of i, dc/dx is the gradient of the concentration in the boundary layer and x is the thickness of the boundary layer. The following boundary conditions apply:

$$x = 0: \quad c = c_m$$

$$x = \delta: \quad c = c_b$$

where c_m is the concentration at the membrane surface and c_b is the concentration in the bulk. Integration of eq. (11.16) results in:

$$\ln \frac{c_m - c_p}{c_b - c_p} = \frac{J * \delta}{D} \tag{11.17}$$

or:

$$\frac{c_m - c_p}{c_b - c_p} = \exp\left(\frac{J * \delta}{D}\right) \tag{11.18}$$

The ratio of the diffusion coefficient, D, and the thickness of the boundary layer, δ, is called the mass transfer coefficient $k = D/\delta$. The ratio of c_m/c_b is called the *concentration polarization modulus*. This ratio increases with increasing flux (c_m will be higher), with increasing retention (higher c_m) and with decreasing mass transfer coefficient, k. When the solute is completely retained by the membrane ($R = 1$ and $c_p = 0$), eq. (11.18) becomes

$$\frac{c_m}{c_b} = \exp\left(\frac{J}{k}\right) \tag{11.19}$$

From this formula, it is clear that the flux, J, and the mass transfer coefficient, k, are the two factors responsible for the concentration polarization. The consequences of concentration polarization can be summarized as follows:

- *retention can be lower*
  Because of the increased solute concentration at the membrane surface (higher c_m), the observed retention, $R_{obs} = 1 - c_p/c_b$, will be lower than the real or intrinsic retention, $R_{int} = 1 - c_p/c_m$. This is generally the case with low molecular weight solutes, such as salts.
- *retention can be higher*
  When macromolecular solutes are being retained, these compounds can form a secondary membrane, resulting in a higher retention of the other solutes present.
- *flux will be lower*
  The flux is proportional to the driving force. When concentration polarization and/or the formation of a secondary membrane is severe, the combined resistance of the boundary layer, the secondary membrane and the membrane itself will be high and, consequently, the flux decline can be quite considerable, whereas in gas separations concentration polarization hardly occurs and consequently, the flux remains at about the same level.

Values of mass transfer coefficients, which are related to the diffusivity by the film thickness, depend on flow regime (laminar or turbulent), module geometry and fluid properties. Empirical relations between Sherwood number, Sh, and Reynolds number, Re, provide a way to estimate mass transfer coefficients. For turbulent flow in a circular tube, mass transfer rate is characterized[5] by

$$Sh \equiv \frac{k_i D_h}{D_{im}} = 0.023 \times Re^{0.8} \times Sc^{0.25} \tag{11.20}$$

where hydraulic diameter, D_h, equals the tube diameter, D_{tube}, D_{im} is the coefficient of molecular diffusion of species i, $Re = \frac{\rho_f v D_{tube}}{\eta_f}$; $Sc = \frac{\eta_f}{\rho_f D_{im}}$, Schmidt number; η_f, ρ_f, v the viscosity, density and velocity of the feed, respectively.

It should be noted that eq. (11.20) applies to *local flow conditions*. Membranes are applied in modules (see next section) and for a proper design the flow patterns in the chosen module must be taken into account. For laminar flow in a circular tube, a similar equation is available:

$$Sh = 1.86 \times Re^{0.33} \times Sc^{0.33} \times \left(\frac{D_h}{L}\right)^{0.33} \tag{11.21}$$

where L = length of tube.

From these equations, it can be seen that the mass transfer coefficient, k, is mainly a function of the velocity of the feed flow (v), the diffusion coefficient of the solute (D), the viscosity, the density and the module shape and dimensions. Of these parameters, the flow velocity and the diffusion coefficient are the most important.

In reverse osmosis, the consequence of concentration polarization is even more severe. The osmotic pressure of the nonpermeating solute is higher at the membrane surface than in the bulk, as can be seen in Figure 11.9. As a consequence, the external pressure must be (much) higher than in the absence of concentration polarization to compensate the increased osmotic pressure. This phenomena limits the applicability of RO in, for example, desalination of seawater. The osmotic pressure of seawater at several salt concentrations is tabulated in Table 11.1, together with the osmotic pressure of two other aqueous solutions. From this table, it appears that the maximum concentration in sea salt solution at the membrane interface amounts to approximately 10 wt%. Higher concentrations would require much too high a value for the external pressure to overcome the osmotic pressure.

---

5 H. Strathmann, Membranes and Membrane Separation Processes, in Ullmann's Encyclopedia of Industrial Chemistry (1990), Vol. 16A, p. 237.

11.6 Membrane modules

So far, we discussed the performance of membranes in terms of material properties, driving forces and local mass transfer rates. Polymer membranes come usually in the form of sheets or hollow fibers, while ceramic membranes are applied as coated tubes or monoliths. A bare membrane cannot be applied without proper connections to direct a feed stream onto the membrane surface and to collect the permeate stream. The housing that enables just all that is called a module.

In this section, some typical membrane modules and module flow patterns will be discussed. In Figures 11.10 to 11.12, four types of common membrane modules are shown. A *plate-and-frame* module, (Figure 11.10) contains flat membrane sheets, supported on plates which direct the permeate flow. Alternatively, two membrane sheets, separated by a porous spacer, can be wound around a perforated tube to form a *spiral wound module*, often used in reverse osmosis for the production of potable water, see Figure 11.10 and 11.11. The permeate flows through the porous spacer to the central collection tube. In such a module, the packing density is usually higher than in a plate-and-frame module. Ceramic membranes are often applied as porous tubes. A *tubular module* is shown in Figure 11.12. Due to the dimensions of ceramic tubes, only a limited number can be packed in a module. Polymer hollow fibers are much thinner and the resulting packing density in a *hollow fiber module* is very high. The main characteristics of these four types are summarized in Tables 11.2 and in 11.3.

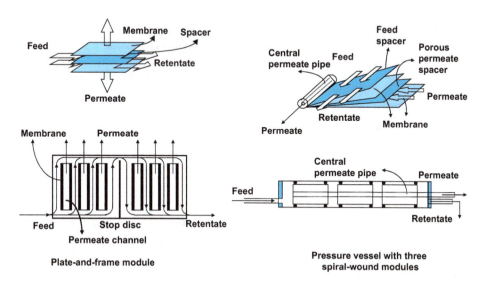

Figure 11.10: Flat sheet and spiral wound modules .

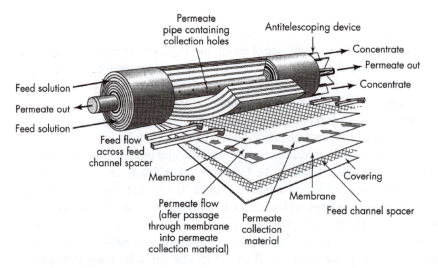

Figure 11.11: Spiral wound module (Reproduced with permission from [15]).

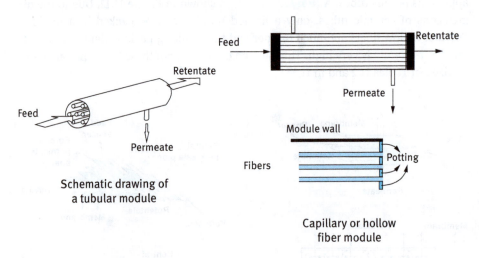

Schematic drawing of
a tubular module

Capillary or hollow
fiber module

Figure 11.12: Tubular and hollow fiber modules.

Unlike the size of distillation columns (or absorption or liquid-liquid extraction columns for that matter), the size of membrane modules is subject to serious limitations. Given a required capacity, the feed and product composition of a certain membrane separation and the driving force, the necessary surface area can be calculated for a chosen membrane type. Usually, however, it is not possible to put the required amount of membrane into just one module. Thus, a number of similar modules have to be build, each of which contains only a fraction of the total surface

Table 11.2: Some characteristics of membrane modules.

	Area/volume m²/m³	Fouling tendency	Prefiltration	Manufacturing costs, $/m²	Area of standard module, m²
Plate and frame	100–500	little	10–25 μm	50–200	5–10
Spiral wound	300–1,000	moderate	10–25 μm	10–50	20–40
Tube >5 mm	100–500	low	none	50–200	5–10
Capillary 0.5–5 mm	500–4,000	moderate	10–25 μm	5–50	50–150
Fiber <0.5 mm	4,000–30,000	high	5–10 μm	2–10	300–600

Table 11.3: Applicability of membrane modules.

	Tubular	Plate-frame	Spiral wound	Capillary	Hollow fiber
MF	++	+		+	
UF	++	+	+	++	
NF	++	+	++	++	+
RO	+	+	++	+	++
Gas sep.			++	+	++
PV/VP	++	++	+	+	
ED		++			

area. Therefore, capacity increase in membrane separations does not follow the well-known 0.6 law, common in many other chemical engineering processes.

A first estimate of the capacity of a single membrane module results from application of eq. (11.3a): the flux is calculated from permeability, membrane thickness and driving force. By definition, the *total permeate flow* follows from the flux and the membrane surface area contained in the module.

Two complications should be considered. Firstly, in a proper approach attention should be paid to possible influence of mass transport limitation. Mass transfer takes place at either surface of a membrane and at high overall permeation rates either side can be subject to mass transport limitation, depending on geometry. Thus, it is appropriate to look at possible *combinations of flow patterns* that govern the mass transfer coefficients, compare the expression for Sherwood numbers in eqs. (11.20) and (11.21). Secondly, the geometry and, in connection with that, flow patterns in a module play an important role, because they determine the driving force as a function of position. Four important combinations of flow patterns can be distinguished:

- cocurrent flow
- counter current flow
- cross flow
- perfect mixing at both sides

These patterns are depicted in Figure 11.13. It is usually far from straightforward to identify the flow patterns in the module types mentioned above. For instance, the flow pattern in a hollow fiber module can be approximated as counter current, but cocurrent flow or cross flow are possible as well. At high permeate rates, in a spiral wound module, the permeate moves away from the membrane surface, characteristic of cross flow. At low permeate rates, on the other hand, the flow type may be described as counter current.

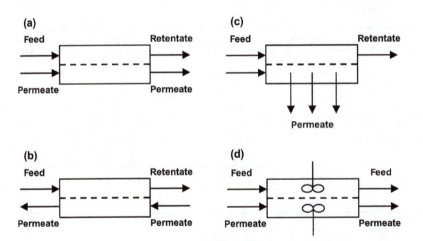

Figure 11.13: Important flow patterns in membrane modules: (a) cocurrent; (b) counter current; (c) cross flow; (d) perfect mixing at both sides.

Here, the design of a membrane module is limited to the situation where the driving force is constant throughout the module: perfect mixing at both sides. In many cases this simplification will give a reasonable estimate. More complicated situations are outside the scope of this book on the fundamentals of separation processes. Interested readers are referred to specialized literature, see references and further reading. The modelling of a simple membrane separator in which perfect mixing can be assumed at both sides of the membrane is illustrated with the separation of air into an oxygen-rich permeate and a retentate that is enriched in nitrogen (oxygen is the faster permeating component). Air is considered a binary mixture of ideal gases and the module (housing, gas phase and membrane) is kept at a constant temperature.

11.6.1 Design procedure

Given the feed flow F, the pressure p at both sides, the molar fraction of oxygen x_F in the feed and the permeance of the chosen membrane for oxygen and nitrogen, how to calculate the unknowns such as oxygen concentration in the permeate and the recovery of oxygen (the fraction of oxygen in the permeate relative to the amount of oxygen in the feed)?

Solution

Like any other unit operation, the model is based on material balances. In this case, the overall and component balances are:

$$F = P + R \tag{11.22}$$

and

$$x_F F = y_P P + x_R R \tag{11.23}$$

where F, P and R indicate the molar flows of the feed, permeate and retentate, respectively; the indices refer to the same flows. The transport of oxygen and nitrogen through the membrane follows from the definition equations of permeance for a pressure driven process:

$$y_P P = A_M \frac{P_{M,O_2}}{\delta_M}(x_R p_R - y_P p_P) \tag{11.24}$$

$$(1 - y_P)P = A_M \frac{P_{M,N_2}}{\delta_M}[(1 - x_R)p_R - (1 - y_P)p_P] \tag{11.25}$$

where A_M is the membrane surface area and $P_{M,i}/\delta_M$ the permeance of component i. So far we have 11 variables (counting the permeance of a component for 1) of which six are given, and only four equations. An additional relationship or value is required to be able to solve the set of equations. A possibility is to lock, for example, the permeate concentration. In distillation calculations, the reflux offers the additional equation. In membrane calculations often the *cut*, the part of the feed that is allowed to permeate, is fixed:

$$cut = \frac{P}{F} \tag{11.26}$$

Now the five equations eqs. (11.22)–(11.26) can be solved. An efficient route to do this is to firstly calculate P and R from eqs. (11.22) and (11.26). Secondly, eliminate the membrane surface area from eqs. (11.24) and (11.25) and replace the ratio of permeances by the selectivity (see section 11.4). Then the molar fractions y_P and x_R can be solved from eqs. (11.23) and (11.24). Subsequently, the eliminated membrane surface area can be calculated from eq. (11.24). Finally, from the values thus obtained, any parameter of interest, such as the separation factor or the oxygen recovery, can be calculated.

Though an analytical solution exists, an example calculation is given in the Appendix to this chapter. It shows a numerical solution as programmed in MathCad. As a final result, it is shown how the required membrane surface area varies with a chosen value of the cut. Obviously, *the higher the cut, the higher the required surface area*.

11.7 Concluding remarks

Membrane processes possess a number of distinct advantages compared to alternative separation methods, such as distillation, absorption and extraction. The usefulness of membrane applications may emerge from the following:

- Numerous separations can be carried out with the use of membranes because of the very broad scale of particles sizes covered, as shown in Figure 11.4.
- Energy requirements of membrane separations are usually low because generally no phase change is involved. An exception is pervaporation, where the feed is in the liquid phase and the permeate in the vapor phase.
- Membrane separations usually function effectively at room temperature.
- The list of suitable membrane materials is long and still expanding. This means that in many cases appropriate membranes exist with a high selectivity for the species to be separated.
- A membrane unit simply comprises a number of membrane modules in parallel and a suitable number of compressors (or pumps).

The above may give the impression that a membrane process is the feasible technology for almost any separation. However, the use of membranes for separation is seriously limited due to the following disadvantages:

- Usually, membrane processes have only a few stages, which requires high selectivities, much higher than would be necessary, for example, in distillation where a lower selectivity (relative volatility) is compensated by a higher number of stages.
- Chemical resistance can be an issue. Polymer membranes may weaken, swell or even dissolve in contact with a variety of organic compounds, such as aromatics and oxygenated solvents.

- Sometimes it is impossible to apply membranes because of lack of thermal stability above the required temperature (dictated by compatibility with connected unit operations). Not only the (polymer) membrane cannot withstand temperatures much above say 80 °C, the same is true for glues and other sealing material normally used in membrane modules. Ceramic-based membranes, on the contrary, can be used to much higher temperatures. However, sealing problems due to differences in thermal expansion in the latter modules prevent their large-scale application, at the time being.
- Scaling up membrane units means simply applying more modules in parallel instead of installing bigger modules. Therefore, the investment of a membrane separation process will increase almost linearly with scale and may be favorable only below a certain maximum capacity as compared to unit operations obeying the power 0.6 law, such as distillation, extraction and absorption.
- Fouling of membranes may greatly restrict permeability of membranes and make them even unsuited for some types of feed streams or expensive regeneration schemes are necessary. Also longer-term flux decline may occur, limiting the membrane's lifetime to only a fraction of the expected time-on-stream.
- In some applications, especially at high osmotic pressure, the cost of compressor energy may be high.

In conclusion: *membrane separations have found their way in chemical industry and its market is still expanding because of ongoing improvement of membrane materials and membrane module design.*

Nomenclature

D	Coefficient of diffusion	$m^2\ s^{-1}$
F, P, R	Feed, product and retentate flow	$mol\ s^{-1}$
J_i	Component flux	$mol\ s^{-1}\ m^{-2}$
K_i	Distribution coefficient of component i	–
$P_{M,i}$	Permeability of component i, see eqs. (11.3a) and (11.3c)	
R_i	Retention of i, eq. (11.14)	–
$\bar{v}$	Mean speed	$m\ s^{-1}$
x, y	Mole fractions	–
S	Selectivity, eq. (11.15)	–
δ_M	Membrane thickness	m
λ	Mean free path	m

Exercises

1 Show that permeability can be expressed in $m^2 s^{-1}$.

2 The concentration of a solute at both sides of a membrane with thickness 30 μm is 0.030 kmol m^{-3} and 0.0050 kmol m^{-3}. The equilibrium constant $K = 1.5$ and $D_A = 7.0\ 10^{-11}\ m^2\ s^{-1}$ in the membrane. The mass transfer coefficients at both liquid sides can be considered as infinite.

   Calculate the flux and the concentrations at the membrane interfaces inside the membrane.

3 Repeat the previous exercise in case that the mass transfer coefficient at the product side amounts to $2\ 10^{-5}\ m\ s^{-1}$.

4 Calculate the osmotic pressure of a solution containing 0.1 mol $NaCl/kg\ H_2O$ at 25 °C. The density of pure water at 25 °C amounts to 997.0 kg m^{-3}.
   Compare the result with the experimental value given in Table 11.1.

5 The permeability of a membrane is experimentally determined by measuring the flow of an aqueous NaCl solution in a reverse osmosis module at a given pressure difference. The inlet concentration $c_{in} = 10\ kg/m^3$ solution with a density of 1,004 kg/m³ solution. The product concentration $c_{out} = 0.39\ kg/m^3$ solution with a density of 997 kg/m³ solution.

   The measured flow rate is $1.92 \times 10^{-8}\ m^3$ solution/s at a pressure difference of 54.4 atm, the membrane area $A = 0.0020\ m^2$.
   Calculate the permeability constants for water and sodium chloride.

6 Ultrafiltration is used to remove to remove proteins from whey. For this purpose, dead-end experiments have been performed using a flat plat UF membrane with an area of $25 \times 25\ cm^2$. To determine the clean membrane resistance some clean water ($\eta = 1.10^{-3}$ Pa.s) filtration experiments have been conducted where the pressure is measured for various flowrates. The following data were obtained:

Flowrate (l/h)	3.0	3.5	4.5	5.0
TMP (bar)	0.62	0.75	0.88	1.00

a. Determine the clean membrane resistance R_M.
   A subsequent experiment in which whey is filtered for 20 mins at a constant flux van 50 l/h/m2 20, resulting in the formulation of the following fouling model:

$$\frac{dR}{dt} = kJ,\ R(0) = R_M$$

   where R (1/m), $k = 41.5\ (m^2/m^3)$, J (m/s) and t(s).

b. Calculate the energy consumption of the pump that has been used during this experiment. Pump efficiency $\eta_P = 0.7$.

c. Determine the pump power required for pilot plant scale filtration test that contains a membrane area of 40 m² and operates at an average trans membrane pressure of 0.5 bar. Pump efficiency $\eta_P = 0.7$

7 For an UF membrane clean water experiments have yielded the following results:

Flux (l/h/m2)	30	40	50	60	70
TMP (Pa)	4,100	5,500	7,000	8,000	9,700

a. Calculate the clean membrane resistance.
   In the second experiment, river water is filtered over the membrane at constant flux (60 l/h/m²) during 15 min. (900 sec), providing the following data:

Filtration time (s)	60	180	300	420	540	660	780	900
Total resistance $(m^{-1}) \times$	10^{11}	5.30	5.98	6.75	7.61	8.58	9.67	10.9	12.3

The filtration behavior can be described by an exponential model or a quadratic model (for constant flux filtration):

$$R_t = R_M \exp(kJt)$$
$$R_t = R_M + kJt^2$$

b. Which of the two models would be preferable to use, explain?
c. Calculate the power consumption of the pump for constant flux filtration when $\eta = 0.001$ Pa.s (viscosity (river) water), $A = 0.05$ m² (membrane area) and $\eta_P = 0.7$ (–) (pump efficiency)

Appendix

Calculation of required membrane area as a function of the cut and feed pressure

$F := 50$ $xF := 0.2$ feedlow [mol/s] and mole fraction O_2 in feed

$PO2 := 0.04$ $PN2 := \dfrac{PO2}{4}$ membrane properties, permeances [mol/(s.m^2.bar)]

$+$ $pP := 1.5$ pressure product side [bar]

$yP := xF$ $xR := xF$ initial guesses for solving Eq. 11.19–11.23

given $(1 - yP) \cdot PO2 \cdot (xR \cdot pR - yP \cdot pP) = yP \cdot PN2 \cdot [pR \cdot (1 - xR) - pP$
$\cdot (1 - yP)]$

$xF \cdot F = yP \cdot \text{cut} \cdot F + xR \cdot (1 - \text{cut}) \cdot F$

solve_for (cut, F, pR) := Find (yP, xR)

$j := 4 .. 6$ $pR_j := j$ three feed (= retentate) pressures [bar]

$i := 0 .. 20$ $\text{cut}_i := 0.1 + \dfrac{i}{25}$ range of cuts

$\begin{bmatrix} yP_i \\ xR_i \end{bmatrix}$:= solve_for (cut$_i$, F, pR_j)   calculated product and retentate mole fractions

AM (i, pR_j) membrane area AM as a function of the cut for
feed pressures of 4, 5 and 6 bar

$:= \dfrac{yP_i \cdot \text{cut}_i \cdot F}{PO2 \cdot (xR_i \cdot pR_j - yP_i \cdot pP)}$

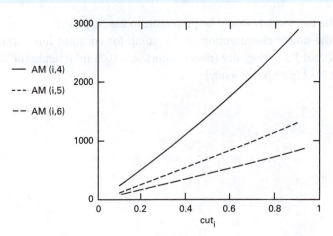

— AM (i,4)

--- AM (i,5)

-- AM (i,6)

Chapter 12
Separation Method Selection

12.1 Introduction

The aim of this chapter is to explore the issue of how to choose the right separation process, or narrow down the number of choices, for a given separation. In most of the cases, however, this selection process is complicated because there is no standard procedure by which separation processes can be chosen. Therefore, we will follow a *heuristic approach* using simple rules of thumb and direction guidelines that can be followed to arrive at the separation process sequence to be adopted in the final plant design. The degree to which this approach reaches the optimum separation scheme depends on the time and money allocated to the development and analysis and on the skills of the process design engineer in applying guidelines to the selection of sequences for detailed evaluation.

When the feed to a separation system is a binary mixture, it may be possible to select a separation method that can accomplish the separation task in just one piece of equipment. In that case, the separation system is relatively simple. In most practical situations, however, the feed mixture involves a complex mixture of many components. For these mixtures a *separation train* is required, consisting of a number of operations in which the separations are sequenced. The separation in each (unit) operation is made between two components designated as key components for that particular separation unit. For example, in a multistage distillation, the heavy key component is the lightest component (with the lowest boiling point) that still is going over the bottom. The light key component is the component with the lowest volatility that still is leaving the distillation with the distillate. Each effluent is either a final product or a feed to another separation device. The synthesis of a multicomponent separation system can be very complex because it involves not only the selection of the separation methods, but also the manner in which the separation operations are sequenced. We will then deal with how a series of separations should be oriented to separate a complex mixture.

12.1.1 Industrial separation processes

The complexity of recovering products in industrial chemical processes is clearly illustrated in Figure 12.1 in which the flow sheet of a naphtha cracker is shown. In the reactor furnaces, naphtha is cracked at a temperature of around 900 °C to a mixture of hundreds of light hydrocarbons, of which the unsaturated hydrocarbons (ethylene, propylene and butylene) are the main products. To recover these products the gaseous reactor effluent is first quenched in an absorption tower with

https://doi.org/10.1515/9783110654806-012

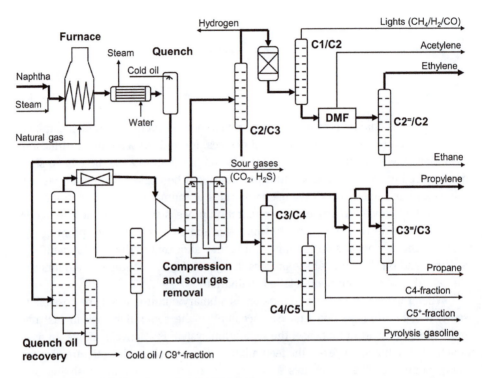

Figure 12.1: Schematic flow sheet of a naphtha steam cracker.

heavy oil to cool the gas stream and condense the heavy oil constituents. In a subsequent operation, some acid gases are removed by reactive absorption to prevent corrosion problems in the final distillation unit. Further separation is accomplished in a series of distillations in which, first, the C2, C3 and C4 hydrocarbons are recovered and finally the saturated and unsaturated hydrocarbons are separated. Traces of acetylene are removed from the C2 hydrocarbon stream by selective absorption in DMF prior to the final distillation.

A second example of a process with a complicated separation train to recover the final product from the reaction mixture is the industrial synthesis of caprolactam, intermediate in the nylon 6 production. The flowsheet of the final step in this process is schematically represented in Figure 12.2. After the reaction in a strongly acidic medium the caprolactam is obtained by neutralization with ammonia, resulting in a two-phase mixture consisting of an ammonium sulphate and a caprolactam rich phase. The caprolactam is recovered from the top layer by extraction with an organic solvent such as benzene, toluene or cyclohexane. A second extraction is used to recover residual caprolactam from the ammonium sulphate rich phase. The caprolactam is recovered from the solvent by back extraction into water and further purified by various techniques such as ion exchange to remove residual contaminants before final evaporation of the water.

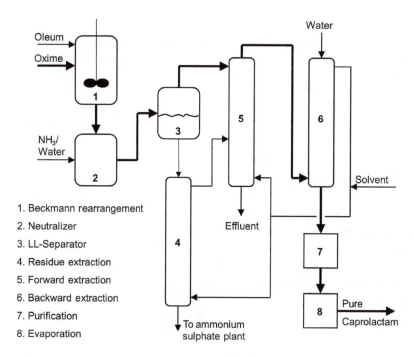

Oleum
Oxime
NH₃/Water
Water
Solvent
Effluent
To ammonium sulphate plant
Pure Caprolactam

1. Beckmann rearrangement
2. Neutralizer
3. LL-Separator
4. Residue extraction
5. Forward extraction
6. Backward extraction
7. Purification
8. Evaporation

Figure 12.2: Schematic representation of the final step of the caprolactam synthesis.

For both processes it is interesting to analyze why the used separation methods have been selected and how it would be possible to improve the process concept by the selection of different separation methods. It is clear that the compositions of the mixtures to separate, scale of operation and the minimization of unnecessary phase changes are important factors that have to be considered. For instance, in a naphtha cracker, absorption is used to selectively remove trace amounts of acid gases without the need of a phase change. Subsequent recovery and purification is done by distillation because this is the most economical large-scale separation technology. The most difficult separation is performed last, being the ethylene/ethane and propylene/propane separations. In the case of caprolactam, extraction is employed because the product has a boiling point in between the water and residual ammonium sulphate. Back-extraction is used to remove nonpolar impurities that were coextracted with caprolactam in the forward-extraction step.

12.1.2 Factors influencing the choice of a separation process

12.1.2.1 Economics
In most situations, the ultimate criterion for the selection of a separation method is economics, of course conditional to technical feasibility. Determining factors are

the **economic value of the products** and the **scale of operation**. The economic criterion is usually also subject to a number of constraints originating from a corporate attitude such as market strategy and timing, reliability, risks associated with innovation and capital allocation. Two extreme cases are very helpful to illustrate the influence of these constraints:

1. The product is of high unit value with a short market life expectancy
2. The product is a high-volume chemical with many producers in a highly competitive market

Case 1, which could present a patented active pharmaceutical ingredient, would highly benefit from a short process development time, and the procedure would probably be to choose the first successful separation method found. The ultimate overall economics of this situation are determined by getting into the market ahead of the competition and selling as much of the product as possible while the market lasts. Clearly many separation processes, which would be unsuitable for substances with a low value, can be considered for the recovery and purification of these high-value products.

In Case 2, a typical case for many products from bulk chemicals to "over the counter pharmaceuticals," the ultimate viability of the plant dictates that the process development and design teams do the most thorough evaluation of various process schemes possible to make the closest approach to the economic optimum design. The desired plant capacity can be an important factor in separation process selection, since some processes are difficult to carry out on a very large scale. The lower the economic value of the product, the more important it will be to select a process with relatively low energy usage and to select a process where the unit cost of any added mass-separating agent is relatively low. A small loss of a costly mass-separating agent can be an important economic penalty to a process. In biorefineries and circular economy processes, where typically the product concentrations are low, losses of mass-separating agents are of high impact when expressed per kg of the (highly) diluted product.

12.1.2.2 Feasibility

First and foremost, any separation process to be considered in a given situation must have the potential of giving the desired result. This *feasibility criterion* is commonly used to make a first selection between separation methods that may work and certainly won't work. Often the question of process feasibility will have to do with the need for *extreme processing conditions*. Although the dividing lines are not easy to draw, the general idea is that a process which requires very high or very low pressures or temperatures will always suffer in comparison with one that does not require extreme conditions.

Another situation where the feasibility condition enters into consideration occurs when a mixture of many components is to be separated into relatively few different products. In such a case it will be necessary for each component to enter the proper product. Since different separation processes accomplish the separation on the basis of different principles, it is quite possible for different processes to divide the various components in different orders between products. Consider, for example, a distillation process that separates a complex mixture with a range in polarity into various cuts. The order depends on relative volatility, while a separation with a mass-separating agent such as a solvent or sorbent can separate the same mixture in the order of the polarity.

12.1.2.3 Product stability

Often the question of avoiding *damage to the product* can be a major consideration in the selection of a separation process. Thermal damage may be manifested through denaturation, formation of unwanted color, polymerization etc. When thermal damage is a factor in separation by distillation, a common approach is to carry out the distillation under vacuum to keep the reboiler temperature as low as possible. Frequently, evaporators and reboilers are given special design to minimize the holdup time at high temperature of material passing through them. Alternatively, a separation sequence may be found with a milder thermal exposure. Also exposure to acidic and basic environments can compromise product stability, or in the case of delicate biological molecules, exposure to organic solvents.

12.1.2.4 Design reliability

On all factors influencing the decision to choose one process in preference to another, design reliability is the most important factor. Regardless of any other considerations, the constructed plant must work properly to produce an acceptable product that can be sold at a profit. The economic consequences of a design failure are too dramatic to accept any potentially malfunctioning separation process design. It is therefore not surprising that those separation operations that are well understood and can be readily designed from first principles are favored in an industrial environment. This is illustrated in Figure 12.3, showing that the degree to which a separation operation is *technologically mature* correlates well with its *commercial use*. Design reliability is not really definable in quantitative terms because it actually relates to the amount of testing and demonstration that must be done before a suitable commercial scale is produced, and it should be noted that even for the most technologically mature separation techniques, the development of a new process involves testing on pilot and demonstration scales before a full scale process can be designed and implemented.

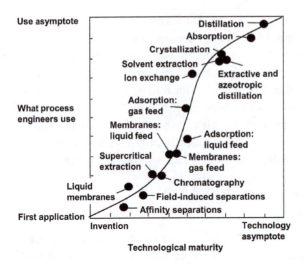

Figure 12.3: Technological and use maturities of separation processes.

12.2 Selection of feasible separation processes

12.2.1 Classes of processes

In the process of selecting a feasible separation process some generalizations can be drawn with respect to the advantages and disadvantages of different classes of separation processes for various processing applications. The distinction between energy-separating-agent, mass-separating agent and rate-governed processes is Important in this classification. In general, the energy usage of processes is proportional to the amount of material that needs to be thermally separated. Typically, when the initial composition is not too extreme (not too dilute) and the relative volatility is large enough to effectively apply a thermal separation, this approach is preferred. The use of mass-separating agents always requires a thermal regeneration to obtain the final product, and this can be energy intensive. Rate-governed processes are typically carried out in much higher dilution, and therefore a multistage *rate-governed separation* process should only be considered when it gives a significantly better separation factor than an equilibration process. In the same way, a *mass-separating-agent* process should only be considered when it provides a better overall separation factor than an *energy-separating-agent* process. The overall energy duty is heavily governed by the initial composition, and when high boiling species are present in high dilution, often it is beneficial to separate them using a mass-separating agent with a high selectivity, instead of distilling off the large quantities of the lower boiling species. The selectivity is an important factor when mass-separating agents are used, because it largely governs the energy duty in the regeneration.

12.2.2 Initial screening

The selection of a best separation process must frequently be made from a number of candidates. When the feed mixture is to be separated into more than two products, a combination of two or more operations may be the best. Even when only two products are to be produced, a hybrid process of more than two or more operations may be required or the most economical. The first step in the selection of a separation process is to define a problem. Table 12.1 gives an overview of four important categories that have to be considered in the *definition of the separation problem* and the selection of feasible separation operations. These factors have to do with feed and product conditions, property differences that can be exploited and certain characteristics of the candidate separation operations.

Table 12.1: Factors that influence the selection of feasible separation operations.

Feed conditions	Composition, particularly concentration of species to be recovered or separated
	Flow rate
	Temperature and pressure
	Phase state (solid, liquid, gas, mixed)
Product conditions	Required product purities
	Required product recovery
	Temperature and pressure
	Phase states
Property differences to be exploited	Molecular
	Thermodynamic
	Transport
Characteristics of separation operations	Ease of scale-up
	Ease of staging
	Temperature, pressure and phase-state requirements
	Physical size limitations
	Energy requirements

The most important *feed conditions* are composition and flow rate, because the other conditions can be altered to fit the required conditions of a candidate separation operation. In general, however, the vaporization of a liquid feed that has a high heat of vaporization, the condensation of a vapor feed with a refrigerant, and/ or the compression of a vapor feed can add significantly to the cost. The most important product conditions are the required purities and recovery specifications because the other conditions can be altered by energy transfer after the separation is achieved. *Product purity specifications* should ultimately be established by customers, although they may not be stated explicitly. *Recovery specifications* are set to assure an economic process. Ideally, the recovery specification should be a variable to be optimized by the process design. In practice, there is usually insufficient time to consider recovery as a variable, and it is specified either arbitrarily or by consensus.

As mentioned before, some separation operations are well understood and can be designed and scaled up to a commercial size from laboratory data without extensive testing on pilot and demonstration scale. Table 12.2 ranks the more common separation operations according to *ease of scale-up*. Operations ranked near the top are frequently designed without the need for any laboratory data or pilot-plant tests, this is particularly valid when similar operations are already in use elsewhere. For operation types near the top that involve new combinations of chemicals, pilots are still often performed. Operations near the middle usually require laboratory data, and more extensive testing on pilot and demonstration scales, while operations near the bottom require extensive laboratory research in combination with pilot plant testing on actual feed mixtures. Also included in the table is an indication of the *ease of providing multiple stages* and to what extent parallel units may be required to handle high capacities. Single-stage operations are utilized only when the separation factor is very large or only a rough or partial separation is needed. When higher product purities are required, either a large difference in certain properties must exist or *efficient countercurrent-flow cascades* of many contacting stages must be provided. Operations based on a barrier are generally more expensive to stage than those based on the use of a solid agent or the creation or addition of a second phase. Some operations are limited to a maximum size. For capacities requiring a larger size, parallel units must be provided.

Table 12.2: Ease of scale-up and staging of the most common separation operations.

Operation in decreasing ease of scale-up	Ease of staging	Need for parallel units
Distillation	Easy	No need
Absorption	Easy	No need
Extractive and azeotropic distillation	Easy	No need
Liquid-liquid extraction	Easy	Sometimes
Membranes	Repressurization between stages	Almost always
Adsorption	Easy	Only for regeneration
Crystallization	Not easy	Sometimes
Drying	Not convenient	Sometimes

12.2.3 Separation factor

An important consideration in the selection of feasible separation methods is the separation factor that can be achieved for the separation between two key components of the feed. This factor has already been extensively discussed in Chapter 1 and is defined for the separation of component 1 from component 2 between phases I and II in a single stage of contacting as:

$$SF = \frac{C_1^I/C_1^{II}}{C_2^I/C_2^{II}} \tag{12.1}$$

where C is a composition variable, such as mole fraction, mass fraction or concentration. If phase I is to be rich in component 1 and phase II is to be rich in component 2, the separation factor is large. The value of the separation factor is limited by *thermodynamic equilibrium*, except in the case of membrane separations that are controlled by relative rates of mass transfer through the membrane. In general, components 1 and 2 are designated in such a manner that the separation factor is larger than unity. Consequently, *the larger the value of the separation factor, the more feasible is the particular separation operation*. In general, operations employing an energy-separating agent are economically feasible at a lower value of SF than those employing a mass-separating agent, which is due to the need to regenerate the mass-separating agent in a second operation. This is illustrated in Figure 12.4, where the required separation factor for considering extractive distillation and extraction are plotted against the separation factor obtained in classical distillation. To interpret the meaning of the graph in Figure 12.4, it can be helpful to consider for binary mixtures to be separated, what would be the required heat duty if it would be done by distillation. Then, obviously, as the relative volatility increases, distillation becomes more effective and can be done with a smaller reflux ratio. Addition of a solvent in extractive distillation or through liquid-liquid extraction may reduce the reflux ratio in the primary distillation column (or replace it by an extraction column)

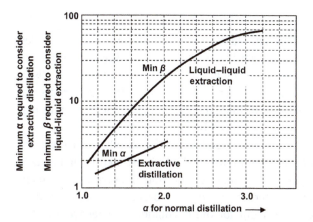

Figure 12.4: Relative separation factors for equal-cost separation operations. (adapted from M. Souders, Chem. Eng. Progr. 60(1964)2, 75–82).

at the cost of a recovery operation. Arguably, when you calculate the effectiveness of a binary distillation, it dramatically increases up to a relative volatility of three. That means with this increasing relative volatility it also becomes much more difficult to develop alternative technology that is even more effective.

In general, mass-separating agents for extractive distillation and liquid-liquid extraction are selected according to their ease of recovery for recycle and to achieve relatively large values of the separation factor. The ease of recovery constraint calls for a solvent that preferably has a boiling point that differs significantly from the species in the mixture to separate, so that thermal regeneration does not pose any problems. The objective of the design of a separation process is to exploit the property differences in the most economical manner to obtain the required separation factors. Processes that emphasize molecular properties in which the components differ to the greatest should be given special attention. This will be illustrated in several examples in the following paragraphs.

12.3 Separation of homogeneous liquid mixtures

At present, distillation is used for over 90% of all separations in the chemical process industry. Considering this dominance, distillation, although not always energy-efficient, is apparently the benchmark against which all newer processes must be compared. Therefore, we will start with an analysis why distillation is so dominant and then proceed to define those separation conditions where the selection of other processes is favored. Finally, the conditions will be defined where one process is preferred over another.

12.3.1 Strengths of distillation

The comparisons between the different classes of separation processes so far led rather strongly toward distillation, in particular when not too dilute mixtures are considered. Economically, if a stream can be easily vaporized or condensed, distillation or a related vapor-liquid separation process is most often the process of choice for separating the components in that stream. There are five important reasons for this dominance, which are summarized below.

- *Small equipment requirements* Compared to virtually all other processes the equipment requirements of distillation are small. As illustrated in Figure 12.5, usually just a column, a reboiler, a condenser and some relatively small ancillary equipment are required. Alternative processes based on mass-separating agents require additional investment for the circulation and recovery of the mass-separating agent. The same holds when distillation is compared to crystallization, another energy-separating-agent equilibration process, involving

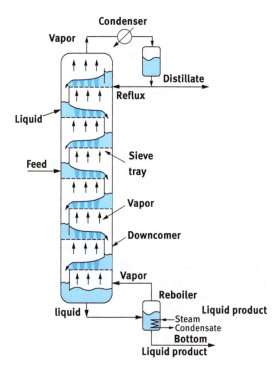

Figure 12.5: Schematic of a distillation column equipped with sieve trays.

solid phases and therefore requiring additional operations (filtration, washing, drying) to obtain the final product. All of these complications often lead to relatively high investments for alternative processes unless separation by distillation is very difficult to effect

- *Easy staging* Distillation can be staged very easily. It is possible to have 100 or more theoretical stages in a distillation column. This means that, even at low relative volatilities down to 1.2, full separations can be achieved. In addition, mass transfer rates between phases are generally high enough to prevent excessive equipment sizes for high degrees of separation.
- *Economics of scale* The economics of scale favor distillation for large-scale separations. Distillation technology scales up by roughly a 0.6 factor on column throughput, which means that doubling a column's capacity increases the column investment only by a factor of about 1.5.

$$\frac{I}{I_0} = \left(\frac{C}{C_0}\right)^{0.6} \tag{12.2}$$

For most alternative processes, the economic improvement of going to very large scales is less pronounced because of their higher scale-up factors. This lowered potential for cost reduction is caused by the more complex flowsheets of the alternative processes and in a number of cases the necessity to place equipment in parallel to accept the whole feed stream. Membrane modules are an excellent example of this problem. To accept large flows, many modules must be aligned in parallel and a considerable part of the savings of going to a larger scale is lost.

– *Energy costs* In most operations, energy costs have not been a dominant consideration. Although this might seem a surprising conclusion at first, energy costs for distillation virtually always run second in impact to capital costs. Therefore, the substitution of an existing distillation process by a more complex, higher-investment but lower-energy-usage process is often difficult to motivate on economic grounds. In addition, in some plants low-pressure steam with little or no value is generated, making energy costs very low. Heat integration of distillation columns can also be practiced on very large scales to help reduce energy costs. Two major examples are olefin plant cryogenic separation systems and cryogenic air separation. Reduction of energy costs has become much more important in recent years due to the correlation between energy usage and CO_2 emission. Therefore, heat integration in distillations and especially for new to build plants, consideration of alternative technologies that are more energy efficient, has received much more attention.

– *Design and scale-up reliability* Distillation is the only separation process that can be designed from the physical properties of the components and vapor-liquid information on important binary-pair components alone. Reliable design methods for distillation have been developed over many years of industrial experience and extensive testing of commercial-scale equipment. For totally new separations that have not been practiced elsewhere before, small-scale testing is required, but scale-up methods for distillation are the most reliable of all the separation methods.

Because of this favorable combination of factors, it is no coincidence that *distillation is the most frequently used separation process in practice*. In fact, a sound approach to the selection of appropriate separation processes for the separation of homogeneous liquid mixtures is to begin by asking: Why not distillation? Unless there is a clear technical reason why distillation is not well suited, distillation will be an important candidate. Next to the technical feasibility, also considerations on energy efficiency are nowadays important initial evaluation criteria for distillation.

12.3.2 Limitations of distillation

In spite of this impressive array of factors, which cause distillation to be favored, other considerations have led to the development and installation of other separation techniques in many applications. Some situations which lead to distillation separation costs which are higher than those of alternative processes, or in which other factors decide against the use of distillation:

- *Low relative volatility* Distillation becomes very expensive when the relative volatility between key components is too low. Seldom will a relative volatility over part or the entire distillation curve of less than about 1.2 give acceptable economics for distillation. Relative volatilities less than 1.2 obviously include azeotrope-forming situations, for which the relative volatilities become 1 at the azeotropic point. In such cases, azeotropic or extractive distillation, pervaporation, solvent extraction and adsorption should be investigated for providing reasonable solutions. Up to a relative volatility of about 3 the energy usage reduces for distillation, whereas above 3, hardly any reduction of energy usage is possible. Considerations on alternative technology reduce at increasing relative volatility.
- *Feed composition* Feed compositions and product purity requirements can present unusual problems. Three problems dominate this category:
  - If the product of interest is a high boiler in low concentration, less than 20 wt%, in the feed, energy costs per unit of desired product to boil up all of the low-boiling material, as well as column capital cost, can become excessive. A typical example of this situation is the liquid phase oxidation of cyclohexane where conversions of only a few percent are used to achieve the desired selectivities. The high boiling product is thus present in small amounts. Also when producing through fermentation, the products often have higher boiling than water, and are present in small amounts.
  - If a small concentration (only a few percent) of high-boiling contaminants must be removed from a desired product, the energy and capital cost for distillation can also become excessive.
  - If the boiling range of one set of components overlaps the boiling range of another set of components from which it must be separated, an extreme amount of distillation columns would be required for a complete separation. In these situations mass-separation-agent based processes have the advantage of being able to separate one set from another in one step based on their difference in chemical structure. Industrial well-known examples are olefin/paraffin and paraffin/aromatic separations.
- *Extreme conditions* If distillation temperatures are less than −40 °C or higher than 250 °C, the cost of refrigeration to condense column overheads in the first case and heat to vaporize in the second escalate rapidly from typical cooling water and steam costs. If required column pressures are less than about

20 mbar, then column size and vacuum pump costs can escalate rapidly. If column pressures higher than about 50 bars are required, column investment can also escalate as a result of the thick vessel walls to meet the pressure requirements.

– *Small capacities* Earlier it was mentioned that distillation has a relatively low scale-up factor of about 0.6. Unfortunately if a process scales up well economically, that same process does not scale down well. That is, its investment does not reduce, as throughput is reduced, as much as other separation techniques with a higher scale-up factor. Thus, alternative processes, which do not compete for large production rates, may compete well at smaller production rates.

– *Product degradation* Various types of undesirable reactions can sometimes occur at column temperatures. Thermally labile components in the feed can undergo reactions, which result in either significant product loss or in the formation of unwanted byproducts that are difficult to separate. In addition, the intrusion of oxygen from small leaks under high vacuum can cause severe product purity problems through unwanted side reactions.

– *Column Fouling* Column fouling rates can be unacceptable. Although product loss rates might be completely acceptable and there are no product quality concerns, precipitation or polymerization reactions can cause unacceptable fast column fouling. Especially when a column cannot be operated for a few days before it must be shut down and cleaned will hardly be economical unless alternative processes are incredibly expensive.

Despite its industrial dominance, this paragraph illustrates that distillation is not always the right answer for a given separation, and especially with revisited considerations on energy usage and CO_2 footprint, considerations of alternative technology are becoming more important. The other processes we have discussed obviously have their places in the scheme of separation trains encountered in industrial processes. In fact, new separation opportunities, since they will often involve more complex and less volatile mixtures than the present spectrum of separations, will be served by these alternative processes. This is certainly the case for biorefineries and bio-based production in fermentation processes.

The following paragraphs have the objective to illustrate how alternative separation methods can be used when the conditions are unfavorable for applying distillation. Three often-encountered situations (low relative volatility, overlapping boiling points and low concentrations) will be analyzed on the basis of industrial examples to clarify the advantages of alternative technologies. Because of its exemplary nature, the approach is not to present an all-covering extensive analysis but merely to show how the *exploitation of different separation principles* can provide alternative solutions to separation problems.

12.3.3 Low relative volatilities

Low relative volatilities are generally encountered when the difference in boiling points of components to be separated is insufficient to achieve a good separation. In principle two distinct situations exist, that both require a different approach to achieve full separation. An example of the first situation is the separation of benzene from cyclohexane or 1-butene from isobutane. A comparison of the properties of these components in Table 12.3, illustrates clearly that the difference in boiling point between cyclohexane and benzene results only in a relative volatility of 1.01, which is far too small to apply distillation. Although the boiling point difference for the 1-butene/isobutane system is larger, non-ideal behavior of the liquid phase is responsible for obtaining only *very low relative volatilities*. In these cases the molecular difference in chemical structure can be exploited to increase the separation factor to a sufficiently high level. This can be done by the addition of a polar solvent (n-methylpyrrolidone (NMP), dimethyl formamide (DMF) sulfolane, furfural) that increases the volatility of the hydrocarbon stronger than the volatility of the aromatic. So introducing a polar solvent in the top of a distillation tower will preferentially force the cyclohexane to the vapor phase and thereby facilitate separation. For these low relative volatility situations *azeotropic and extractive distillation* are often economically favored because of their high mass transfer rates and ease of phase disengagement. In cases of a very asymmetric feed (one of the components in the mixture is highly diluted), where the diluted species is the one to be extracted, liquid extraction can be favorable over extractive distillation. As illustrated in Figure 12.6, a second distillation column is required to separate the benzene from the extractive solvent. The resulting flow sheet is indeed more complex than a distillation flow sheet, but the mass transfer rate and the phase disengagement ease of distillation are also present, and the much higher energy efficiency make this process economically favored in a number of low relative volatility situations.

An example of the second situation is the separation of xylene isomers and ethylbenzene. Xylenes are obtained commercially from the mixed hydrocarbon stream manufactured in naphtha reforming units in oil refineries. p-Xylene is the desired product for the manufacture of phthalic acid and dimethyl terephthalate. As shown

Table 12.3: Some pure component properties of benzene, cyclohexane, 1-butene and isobutane.

Property	Benzene	Cyclohexane	1-Butene	Isobutane
Molecular weight	78.1	84.2	56.1	58.1
Boiling point (°C)	80.1	80.7	−6.2	−11.7
Melting point (°C)	5.6	6.5	−185.3	−159.5
Dipole moment (debye)	0.0	0.3	0.3	0.1
Polarizability (10^{-31} m^3)	103	110	80	82

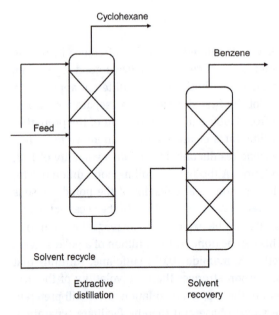

Figure 12.6: Schematic of an extractive distillation process.

in Table 12.4, the boiling points of p-xylene, m-xylene and ethylbenzene differ only marginally. With a separation factor of only 1.02 ordinary distillation would require about 1,000 stages and a reflux ratio of more than 100. Because the chemical nature of all four constituents is almost the same, the addition of an extractive solvent will act approximately the same on all isomers. The only *macroscopic property* that differs considerably is the melting point, allowing selective recovery of the p-xylene from the mixture by crystallization. For crystallization the separation factor is nearly infinite due to the additional advantage of shape selectivity in crystal growth. The resulting flowsheet for industrial p-xylene recovery is shown in Figure 12.7.

Table 12.4: Some pure component properties of xylenes and ethylbenzene.

Property	o-Xylene	m-Xylene	p-Xylene	Ethylbenzene
Molecular weight	106.2	106.2	106.2	106.2
Boiling point (°C)	144.2	139.5	138.7	136.5
Melting point (°C)	−25.0	−47.7	13.5	−94.7
Dipole moment (debye)	0.21	0.10	0.00	0.20
Polarizability (10^{-31} m^3)	141	142	142	135
Molecular volume (cm^3/mol)	121	123	124	123
Kinetic diameter (Å)	6.8	6.8	5.8	6.3

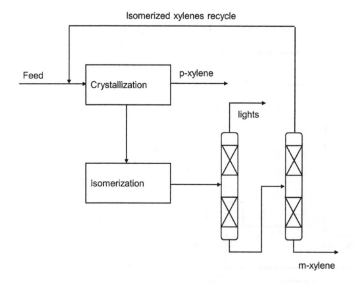

Figure 12.7: Schematic of process for p-xylene manufacture by crystallization and isomerization.

Other properties that may be successfully exploited in the separation of isomer mixtures is the difference in kinetic diameter and complexation behavior resulting from the difference in isomer structures. In the case of p-xylene separation, this is used in a large-scale continuous adsorption process where shape selective zeolites are used as adsorbents to achieve full separation. One of the properties they use is the relative linearity of the p-xylene molecule compared to the other xylene isomers. As a result, the p-xylene kinetic diameter is smaller and the molecules fit better into the zeolite pores. An increase in the separation factor to values exceeding 3 is obtained.

12.3.4 Overlapping boiling points

Overlapping boiling points are generally encountered when it is attempted to *separate one class of components from another class*. Industrial examples are the alifatics/aromatics, olefins/paraffins and paraffins/isoparaffins separation. In these cases the strategy for obtaining a suitable separation factor is roughly the same as with low relative volatilities. The main difference, however, is that vapor-liquid equilibria are avoided because it is almost impossible to overcome the large differences in vapor pressure by enhancement of the separation factor.

In the case of *alifatics/aromatics*, the desired separation factor is obtained by introducing a highly polar solvent to form a nonideal solution that will split into two phases. The used solvents such as sulfolane, NMP, DMF and tetraethylene glycol have a purely physical interaction with the hydrocarbon phase. The alifatics are

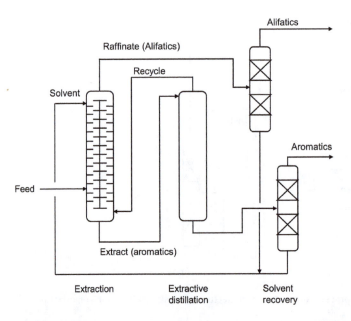

Figure 12.8: Schematic of extraction processes for alifatics/aromatics separation.

repelled stronger by the polar solvent than the aromatics, generating the desired separation factor. Unique in these processes is that the extraction columns are operated with a partial recycle of the extract in order to obtain full separation of the alifatics and aromatics. As shown in Figure 12.8, both product streams extract and raffinate have to be distilled to recover the solvent. Alternatively, water washing of the raffinate to wash out the polar solvent is practiced, and in that situation the water needs to be separated by distillation from the solvent before reuse. These additional distillations make an extraction process in general economically unattractive, except for situations where a large amount of distillation operations would be required to obtain the same separation.

12.3.5 Low concentrations

If only a small amount of one component is to be removed from a large amount of one or more other components, changing the phase of the latter components should be avoided. In such a case, stripping, extraction or selective adsorption are the preferred techniques to remove the minor component.

Stripping is often used to remove small amounts of volatile chemicals from aqueous systems. It is particularly good for components that are immiscible with water and tend to have extremely high volatilities such as benzene, toluene and other hydrocarbons. Stripping of low volatility components that are completely miscible with water is generally uneconomical because of the large amounts of stripping gas or steam that are required.

Extraction is generally favored for the removal or recovery of reasonable amounts of a desired component with a low volatility from reaction mixtures or waste streams. Such examples are often encountered in biotechnological processes where the product stream contains typically up to 10 wt% of the desired product (organic acids, amino acids, penicillins, alcohols), which is recovered by extraction with solvents that exhibit a significant miscibility gap with water such as heptane, methylethylketone, but also higher alcohols and composite solvents based on these. Composite solvents contain an extractant that specifically addresses the solute that is desired to increase the distribution ratio of that solute. The extractant is diluted in the second solvent component, the diluent, which should be selected such that the miscibility gap is large, the viscosity of the solvent phase not too high, and preferably stabilizes the complexes formed between the extractant and the solute, enhancing the distribution further.

At very low concentrations (<1,000 ppm), extraction usually becomes highly uneconomical because the removal of solvent residues from the raffinate and recovery of solvent from the extract stream becomes too expensive. The same holds for stripping because the stripped components have to be removed from the gaseous stream. In those cases, adsorption is often the technology of choice. Typical examples are the removal of contaminant by adsorption on activated coal or the removal of trace amounts of water to prevent catalyst deactivation from monomer and solvent streams in polymer processes.

12.3.6 Dissolved solids

The *final product* from many industrial chemical processes is a *solid material*. For these systems of dissolved solids, such as inorganic salts in water or essentially nonvolatile or high melting organics, crystallization is the predominant separation method. Even when the final product is not a solid, solid-liquid or solid-gas separation operations may be involved. An already discussed example is the separation of xylene isomers. Crystallization is induced commonly by one of four mechanisms: cooling, solvent evaporation, reaction and the addition of an anti-solvent. Evaporation of solvent is the preferred mode for systems in which solubility shows a large decrease with increasing temperature, or when the solubility shows little or no temperature dependence. In cases of significantly increasing solubility with increasing temperature, cooling crystallization may be an option.

A common flowsheet for the separation section of a process for the manufacture of solid products is shown in Figure 12.9. In these processes, a crystallizer is used to produce the solid product, always followed by a solid-liquid phase separation to remove as much of the solvent as possible without the need of evaporation.

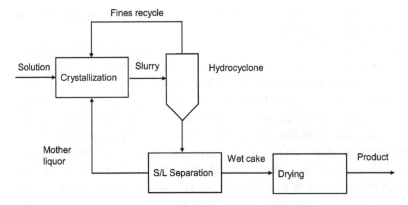

Figure 12.9: Schematic of a general solids production process.

In many crystallization processes, a hydrocyclone is used for control of the particle size by separation of fines that are recycled to the crystallizer. The remaining solvent residues are finally removed in a drying step. The physical nature of the material to be handled is the primary item for consideration in dryer selection. A slurry will demand a different type of dryer from that required by a coarse crystalline solid.

As discussed before, crystallization is an alternative to other separation techniques for liquid mixtures. However, it should always be kept in mind that *separation processes involving handling of a solid have a disadvantage in continuous operation relative to processes in which all phases are fluid*. This disadvantage stems from the difficulty of handling solids in continuous flow. Therefore, crystallization should only be considered when other techniques are not feasible or when the desired final state of the product is solid and a crystallization operation is required anyhow.

12.4 Separaton systems for gas mixtures

12.4.1 General selection considerations

In the previous section, we have primarily dealt with the separation of liquid mixture feeds. The primary techniques are ordinary and enhanced distillation. If the feed consists of a vapor mixture in equilibrium with a liquid mixture, the same techniques can often be employed. If the feed is a gas mixture and a wide gap in volatility exists between two groups of chemicals in the mixture, it is often preferable to partially condense the mixture, separate the phases, and send the liquid and gas phases to separate separation systems. Whereas ordinary distillation is the dominant method for the separation of liquid mixtures, no method is dominant for gas

mixtures. The separation of gas mixtures is further complicated by the fact that whereas most liquid mixtures are separated into nearly pure components, the separation of gas mixtures falls into the following three categories:
1. Enrichment to increase the concentration of one or more species
2. Sharp splits to produce nearly pure products
3. Purification to remove one or more low-concentration impurities

Enrichment is defined as a separation process that results in the increase in concentration of one or more species in one product stream and the depletion of the same species in the other product stream. Neither high purity nor high recovery of any components is achieved. *Gas enrichment* can be accomplished with a wide variety of separation methods such as physical absorption, adsorption, condensation and membrane permeation.

Sharp separations are often referred to as *bulk separation* and are designed to produce two high-purity product streams at high recovery. Separations in this category can be difficult to achieve for gas mixtures. The separation methods that can potentially obtain a sharp separation in a single step are physical absorption, adsorption and cryogenic distillation. Chemical adsorption is often used to achieve sharp separations, but generally limited to situations where components to be removed are present in low concentrations. Also reactive absorption can be applied, for example, in carbon capture from flue gas or removal of sour gases (CO_2 and H_2S) from natural gas. When the initial concentration of these gases is high, and the desired separation sharp, a combination of a physical absorption and a reactive absorption may be considered. In such tandem operation the bulk of the sour gases is removed in a physical absorption process, which does not have a high energy penalty in the solvent recovery, and the polishing step involves a reactive absorption which exhibits a much higher capacity at the cost of higher regeneration energy costs.

The special case involving the removal of a low (<5 mol%) mole fraction impurity at high (>99%) recovery is called *purification separation*. Purification typically results in one product of very high purity. It may or may not be desirable to recover the impurity in the other product. The separation methods applicable to purification separation include adsorption, chemical absorption and catalytic conversion. Physical absorption is not included in this list as this method typically cannot achieve high purities.

Besides the separation category and separation factor, the *production scale of the process* is a major factor in determining the optimal separation method for the separation of gas mixtures because economies of scale are most pronounced for cryogenic distillation and absorption. This effect of economics of scale is perfectly illustrated in air separation. As Figure 12.10 shows, membrane separations are most economical at low production rates, adsorption at moderate rates and cryogenic distillation at large production scales.

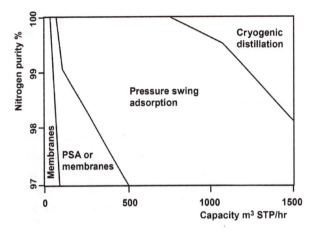

Figure 12.10: Comparison of the economics of operation of various techniques for air separation.

12.4.2 Comparison of gas separation techniques

12.4.2.1 Absorption

Physical absorption is typically used for applications for the recovery of components that are present in reasonable concentrations. The absorption liquid should be selected to have a separation factor of three or higher for economical operation. Unless values of the separation factor are about 10 or higher, absorption and stripping operations cannot achieve sharp separation between two components. Nevertheless, these operations are used widely for *preliminary or partial separations* where the separation of one key component may be sharp, but only partial separation of the other key component is required. If the composition of the absorbate in the gas is to be reduced below 100 ppm in the lean gas product (such as sour gases in natural gas) it should be considered to combine physical absorption with chemical absorption or adsorption. An important advantage of absorption is its favorable scale-factor, which is comparable to distillation. Therefore, absorption is particularly suited for large-scale applications and situations where the product should be obtained in the liquid phase for further processing. Typical examples where the final product is in the liquid state are the recovery of oxidation products from partially converted gas streams and the absorption of NO_x and SO_2 for the production of nitric acid and sulfuric acid. In these acid-production applications, absorption in a liquid is essential to obtain the desired product state and has the additional advantage of allowing interstage cooling in the absorption tower.

When considering chemical absorption, the impurities must contain acid-base functionalities or other reactive groups. The reactivities of the impurities must be much higher than the bulk stream components to allow selective impurity removal. Chemical absorption is typically favored when desired removal is high and the partial pressure of the components to be removed from the bulk stream is low (<1%). For the removal of acidic (sour) gases (CO_2 and H_2S), one of the most widely used applications of chemical absorption, typically amine group containing reactive solvents are applied.

12.4.2.2 Adsorption

Adsorption should be favored for processes that require essentially complete removal of water vapor and for small-scale desiccation operations. Adsorption is a very versatile separation technique that can be used for all three separation categories. Selectivity in adsorption is controlled by molecular sieving or adsorption equilibrium. When solutes differ significantly in molecular size and/or shape, zeolites and carbon molecular-sieve adsorbents with very small pores can be used to advantage. These adsorbents have very narrow pore-size distributions that are capable of very sharp separations based on differences in the molecular kinetic diameter. Adsorbents made of activated alumina, activated carbon and silica gel separate by differences in adsorption equilibria, which must be determined experimentally. *To be economical, the adsorbent must be regenerable*. This requirement excludes the processing of gas mixtures that contain:

1. High-boiling organic compounds because they are difficult to remove
2. Lower-boiling organic compounds that may polymerize on the adsorbent surface
3. Highly acidic or basic compounds that may react with the adsorbent surface

Molecular sieves are excellent adsorbents for selective water removal. Solid-phase desiccant systems are relatively simple to design and operate. Generally, they are the lowest cost alternative for processing small quantities of gas. For bulk separations, the more strongly adsorbed component should be in minority in the feed. Some large-scale enrichments examples include hydrogen recovery, methane enrichment and oxygen enrichment from air. In purification separations adsorption should only be considered for dilute solutions. *Activated carbon adsorption is generally uneconomical for the removal of >1,000 ppm contaminants*. This is due to the fact that the costs of regeneration become too excessive in those cases. Regeneration can be performed by steam stripping or burning. During burning part of the active coal is lost, resulting in excessive cost when large amounts of contaminants have to be handled.

12.4.2.3 Membrane separation

The separation factor defined earlier can be applied to gas permeation if the relative volatility is replaced by the permselectivity, which is the ratio of the membrane permeabilities for the two key components of the feed-gas mixture. This permeability depends on both solubility and diffusivity through the membrane at the conditions of temperature and pressure. Permeabilities and permselectivities are best determined by laboratory experiments. In general, gas permeation is commercially feasible when the permselectivity between the two components is greater than 15. Commercial applications include the recovery of carbon dioxide from hydrocarbons, the adjustment of the hydrogen–carbon monoxide ratio in synthesis gas, the recovery of hydrocarbons from hydrogen and the separation of air into nitrogen- and oxygen-enriched streams.

12.5 Separation methods for solid-liquid mixtures

Solid-liquid separation equipment selection is dependent on such a huge number of different factors and the choice is so wide, that it is impossible to present a recipe that can be followed to select the right piece of equipment for each application. When developing a solid-liquid process to separate two or more solids from each other, many combinations of machine and technique are possible. Typical for solid-liquid separations is that *several of these combinations will result in an adequate solution to the problem*, while obtaining the optimal solution would inevitably too expensive, if not impossible in an industrial situation.

The first step in the selection process is to choose the most appropriate technology from sedimentation, filtration or a combination of these two operations. The use of gravity sedimentation systems would generally be considered first because they tend to be relatively cheap, and are ideally suited for large continuous liquid flows and automatic operation. Settlers can also be used for classifying particles by size or density, which is usually not possible with filtration. However, *settling does not give a complete separation*. One product is a concentrated suspension and the other is a liquid, which may contain fine particles of suspended solids. In spite of this limitation it is important to note that even if sedimentation does not achieve exactly the separation required, it may still be a valuable first step in a process and may be followed by filtration. Preconcentration of the solids will reduce the quantity of liquid to be filtered and the concentrated suspension can then be filtered with smaller equipment than would be needed to filter the original dilute suspension. However, too small a difference in density between the solid and fluid phases would eliminate sedimentation as a possibility, unless this density difference can be enhanced, or the gravity force field increased by centrifugal action. Having decided on sedimentation, the next stage is to consider the different equipment available and whether the separation is to be effected continuously or discontinuously. Various attempts have been made to provide a basis for the selection of sedimentation equipment. Table 12.5 contains some of this information, which points to the principal factors influencing such selections.

Opposite to sedimentation, filtration provides an almost unlimited amount of design control since it relies on the filter material, a man-made septum chosen to retain exactly those particles that must be separated. Filtration systems, however,

Table 12.5: Selection of sedimentation machinery.

Selection criterion	Gravity sedimentation	Sedimentation centrifuges			Hydrocyclones
		Tubular bowl	Disc	Scroll	
Process scale (m^3/hr)	1–>100	1–td32#10	1–>100	1–>100	1–>100
Settling rate (cm/s)	0.1–>5	<0.1–5	<0.1–5	<0.1–>0.5	0.1–>0.5
Operation	batch or continuous	batch	batch or continuous	batch or continuous	continuous
Objectives:					
clarified liquid	x	x	x	x	x
concentrated solids	x	x	x	x	x
washed solids	x				x

are far less suitable for continuous production and tend to be more expensive per volume treated than sedimentation systems. Therefore, filtration is mostly applied when *one wants to obtain the solids* in the form of a cake, either as the desired end product or in removing undesired particles from liquid products. Although even more complicated, several attempts have been made to develop a methodology for filtration equipment selection. For example, the cake build-up rate criterion relates the rate of filtration to equipment types, as shown in Table 12.6. However, it must be realized that all such generalizations are merely tentative suggestions on filtration equipment selection. Many types of filters are capable of similar performance and could be interchanged.

Table 12.6: Equipment selection and rate of filtration.

System characteristic	Cake thickness build-up rate	Equipment
Rapid separation	cm/s	Gravity pan screens Top-feed vacuum filters Filtering centrifuge Vibratory conical bowl
Medium separation	cm/min	Vacuum drum, disc, belt Solid-bowl centrifuge
Slow separation	cm/h	Pressure filters Disc and tubular centrifuge
Clarification	Negligible cake	Cartridges Granular beds Filter aids, precoat filters

For the separation of very fine particles down to molecular size, membrane filtration techniques such as microfiltration, ultrafiltration, nanofiltration and reverse osmosis should be considered. The range of application of these four pressure driven membrane separation processes is summarized in Figure 12.11. Ultrafiltration and microfiltration are basically similar in that the mode of separation is molecular sieving through increasingly fine pores. Microfiltration membranes filter colloidal particles and bacteria from 0.1 to 10 μm in diameter. Ultrafiltration membranes can be used to filter dissolved macromolecules, such as proteins, from solutions. The mechanism of separation by reverse osmosis membranes is quite different. In reverse osmosis membranes the membrane pores are so small that they are within the range of thermal motion of the polymer chains that form the membrane. Nanofiltration membranes fall into a transition region between pure reverse osmosis and pure ultrafiltration. Both techniques are used when low molecular weight solutes such as inorganic salts or small organic molecules have to be separated from a solvent. The most important application of reverse osmosis is the desalination of brackish groundwater or seawater.

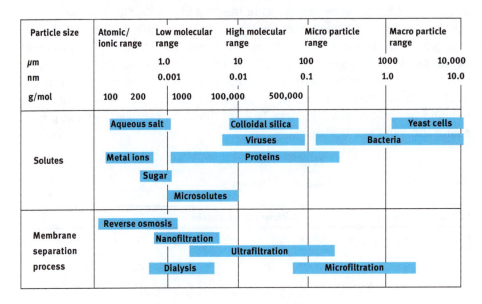

Figure 12.11: Application range of various membrane filtration processes.

Excercises

Exercise 1

The vent gas from several polymer plants contains air, paraffins and olefins (monomers). The composition is given by (1 bar, 25 °C):

Air	91	mol%			
Butane	1.5	mol%	1-Butene	3	mol%
Pentane	1.0	mol%	1-Pentene	2	mol%
Hexane	0.5	mol%	1-Hexene	1	mol%

Within the R&D department of the company, it is explored, which techniques may be technically and economically feasible to recover the three olefins (1-butene, 1-pentene en 1-hexene) in pure state. The company has decided to approach you as a consultant and prepared several questions for the first meeting. As a first step they are considering the purification of the air by simultaneous recovery of the olefins and the paraffins.

a. What separation methods are in your opinion applicable for this simultaneous olefin and paraffin recovery from the vent gas.

After recovery the olefin/paraffin needs to be separated in two fractions, the paraffins (butane, pentane, hexane) and the olefins. This is in principle possible by selective absorption of the olefins.

b. Can you determine which two key components need to be separated (selectivity > 1) in order to achieve a full olefin/paraffin separation.

c. What is the minimal required difference in affinity (activity coefficients) of the absorption liquid towards both key components to achieve this full olefin/paraffin separation. In other words, achieve a selectivity = $(x_1/y_1)/(x_2/y_2) > 1$.

	C_4	$C_4^=$	C_5	$C_5^=$	C_6	$C_6^=$
Vapor pressure (25 °C, bar)	2.4	3.0	0.68	0.85	0.20	0.24

A possible alternative for absorption is extraction with a solvent that has a higher affinity for olefins compared to paraffins.

d. What are in your eyes the two key components in order to obtain a full olefin/paraffin separation when extraction with a high polar solvent is applied?

e. Which of the below solvents are in principle suitable (selectivity > 1) for a complete separation of the olefin and paraffin fractions.

Distribution coefficient
(concentration in solvent/concentration in feed)

NMP	0.10	0.12	0.08	0.08	0.06	0.05
DMSO	0.05	0.15	0.04	0.12	0.03	0.10
DEG	0.02	0.25	0.01	0.20	0.01	0.15

Exercise 2

In a large chemical plant methylcyclohexane is separated from toluene by distillation. For the design of a new plant both the use of extractive distillation and extraction are considered. A detailed evaluation requires first the characterization of the current separation between methylcyclohexane and toluene.

a. Calculate from the given data the relative volatility for the separation of methylcyclohexane and toluene by distillation. Consider the mixture to behave almost ideal.

b. Which of the given solvents might be feasible to enhance the relative volatility sufficiently to make extractive distillation economically attractive.

c. The same question for extraction. Assume that the solvents and the methylcyclohexane/toluene mixture are almost immiscible.

   Besides methylcyclohexane (45%) and toluene (45%) the feed also contains 5% of cyclohexane as well as benzene.

d. What would be the optimal arrangement of separation steps for distillation, extractive distillation and extraction (for each one) to separate this mixture into its four pure constituents? What distillative separation might become problematic and why?

e. Which of the arrangements seems, considering the number of operations (columns), the most economical solution.

f. Why is the counting of the number of operations sufficient to get a first impression about the most economical route? What effects are not taken into account?

	Benzene	Cyclo-hexane	Methyl-cyclohexane	Toluene
Vapor pressure (80 °C, bar)	1.01	0.99	0.54	0.39

Activity coefficient at infinite dilution:

NMP	1.5	7	8	2
DMSO	4.0	25	30	4
DEG	7.0	41	60	10
NMF	1.5	10	10	2.5

Appendix
Answers to Exercises

1 Introduction

2. $\alpha = 2.26$

2 Evaporation and distillation

1. $A = 9.2082; B = 2755.6; C = 219.16 \,°C$

2. $P^0_{benzene} = 1.013$ bar

3. a. 100 °C and 1.0 atm: $x = 0.257$ $y = 0.456$
   b. 100 °C and 1.5 atm: $x = 0.736$ $y = 0.871$

4. a. $T_{feed} = 369.9$ K
   b. $V/L = 0.296$

5. b. $x = 0.318$, $y = 0.538$
   c. $x = 0.375$, $L = 466.7$ mol/h, $V = 233.3$ mol/h
   d. $x_1 = 0.36$, $y_1 = 0.585$
       $x_2 = 0.295$, $y_2 = 0.511$
       $V_1 = 0.4\,F$, $V_2 = 0.18\,F$, $L_2 = 0.42\,F$

6. a. 327.5 K
   b. $V'' = 101.4$ mol/s
   c. 3042 kJ/s
   f. $N_{ts} = 11.5$ including reboiler
   g. ≈ 6 m

7. a. $N_{min} = 5.4$ (Fenske)
       $R_{min} = 0.77$
   b. $N = 16 +$ reboiler
       Feedstage = tray 4 from bottom

8. a. $R_{min} = 1.08$ (Underwood)
   b. $N_{min} = 6.4$ (Fenske)
   c. $N_{ts} = 11 +$ reboiler
   d. $B = 487.3$ kg/h
       $D = 420$ kg/h
   e. 255 kg/h

https://doi.org/10.1515/9783110654806-013

9. a. $B = 55$ kmol/hr, $D = 45$ kmol/hr
   b. $N_{min} = 8$
   c. $R = 2.45$
   d. $V''/B = 2$

10. a. $D = 402$ kmol/h, $B = 148$ kmol/h
    b. $N_{min} = 12$
    c. $N = 17$
    d. Tray 10
    e. Slope feed line = 0, saturated vapor

11. a. 85%
    b. $F = 119$ kmol/h, $B = 69$ kmol/h
    c. –
    d. $R_{min} = 1.33$
    e. $V''/B = 2.5$
    f. –

3 Absorption and stripping

1. $x_{out} = 6.44 \times 10^{-6}$

2. $K = y\dfrac{p^0}{P_{tot}} = 0.682$
   For $L/G = 1.5*(L/G)_{min} = 1.5 \cdot 0.648 = 0.97$, $N_{ts} \approx 5.4$

3. $(L/G)_{min} = 80.5$
   at $L/G = 120.8$, $N_{ts} \approx 4.4$

4. a. Weight fraction base, $(L/G)_{min} = 0.471$
   b. $N_{ts} \approx 8$

5. a. $K = 5 \times 10^{-3}$
   b. Use conventional mole fractions; $N = 2.25$
   c. For $L/G = 0.005$, $N = 8$
      For $L/G = 0.004$ the desired separation cannot be achieved because the minimum L/G ratio = 4.44×10^{-3}
   d. N_{ts}(stripping) = 2.8
   e. N_{ts}(stripping) = 2.8; N_{ts}(absorption) = 2.3
   f. $0.1 < x_{in}$ (absorber) < 0.2

4 G/L-Contactors

3. a. 2.7 m
   b. 0.14
   c. 6.0 m

4. a. $N_{ts} \approx 3.7$
   b. $E_{MV} = 0.69$
   c. $N_s = 4.4$
   d. $x_{out} = 0.16 \cdot 10^{-6}$
   e. $Q_{max} = 3.4 \text{ m}^3 \text{ s}^{-1}$; $Q_{min} = 0.1 \text{ m}^3 \text{ s}^{-1}$

5. $F_p = 540 \text{ m}^{-1}$ $u_{flood} = 1.18 \text{ m s}^{-1}$
            $F_p = 125 \text{ m}^{-1}$ $u_{flood} = 2.46 \text{ m s}^{-1}$

6. a. 20%
   b. $N_{OV} = 3.62$
   c. $k_{OV} = 5.68 \cdot 10^{-3} \text{ m s}^{-1}$
   d. $u_{flood} = 0.92 \text{ m s}^{-1}$
   e. 20% higher

7. a. $N_{ts} = 14.34$
   b. $EMV = 0.52$
   c. $N_S = 25.1 < 30$
   d. 4.9 ppb
   e. $Q_{max} = 0.969 \text{ m}^3 \text{ s}^{-1}$

8. a. $N_{ov} = 9.38$
   b. $u_v = 0.445 \text{ m s}^{-1}$
   c. $Ah = 813 \text{ m}^2 \text{ m}^{-1}$
   d. $N_{TU} = 12,9 > 9.38$
   e. $f_{abs} = 0.9976$
   f. %flood = 66%

9. a. $Nov = 27.4$
   b. $kov = 6.2 \ 10^{-3} \text{ m s}^{-1}$
   c. $f_s = 1$
   d. 1620 kg hr^{-1}
   e. $e = 0.64$

5 Liquid–liquid extraction

1.   Graphical $N_{ts} = 4.1$
          Kremser $N_{ts} = 4.13$

2. a. $y_1 = y_{out} = 9.26 \times 10^{-4}$
          $y_{N+1} = y_{in} = 6 \times 10^{-6}$
    b. $R' = 51.6 (a.u.)$
          $x'_N = 0.0178$

3. a. $S/F = 495$
    b. $S/F = 90$
    c. $S/F = 47.3$
    d. $S/F = 22$
    e. $S/F = 4.95$

4. a. $X_{in} = 0.13; X_{out} = 5.025 \cdot 10^{-3}; Y_{out} = 0.105$
    c. $F/S = 0.84$
    d. $N_{ts} = 3.9$

5. a. $S_{min} = 232$
    b. $N = 4.5$
    c.   Approximate solution:

Extract	Ac	H_2O	EA	kg/h	Raffinate	Ac	H_2O	EA	kg/h
E1	0.51	0.04	0.45	769	R1	0.38	0.59	0.03	914
E2	0.43	0.03	0.54	673	R2	0.32	0.66	0.02	809
E3	0.34	0.025	0.635	590	R3	0.24	0.74	0.02	714
E4	0.24	0.02	0.74	473	R4	0.15	0.84	0.01	622
E5	0.10	0.01	0.89	350	R5	0.05	0.945	0.005	500

6. a. $y_{FA} = 0.16$
    b. $S/F_{min} = 0.26$
    c.   3 stages

7. a. $S_{min} = 27.3$ ton/h
    b. $x_C = 0.22$
    c. $x_C = 0.46$

9. a. $S_{min} = 0.073$ kg s^{-1}
    b. $E_1 = 0.17$ kg s^{-1}
    c. $N_{ts} \approx 3$
    d. $x_{E2pyr} = 0.10; E_2 = 0.12$ kg s^{-1}

10. a. water and biodiesel are almost completely immiscible
    furthermore the tielines are favorable
    b. 5.1 t/h
    c. 3 stages
    d. distillation

6 Adsorption and ion exchange

3.   14.8 kg/m^3

4. a1. $W/V = 0.180$ kg/m^3
   a2. $V = 1003$ m^3
   a3. $6.15 \cdot 10^6$ s (71 days)
   b1. 2.14 kg/m^3

6. b. 200 s
   d. 222 s

7. a. $q_{m1} = 1.301$ $q_{m2} = 1.800$
   b. $q_1 = 0.751$ mol kg^{-1}
      $q_2 = 0.582$ mol kg^{-1}
   c. $v_1/v_2 = 0.912$

8.   5.33 eq kg^{-1} dry resin

9. b. $V_{min} = 2.333 \times 10^{-3}$ m^3 s^{-1}

7 Drying of solids

1.   0.34 g/s m^2

2.   3.1 h

3.   0.17 g/s ($h = 54.1$ W/m^{2}K)

4. $p_{sat}/p_{sat}^\infty = 1.011$

5. $T > 332.6$ K

6. a. 785 kg
    c. 50.4 h
    d. 175 kg
    e. 3.7 h

7. a. $H_{rel} = 10\%$
    b. $r_c = 2.7 \times 10^{-4}$ kg m^{-2} s^{-1}
    c. $t_c = 926$ s

8. a. $r_c = 7.64 \times 10^{-4}$ kg m^{-2} s^{-1}
    b. $h = 56.7$ W m^{-2} K^{-1}
    c. $v = 5.3–5.4$ m s^{-1}

9. a. $r_{vap} = 0.0244$ kg s^{-1}
    b. $N = 15$
    c. 36–37 °C

8 Crystallization

1. a. 0.436
    b. 0.718

2. a. maximum production rate 3182 kg/h
    b. maximum yield 0.795

3.   63.6 kg/h crystals/h

4.   195 kg/h

5. a. 0.142 kg Na_2SO_4.10aq per kg free water fed
    b. 0.187 kg Na_2SO_4.10aq per kg free water fed
    c. 0.343 kg Na_2SO_4.10aq per kg free water fed
    d. 0.443 kg Na_2SO_4.10aq per kg free water fed

6. $\Delta E = 8.2$ kJ/mol

7. $s = 0.006$

8. a. $G = 0.097 \ \mu m/s$
    b. $B^0 = 3.7 \times 10^4$ no/s kg

9. a. 1.9 mm
    b. $G = 0.082$ mm/h

10. $\tau = 9740$ s

    $V_{crystallizer} = V_{solution} + V_{crystals} = 27.6 + 1.02 = 29 \ m^3$

11. –

12. a. $0.02 \ mol \ kg^{-1}$
    b. $293 \ kg \ hr^{-1}$
    c. 0.9 vol%

13. a. 0.60 (60%)
    b. $31250 \ kg \ hr^{-1}$
    c. $208490 \ kg \ brine \ hr^{-1}$
    d. 1500 s
    e. $52.7 \ m^3$
    f. $2 \ m^3$

9 Sedimentation

1. $Re_p = 0.3$
    $V_\infty = 4.27 \times 10^{-3} \ m/s$

2. $A = 11.8 \ m^2$

3. $V_\infty = 0.0.034 \ m/s$
    $A = 8.7 \ m^2$

4. $A = 2.9 \ m^2$
    $D = 1.9$ m (cylindrical tank)

5. 12.6 μm (check Re = 0.055)
    $L_1 + L_2 = 28.4$ m

6. $V_{g\infty} = 1.75 \times 10^{-6} \ m/s$
    $V_{c\infty} = 0.98 \times 10^{-2} \ m/s$

7.   49 Hz (98π rad/s)

8. $\Sigma_{process} = 918$ m^2
        $\Sigma_{machine} = 3580$ m^2
        efficiency = 25%

9. $Q = 4.47 \times 10^{-4}$ m^3/s

10.   a. $9.05 \cdot 10^{-5}$ m^3 s^{-1}
     b.   0.15 µm (Re < 1)

11. $D = 0.75$ m
        4 hydrocyclones in parallel

12.   a. $d = 24.73$ µm
     b. $A = 599$ m^2
     c. $\omega = 840$ rpm
     d. $Q_{out} = 0.001$ m^3 s^{-1}

10 Filtration

1       $\alpha = 4.67 \times 10^{12}$ m/kg, $R_m = 2.45 \times 10^{12}$ 1/m
2       $\alpha = 5.23 \times 10^{11}$ m/kg
3. $\alpha = 5.97 \times 10^{11}$ m/kg, $R_m = 9.04 \times 10^{11}$ 1/m
4. $\alpha_0 = 8.11 \times 10^8$ m/kg, $n = 0.38$
5. $\alpha_0 = 2.36 \times 10^8$ m/kg, $n = 0.574$
6.   filtrate production rate = $3.15 \cdot 10^{-5}$ m s^{-1}
       dry solid production rate = $3.15 \cdot 10^{-3}$ kg m^{-2} s^{-1}
       cake thickness = 0.90 mm
7.   9 m^2
8.   0.627 kg s^{-1}
9.   no single set of answers
10.   0.66 kg s^{-1}
11.   3.77 h
12.   a.   31.1 s
     b.   2.75 times faster
13.   a. $\alpha = 3.3 \times 10^{11}$ m kg^{-1}
     b. $w_r = 153$ kg hr^{-1}

14.　a.　$\alpha = 8.5 \times 10^{10}$ m kg^{-1}
　　　　$R_m = 2.22 \times 10^{11}$ m^{-1}
　　b.　$(dV/dt) = 1.081 \times 10^{-3}$ m^3 s^{-1}
　　c.　$V = 0.234$ m^{-3}
　　d.　$t = 216$ s

11 Membrane filtration

2.　　　8.75 10^{-8} kmol/m^{2}s
　　　　0.0075 and 0.045 kmol/m^3

3.　　　7.45 10^{-8} kmol/m^{2}s
　　　　13 and 8.7 mol/m^3

4.　　　4.88 atm

5.　　　$P_{M,water}/\delta_M = 2.07\ 10^{-4}$ kg/m^2.s.atm
　　　　$P_{M,salt}/\delta_M = 3.90\ 10^{-7}$ m/s

6.　a.　$R_M = 4.6.10^{12}$ (m^{-1})
　　b.　$E = 7000$ Nm
　　c.　$P = 80$ Nm s^{-1}

7.　a.　$R_M = 5.10^{11}$ (m^{-1})
　　b.　exponential model
　　c.　$E = 2.9.10^{11}$ Nm

12 Separation method selection

1.　a.　Adsorption, absorption, condensation
　　b.　1-butene and hexane
　　c.　$S = >1 \Rightarrow \dfrac{\gamma_{hexaan}}{\gamma_{1-buteen}} > 15$
　　d.　1-hexene and butane
　　e.　$S > 1$ 　　　　　　NMP　　　　　$S = 0.5$ (not suited)
　　　　　　　　　　　　　DMSO　　　　$S = 2.0$ (OK)
　　　　　　　　　　　　　DEG　　　　　$S = 7.5$ (OK)

2. a. $\alpha_{12} = \dfrac{P_1^*}{P_2^*} = \dfrac{0.54}{0.39} = 1.38$

    b.
NMP	$\alpha_{12} = 5.52$
DMSO	$\alpha_{12} = 10.4$
DEG	$\alpha_{12} = 8.28$
NMF	$\alpha_{12} = 5.52$

For $\alpha_D = 1.38$ economics extractive vs normal distillation demand $\alpha_{ED} \geq 2$
Conclusion: all four solvents should be evaluated

    c.
NMP	$\beta_{12} = 4.0$
DMSO	$\beta_{12} = 7.5$
DEG	$\beta_{12} = 6.0$
NMF	$\beta_{12} = 4.0$

For $\alpha_D = 1.38$ economics extraction vs distillation demand
$\alpha_E \geq 5$
Conclusion: only DMSO and DEG interesting for evaluation

d.

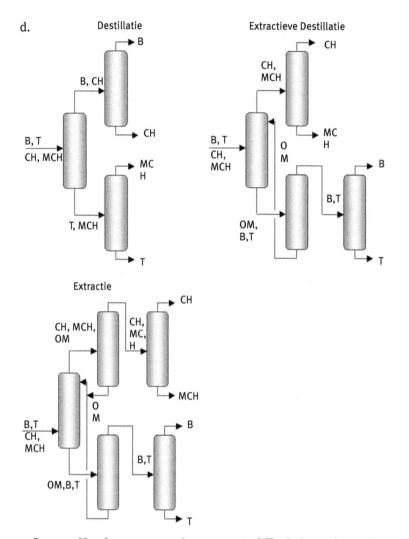

Destillatie

Extractieve Destillatie

Extractie

Benzene/Cycohexane separation $\alpha_{12} = 1.02$, difficult, large chance for azeotrope

e. Distillation

References and Further Reading

General

[1]   C.J. Geankoplis, *Transport Processes and Separation Process Principles*, 5th Edition, Prentice Hall, Englewood Cliffs NJ, **2018**.
[2]   R.E. Treybal, *Mass-Transfer Operations*, 3rd Edition, McGraw-Hill, New York, **1980**.
[3]   J.L. Humphrey and G.E. Keller II, *Separation Process Technology*, McGraw-Hill, New York, **1997**.
[4]   F.M. Khoury, *Multistage Separation Processes*, 4th Edition, CRC Press, Boca Raton, **2015**.
[5]   C.J. King, *Separation Processes*, 2nd Edition, McGraw-Hill, New York, **2013**.
[6]   W.L. McCabe, J.C. Smith and P. Harriot, *Unit Operations of Chemical Engineering*, 7th Edition, McGraw-Hill, New York, **2005**.
[7]   R.D. Noble and P.A. Terry, *Principles of Chemical Separations with Environmental Applications*, Cambridge University Press, New York, **2004**.
[8]   K. Sattler and H.J. Feindt, *Thermal Separation Processes, Principles and Design*, VCH, Weinheim, **1995**.
[9]   S.J. Setford, *A Basic Introduction to Separation Science*, Rapra Technology, Shawbury, **1995**.
[10] E.J. Henley, J.D. Seader and D.K. Roper, *Separation Process Principles*, International Student Version, 4th Edition, John Wiley & Sons, Hoboken, NJ, **2016**.
[11] P.C. Wankat, *Rate-Controlled Separations*, Springer, Dordrecht, **1994**.
[12] P.C. Wankat, *Separation Process Engineering: Includes Mass Transfer Analysis*, 4th Edition, Prentice Hall, Upper Saddle River NJ, **2017**.
[13] R.P. Chhabra and B. Gurappa, Coulson and Richardson's Chemical Engineering, Volume 2, *Particle Technology and Separation Processes*, 6th Edition, Butterworth-Heinemann, Oxford, **2019**.
[14] R.K. Sinnot and G. Towler, Coulson and Richardson's Chemical Engineering, Volume 6, *Chemical Engineering Design*, 6th Edition, Butterworth-Heinemann, Oxford, **2019**.
[15] *Handbook of Separation Process Technology*, (R.W. Rousseau, Editor), John Wiley & Sons, New York, **1987**.
[16] *Perry's Chemical Engineers' Handbook*, 9th Edition (R.H. Perry and D.W. Green, Editors), McGraw-Hill, New York, **2018**.
[17] *Encyclopedia of Separation Technology* (D.M. Ruthven, Editor), John Wiley & Sons, New York, **1997**.
[18] *Handbook of Separation Techniques for Chemical Engineers*, 3rd Edition (P.A. Schweitzer, Editor), McGraw-Hill, New York, **1997**.
[19] *Handbook of Chemical Engineering Calculations*, 4th Edition (T.G. Hicks and N.P. Chopey, Editors), McGraw-Hill, New York, **2012**.
[20] *Ullmann's Encyclopedia of Industrial Chemistry*, 7th Edition, Wiley-Interscience, New York, **2002**.
[21] *Kirk-Othmer Encyclopedia of Chemical Technology*, 5th Edition (A. Seidel, Editor), John Wiley & Sons, Hoboken, **2004**.
[22] *Kirk-Othmer Separation Technology*, 2nd Edition (A. Seidel, Editor), John Wiley & Sons, Hoboken, **2008**.

Chapter 2 and 4

[23] A. Vogelpohl, *Distillation the Theory*, De Gruyter, Berlin, **2015**.

https://doi.org/10.1515/9783110654806-014

[24] A. Gorak, E. Sorensen, *Distillation Fundamentals and Principles*, Academic Press, London, **2014**.
[25] A. Gorak, H. Schoenmakers, *Distillation Operation and Applications*, Academic Press, London, **2014**.
[26] J.G. Stichlmair, J.R. Fair, *Distillation: Principles and Practice*, Wiley-VCH, New York, **1998**.
[27] A. Gorak, Z. Olujic, *Distillation: Equipment and Processes*, Academic Press, London, **2014**.
[28] M.J. Lockett, Distillation Tray Fundamentals, Cambridge University Press, Cambridge, **2009**.
[29] R.F. Strigle, *Packed Tower Design and Applications: Random and Structured Packings*, 2nd Edition, Gulf Publishing Company, Houston, **1994**.
[30] J. Maćkowiak, *Fluid Dynamics of Packed Columns*, Springer, Heidelberg, **2010**.

Among many other reference data books, a lot of vapor-liquid data can be found in the following:
[31] B.E. Poling, J.M. Prausnitz and J.P. O'Connell, *The Properties of Gases and Liquid*, 5th Edition, McGraw-Hill, New York, **2000**.
[32] C.L. Yaws, *The Yaws Handbook of Vapor Pressure: Antoine Coefficients*, Elsevier, Amsterdam, **2015**.
[33] J. Gmehling and U. Onken, *Vapor-Liquid Equilibrium Data Collection, Chemistry Data Series*, Dechema, Frankfurt aM., **2003**.
[34] C.L. Yaws, *The Yaws Handbook of Thermodynamic Properties for Hydrocarbons and Chemicals*, Gulf Publishing Co., Houston, **2007**.

Chapter 3

[35] R. Zarzycki and A. Chacuk, *Absorption: Fundamentals & Applications*, Pergamon Press, Oxford, **1993**.
[36] P. Chattopadhyay, *Absorption and Stripping*, Asian Books, **2007**.
[37] D.A. Eimer, *Gas Treating: Absorption Theory and Practice*, John Wiley & Sons, Hoboken, **2019**.

Chapter 5

[38] R.E. Treybal, *Liquid Extraction*, 2nd Edition, McGraw-Hill, **1963**.
[39] *Handbook of Solvent Extraction* (T.C. Lo, M.H.I. Baird and C. Hanson, Editors), John Wiley & Sons, New York, **1983** (reprinted 1991).
[40] *Science and Practice of Liquid-Liquid Extraction* (J.D. Thornton, Editor), Volume 1&2, Oxford Science Publications, Oxford, **1992**.
[41] *Liquid-Liquid Extraction Equipment* (J.C. Godfrey and M.J. Slater, Editors), John Wiley & Sons, New York, **1994**.
[42] *Solvent Extraction Principles and Practice, Revised and Expanded* (J. Rydberg, M. Cox, C. Musikas and G.R. Choppin, Editors), 2nd Edition, CRC Press, **2004**.
[43] V. Kislik, *Solvent Extraction: Classical and Novel Approaches*, Elsevier, Amsterdam, **2011**.
[44] C. Poole, *Liquid Phase Extraction*, Elsevier, Amsterdam, **2019**.

Chapter 6

[45] E. Worch, *Adsorption Technology in Water Treatment: Fundamentals, Processing and Modeling*, De Gruyter, Berlin, **2012**.
[46] D.M. Ruthven, *Principles of Adsorption and Adsorption Processes*, Wiley, New York, **1984**.

[47] C. Tien, *Introduction to Adsorption*, Elsevier, Amsterdam, **2018**.

[48] R.T. Yang, *Adsorbents: Fundamentals and Applications*, Wiley Interscience, **2007**.

[49] D.M. Ruthven, S. Farooq and K.S. Knaebel, *Pressure Swing Adsorption*, Wiley-VCH, New York, **1993**.

[50] F. Rouquerol, J. Rouquerol and K. Sing, *Adsorption by Powders & Porous Solids: Principles, Methods and Applications*, 2nd Edition, Academic Press, San Diego, **2013**.

[51] A.K. Sengupta, *Ion Exchange in Environmental Processes*, John Wiley & Sons, Hoboken, **2017**.

[52] *Ion Exchange Technology* (Inamuddin and M. Luqman, Editors), Spinger, Heidelberg, **2012**.

[53] C.E. Harland, *Ion Exchange: Theory & Practice*, 2nd Edition, RSC Publishing, **1994**.

Chapter 7

[54] C. Strumillo and T. Kudra, *Drying: Principles, Applications and Design*, Topics in Chemical Engineering, Volume 3, Gordon & Breach Science Publishers, New York, **1986**.

[55] *Handbook of Industrial Drying*, 4th Edition (A.S. Mujumdar, Editor), CRC Press, Boca Raton Fl, **2014**.

[56] *Modern Drying Technology* (E. Tsotsas and A.S. Mujumdar, Editor), Wiley, New York **2007**.

[57] I. Dincer and D. Zamfirescu, *Drying Phenomena: Theory and Applications*, Wiley, New York, **2016**.

[58] R.B Keey, *Introduction to Industrial Drying Operations*, Pergamon Press, Oxford, **1978**.

[59] C.M. van't Land, *Industrial Drying Equipment*, Marcel Dekker, New York, **1991**.

Chapter 8

[60] N.S. Tavare, *Industrial Crystallization, Process Simulation Analysis and Design*, Plenum Press, New York, **1995**.

[61] *Crystallization Processes* (H. Ohtaki, Editor), Wiley, Chichester, **1997**.

[62] *Crystallization Technology Handbook*, 2nd Edition (A. Mersmann, Editor) Marcel Dekker, New York, **2001**.

[63] J.W. Mullin, *Crystallization*, 4th Edition, Butterworth Heinemann, New York, **2001**.

[64] *Handbook of Industrial Crystallization*, (A.S. Myerson, Editor), 3rd Edition, Cambridge University Press, Cambridge, **2019**.

[65] A. Lewis, M. Seckler, H. Kramer and G. Rosmalen, *Industrial Crystallization: Fundamentals and Applications*, Cambridge University Press, **2015**.

[66] Alan G. Jones, *Crystallization Process Systems*, Butterworth Heinemann, New York, **2002**.

[67] W. Beckmann, *Crystallization: Basic Concepts and Industrial Applications*, John Wiley & Sons, **2013**.

[68] H. Kramer, G. Rosmalen and J. ter Horst, *Basic Process Design for Crystallization Processes*, Online reader available at www.erallab.com.

[69] D.Kashchiev, *Nucleation*, Butterworth Heinemann, **2000**.

Chapter 9 & 10

[70] A. Bürkholz, *Droplet Separation*, VCH-Verlag, Weinheim, **1989**.

[71] P.N. Cheremisinoff, *Solids/Liquids Separation*, 2nd Edition, Butterworths Heinemann, Boston, **1998**.

[72] Rushton, A.S. Ward and R.G. Holdich, *Solid-Liquid Filtration and Separation Technology*, 2nd Edition, Wiley-VCH, **2000**.

[73] L. Svarovsky, *Solid-Liquid Separation*, 4th Edition, Butterworths, London, **2001**.

[74] W. Strauss, *Industrial Gas Cleaning*, Pergamon Press, Oxford, **1975**.

[75] R.J. Wakeman and E.S. Tarleton, *Solid Liquid Separation: Principles of Industrial Filtration*, Elsevier Science & Technology, Oxford, **2005**.

[76] R.J. Wakeman and E.S. Tarleton, *Solid Liquid Separation: Equipment Selection and Process Design*, Elsevier Science & Technology, Oxford, **2005**.

[77] R.J. Wakeman and E.S. Tarleton, *Solid Liquid Separation: Scale Up of Industrial Equipment*, Elsevier Science & Technology, Oxford, **2005**.

[78] T. Sparks, *Solid Liquid Filtration*, Butterworth Heineman, Oxford, **2011**.

[79] T. Sparks, *Filters and Filtration Handbook*, 6th Edition, Butterworth Heineman, Oxford, **2013**.

Chapter 11

[80] E. Drioli, L. Giorno and F. Macedonio, *Membrane Engineering*, De Gruyter, Berlin, **2018**.

[81] M. Mulder, *Basic Principles of Membrane Technology*, 2nd Edition; Kluwer Academic Publishers, Dordrecht, **1996**.

[82] *Membrane Technology in the Chemical Industry* (S. Pereira Nunes and K.-V. Peinemann, Editors), Wiley-VCH, Weinheim, **2006**.

[83] K. Mohanty and M.K. Purkait, *Membrane Technologies and Applications*, CRC Press, Boca Raton, **2011**.

[84] A.K. Pabby, S.S.H. Rizvi and A.M. Sastre, *Handbook of Membrane Separations: Chemical, Pharmaceutical, Food and Biotechnological Applications*, 2nd Edition, CRC Press, Boca Raton, **2015**.

[85] R.W. Baker, *Membrane Technology and Applications*, 3rd Edition, Wiley, **2012**.

Chapter 12

[86] H.R. Null, *Selection of a Separation Process*, in *Handbook of Separation Technology*, (R.W. Rousseau, Editor), John Wiley & Sons, New York, **1987**, Chapter 22.

[87] S.D. Barnicki and J.J. Siirola, *Separations Process Synthesis*, in *Kirk-Othmer Encyclopedia of Chemical Technology*, 5th Edition (A. Seidel, Editor), John Wiley & Sons, Hoboken, **2004**.

[88] W.D. Seider, D.R. Lewin, S. Widagdo and J.D. Seader, *Product and Process Design Principles: Synthesis, Analysis and Evaluation*, 4th Edition, John Wiley & Sons, New York, **2017**.

Index

https://doi.org/10.1515/9783110654806-015

Printed in the USA
CPSIA information can be obtained
at www.ICGtesting.com
CBHW061537200824
13467CB00012B/477